APPLIED
ALGEBRA I

APPLIED ALGEBRA I

THOMAS J. McHALE
PAUL T. WITZKE
Milwaukee Area Technical College, Milwaukee, Wisconsin

ADDISON-WESLEY PUBLISHING COMPANY
Reading, Massachusetts
Menlo Park, California · London · Amsterdam · Don Mills, Ontario · Sydney

OTHER BOOKS OF INTEREST

INTRODUCTORY ALGEBRA – Programmed
INTRODUCTORY ALGEBRA
CALCULATION AND CALCULATORS

Reproduced by Addison-Wesley from camera-ready copy supplied by the authors.

ISBN 0-201-04767-5
    19 20 MU 95949392

PREFACE

APPLIED ALGEBRA I and APPLIED ALGEBRA II are designed to teach the algebraic concepts and skills needed in basic science, technology, pre-engineering, and mathematics itself. The word "applied" in the title does not mean that major emphasis is given to applied problems, though many basic types of applied problems are included. Rather, the word "applied" means that emphasis is given to topics that are broadly useful, including manipulations with formulas and problem-solving based on formulas. Written at an intermediate level, the texts presuppose one year of high school algebra or the equivalent. For students without that prerequisite or students with serious remedial problems, INTRODUCTORY ALGEBRA can be used as a preparatory text.

The content of APPLIED ALGEBRA I is presented in a programmed format with assignment self-tests with answers within each chapter and supplementary problems for each assignment at the end of each chapter. Answers for all supplementary problems are given in the back of the text.

APPLIED ALGEBRA I is accompanied by the book TESTS FOR APPLIED ALGEBRA I which contains a diagnostic test, thirty-five assignment tests, ten chapter tests, five multi-chapter tests (every two chapters), and a comprehensive test. Three parallel forms are provided for chapter tests, multi-chapter tests, and the comprehensive test. A full set of answer keys for the tests is included. Because of the large number of tests provided, various options are possible in using them. For example, an instructor can use the chapter tests alone, the multi-chapter tests alone, or some combination of the two types.

> Note: The test book is provided only to teachers. Copies of the tests for student use must be made by some copying process.

The following features make the instruction effective and efficient for students:

1. The instruction, which is based on a task analysis, contains examples of all types of problems that appear in the tests.

2. The full and flexible set of tests can be used as a teaching tool to identify learning difficulties which can be remedied by tutoring or class discussion.

3. Because of the programmed format and the full and flexible set of tests provided, the text is ideally suited for individualized instruction as well as more traditional methods.

The authors wish to thank Ms. Marylou Hansen and Ms. Arleen D'Amore who typed and proofread the camera-ready copy, Ms. Joan Szulczewski who prepared the graphs, Mr. Gail Davis who did the final proofreading, and Mr. Allan Christenson who prepared the Index.

HOW TO USE THE TEXT AND TESTS

This text and the tests available in TESTS FOR APPLIED ALGEBRA I can be used in various instructional strategies ranging from paced instruction with all students taking the same content to totally individualized instruction. The general procedure for using the materials is outlined below.

1. The diagnostic test can be administered either to simply get a measure of the entry skills of the students or as a basis for prescribing an individualized program.

2. Each chapter is covered in a number of assignments (see below). After the students have completed each assignment and the assignment self-test (in the text), the assignment test (from the test book) can be administered, corrected, and used as a basis for tutoring. The assignment tests are simply a teaching tool and need not be graded. The supplementary problems at the end of the chapter can be used at the instructor's discretion for students who need further practice.

   > Note: Instead of using the assignment self-tests (in the text) as an integral part of the assignments, they can be used at the completion of a chapter as a chapter review exercise.

3. After the appropriate assignments are completed, either a chapter test or a multi-chapter test can be administered. Ordinarily, these tests should be graded. Parallel forms are provided to facilitate the test administration, including the retesting of students who do not achieve a satisfactory score.

4. After all desired chapters are completed in the manner above, the comprehensive test can be administered. Since the comprehensive test is a parallel form of the diagnostic test, the difference score can be used as a measure of each student's improvement.

ASSIGNMENTS FOR APPLIED ALGEBRA I

Ch. 1:	#1 (pp. 1-10)	Ch. 4:	#13 (pp. 140-150)	Ch. 8:	#26 (pp. 325-338)
	#2 (pp. 11-19)		#14 (pp. 151-161)		#27 (pp. 339-349)
	#3 (pp. 20-32)		#15 (pp. 161-173)		#28 (pp. 350-362)
	#4 (pp. 33-42)				#29 (pp. 363-374)
		Ch. 5:	#16 (pp. 176-192)		
Ch. 2:	#5 (pp. 45-54)		#17 (pp. 192-203)	Ch. 9:	#30 (pp. 377-384)
	#6 (pp. 55-66)		#18 (pp. 204-219)		#31 (pp. 385-398)
	#7 (pp. 66-79)				#32 (pp. 398-413)
	#8 (pp. 79-91)	Ch. 6:	#19 (pp. 223-239)		
			#20 (pp. 240-257)	Ch. 10:	#33 (pp. 416-425)
Ch. 3:	#9 (pp. 94-103)		#21 (pp. 258-269)		#34 (pp. 425-436)
	#10 (pp. 104-113)				#35 (pp. 437-449)
	#11 (pp. 113-125)	Ch. 7:	#22 (pp. 273-286)		
	#12 (pp. 125-136)		#23 (pp. 286-299)		
			#24 (pp. 299-309)		
			#25 (pp. 309-321)		

C O N T E N T S

Chapter 1 SIGNED NUMBERS

In this chapter, we will discuss adding, subtracting, multiplying, and dividing signed whole numbers, decimal numbers, and fractions. Though the operations with signed whole numbers and decimal numbers are presented together, the operations with signed fractions are presented in separate sections. Some combined operations are also included.

1-1   NUMBER LINE, ORDER, ABSOLUTE VALUE

In this section, we will discuss the number line, the meaning of "is greater than", and "is less than", and the meaning of "absolute value".

---

1.   Numbers with a "+" sign are called <u>positive</u> numbers. Numbers with a "-" sign are called <u>negative</u> numbers. For example:

$+3$, $+12.5$, and $+\frac{4}{5}$ are <u>positive</u> numbers.

-7, -16.8, and $-\frac{1}{2}$ are <u>negative</u> numbers.

On the number line, <u>positive</u> numbers are <u>to the right</u> of "0"; <u>negative</u> numbers are <u>to the left</u> of "0".

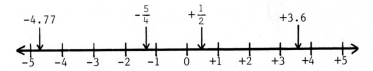

Since "0" separates the positive and negative numbers, "0" is neither positive nor negative. Would it make sense to write either +0 or -0? _____

---

2.   The definition of "<u>is greater than</u>" is:

| A first number <u>is greater than</u> a second number if the first is <u>to the right</u> of the second on the number line. |

Which number in each pair is the <u>greater</u> of the two?

a) 0 or +5   +5

b) +7 or -8   +7

c) -1 or -6   -1

d) -2.7 or -1.3   -1.3

No

3.   The definition of "is less than" is:

> A first number is less than a second number if the first
> is to the left of the second on the number line.

Which number in each pair is the lesser of the two?

a) 0 or -4   ~~0~~ -4   c) -6 or -8   -8

b) -9 or +7   ~~+7~~ -9   d) -5.75 or -5.25   -5.75

a) +5	c) -1
b) +7	d) -1.3

---

4.   We use the symbols ">" and "<" for "is greater than" and "is less than".

> ">" means "is greater than".
> "<" means "is less than".

In each blank, write either ">" or "<".

a) +7 __>__ -2   c) -10 __<__ -1

b) 0 __<__ +5   d) -7.5 __>__ -9.5

a) -4	c) -8
b) -9	d) -5.75

---

5.   Though we always write a "−" before a negative number, we frequently do not write a "+" before a positive number. That is:

$$4, \ 6.8, \ \text{and} \ \frac{7}{8} \ \text{mean} \ +4, \ +6.8, \ \text{and} \ +\frac{7}{8}.$$

In each blank, write a ">" or "<".

a) -1 __<__ 1   c) -5 __>__ -10

b) $\frac{5}{4}$ __>__ $-\frac{5}{4}$ d) $-\frac{7}{8}$ __<__ $-\frac{5}{8}$

a) >	c) <
b) <	d) >

---

6.   The "absolute value" of a number is its distance from "0" on the number line.

Since +7 is 7 units from "0", the absolute value of +7 is 7.

Since -5 is 5 units from "0", the absolute value of -5 is 5.

For practical purposes, the absolute value of a number is its numerical value with the sign ignored. Write the absolute value of each number.

a) +9 __9__   b) -12 __12__   c) +8.8 __8.8__   d) $-\frac{1}{2}$ __1/2__

a) <	c) >
b) >	d) <

---

7.   We use | | as the symbol for absolute value. That is:

|+8| means "the absolute value of +8".

|-5| means "the absolute value of -5".

Therefore: |+8| = 8 |-5| = 5 |-3.7| = __3.7__

a) 9	c) 8.8
b) 12	d) $\frac{1}{2}$

8. Pairs of numbers that are the same distance from "0" on the number line have the same absolute value.

$$|+4| = 4 \quad \text{and} \quad |-4| = 4 \qquad |+2.5| = 2.5 \quad \text{and} \quad |-2.5| = 2.5$$

Name the pair of numbers whose absolute value is $\frac{3}{8}$.

+3/8   and   −3/8

3.7

9. Since "0" is 0 units from "0" on the number line, the absolute value of "0" is 0.

Complete: a) $|-1| = $ __1__   c) $|0| = $ __0__

b) $\left|+\frac{7}{5}\right| = $ __7/5__   d) $\left|-\frac{1}{3}\right| = $ __1/3__

$+\frac{3}{8}$ and $-\frac{3}{8}$

a) 1   b) $\frac{7}{5}$ c) 0   d) $\frac{1}{3}$

1-2   ADDING SIGNED NUMBERS

In this section, we will discuss the "rules" for adding signed numbers.

10. In an addition, the numbers added are called the "terms"; the answer is called the "sum". For example:

In $(+3) + (+2) = +5$:   "+3" and "+2" are the terms.

"+5" is the sum.

In $(-1) + (-6) = -7$:   a) "-1" and "-6" are the terms.

b) "-7" is the sum.

11. We did $(+4) + (+2) = +6$ on the number line below. Notice that the arrows for positive terms go to the right.

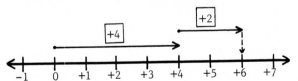

a) terms   b) sum

From the example, you can see the following "rule":

> When adding two positive numbers:
> 1. The sign of the sum is positive.
> 2. The absolute value of the sum is the sum of the absolute values of the two terms.

Use the rule for these:

a) $(+10) + (+5) = $ +15   b) $(+1.3) + (+4.2) = $ +5.5

12.   We did (−1) + (−6) = −7 on the number line below.   Notice that the arrows for <u>negative</u> terms go <u>to the left</u>.

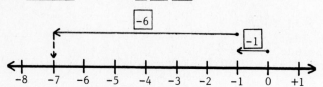

From the example, you can see the following "rule":

> When adding <u>two</u> <u>negative</u> <u>numbers</u>:
>
> 1. The <u>sign of the sum</u> is <u>negative</u>.
>
> 2. The <u>absolute value of the sum</u> is the sum of the absolute values of the two terms.

Use the rule for these:

a) (−8) + (−7) = __−15__   b) (−6.6) + (−9.9) = __−16.5__

a) +15   b) +5.5

---

13.   We did (+5) + (−2) = +3 and (−6) + (+4) = −2 on the number lines below.

a) −15   b) −16.5

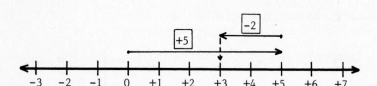

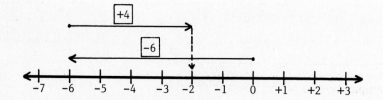

From the examples, you can see the following "rule":

> When <u>adding</u> <u>one</u> <u>positive</u> <u>and</u> <u>one</u> <u>negative</u> <u>number</u>:
>
> 1. The <u>sign of the sum</u> is the same as the sign of the term with the larger absolute value.
>
> 2. The <u>absolute value of the sum</u> is obtained by subtracting the absolute values of the terms.

Use the rule for these:

a) (+3) + (−9) = __−6__   c) 10 + (−7) = __3__

b) (−1) + (+8) = __+7__   d) (−5) + 2 = __−3__

14.   Complete: a) $(-30) + 10 = -20$ c) $(-4.2) + (-5.5) = -9.7$

b) $70 + (-45) = 25$ d) $(-10.5) + 12.7 = 2.2$

a) -6	c) +3
b) +7	d) -3

15.   When one term is "0", the sum is <u>identical</u> to the other term.
      For example:

$(+10) + 0 = +10$ a) $0 + (+3.75) = 3.75$

$0 + (-1.5) = -1.5$ b) $(-60) + 0 = -60$

a) -20	c) -9.7
b) +25	d) +2.2

16.   To add three or more signed numbers, we can add "two at a time" from
      left to right as we have done below.

$$2 + (-7) + 6 + (-5)$$
$$\downarrow$$
$$(-5) \quad + 6 + (-5)$$
$$\downarrow$$
$$1 \quad + (-5) = -4$$

      By adding "two at a time", find each sum.

a) $(-9) + 6 + (-1) = -4$

b) $10 + (-6) + (-7) + 5 = 2$

a) +3.75	b) -60

17.   By adding "two at a time", find each sum.

a) $(-4) + 7 + (-2) = 1$

b) $12 + (-5) + (-8) + 4 + (-7) = -4$

a) -4	b) +2

a) +1	b) -4

1-3   ORDER PROPERTIES OF ADDITION

In this section, we will discuss the two "order" properties of addition. They are the <u>commutative</u> property
and the <u>associative</u> property.

18.   If we interchange the two terms in an addition, we get the same sum.
      For example:

Both $(-6) + (-2)$ and $(-2) + (-6)$ equal -8.

Both $10 + (-15)$ and $(-15) + 10$ equal -5.

The property above is called the "<u>commutative</u>" property of addition.
The word "commutative" means to "change around" or "interchange".

Using the <u>commutative</u> property of addition, complete these:

a) $(-8) + 7 = 7 + (-8)$ b) $12 + (-5) = (-5) + 12$

19. In the expression below, 5 + 7 is enclosed in parentheses ( ). When parentheses contain an addition, they are "grouping" symbols. The operation within the grouping is always performed first. For example:

$$3 + (5 + 7)$$
$$\downarrow$$
$$3 + 12 = 15$$

When the grouping contains negative numbers, we use brackets [ ] as the grouping symbols to avoid a series of parentheses. For example:

$$[(-9) + 7] + (-6)$$
$$\downarrow$$
$$(-2) + (-6) = -8$$

Complete each addition by performing the operation within the grouping first.

a) $(1 + 4) + (-10)$
    $\downarrow$
    $(5) + (-10) = \underline{-5}$

b) $(-4) + [(-1) + (-5)]$
    $\downarrow$
    $(-4) + (-6) = \underline{-10}$

a) 7 + (-8)
b) (-5) + 12

---

20. To perform an addition with three or more terms, we add "two at a time". As the groupings in the examples below show, we can add the terms in any order and get the same sum.

$$(5 + 3) + 6 \qquad\qquad 5 + (3 + 6)$$
$$\downarrow \qquad\qquad\qquad\qquad \downarrow$$
$$8 + 6 = 14 \qquad\quad 5 + 9 = 14$$

The property above is called the "associative" property of addition. The word "associative" means to "group".

The additions below also illustrate the associative property of addition. Complete these.

a) $[(-6) + 4] + (-7)$
    $\downarrow$
    $(-2) + (-7) = \underline{-9}$

b) $(-6) + [4 + (-7)]$
    $\downarrow$
    $(-6) + (-3) = \underline{-9}$

a) $\underline{5} + (-10) = \underline{-5}$
b) $(-4) + \underline{(-6)} = \underline{-10}$

---

21. The important point about the commutative and associative properties of addition is this: the terms can be written in any order or added in any order without changing the sum.

Because of the fact above, there is a shortcut that can be used to add three or more signed numbers. In the shortcut, we add all the positive numbers together and all the negative numbers together. For example:

Original addition: $8 + (-5) + (-7) + 3 + 4 + (-6)$

Rearranged addition: $(8 + 3 + 4) + [(-5) + (-7) + (-6)]$
$$\downarrow \qquad\qquad\qquad\qquad \downarrow$$
$$15 \qquad + \qquad (-18) \qquad = -3$$

a) $\underline{(-2)} + (-7) = \underline{-9}$
b) $(-6) + \underline{(-3)} = \underline{-9}$

(Continued on following page.)

21.   Continued

We can add the positive numbers and negative numbers mentally without rewriting the terms.

$$(-5) + 10 + 4 + (-9) + (-1) + 3$$

Do the addition above mentally:

a) The sum of the positive numbers is __17__ .

b) The sum of the negative numbers is __15__ .

c) The total sum is __2__ .

---

22.   Use the mental shortcut for these:

a)   3 + (-5) + 6 = __4__

b)   (-10) + 7 + (-8) + 3 = __-8__

c)   (-4) + 1 + (-6) + 3 + (-5) = __-11__

d)   (-7) + 8 + (-6) + 2 + (-7) + 4 = __-6__

a) 17

b) -15

c) 17 + (-15) = 2

---

a) 4   b) -8   c) -11   d) -6

1-4   OPPOSITES

In this section, we will discuss the meaning of "opposites".   "Opposites" are also called "additive inverses".

---

23.   Two numbers with the same absolute value but different signs are called "opposites". They are called "opposites" because they are on opposite sides of "0" on the number line.

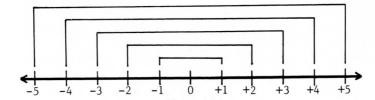

From the number line above, you can see these facts:

+2 and -2 are "opposites".

+5 and -5 are "opposites".

---

24.   Since + 8 and -8 are a pair of opposites, we say:

The opposite of +8 is -8.

The opposite of -8 is +8.

Name the opposite of each number.

a) -10 __10__   b) 3.5 __-3.5__   c) -2.66 __2.66__

-5

25.   When two opposites are added, the sum is "0".   That is:

$$4 + (-4) = 0 \qquad\qquad (-35) + 35 = \underline{0}$$

a) +10   c) +2.66
b) -3.5

26.   Use the mental shortcut for these:

   a) $(-7) + 7 + (-5) = \underline{-5}$

   b) $3 + (-6) + 7 + (-4) = \underline{0}$

0

a) -5   b) 0

1-5   SUBTRACTING SIGNED NUMBERS

Any subtraction of signed numbers can be performed by converting the subtraction to an equivalent addition. We will discuss the method in this section.

27.   Any subtraction of signed numbers can be converted to an equivalent addition by ADDING THE OPPOSITE OF THE SECOND TERM TO THE FIRST TERM.   That is:

$$7 - (-2) = 7 + \text{(the opposite of -2)}$$
$$= 7 + (+2)$$

$$(-9) - (+8) = (-9) + \text{(the opposite of +8)}$$
$$= (-9) + (-8)$$

28.   Any subtraction can be converted to an equivalent addition by using the following definition:

The subtraction symbol "-" means:
ADD THE OPPOSITE OF THE SECOND TERM

Complete the following conversions to additions.

   a) $5 - (+10) = 5 + (-10)$ c) $(-9) - (+3) = (-9) + (-3)$

   b) $5 - (-10) = 5 + (10)$ d) $(-9) - (-3) = (-9) + (3)$

(-9) + (-8)

29.   When the second term of a subtraction is positive, the positive sign is not usually shown.   In such cases, remember that the second term is positive when converting to addition.   For example:

$$2 - 7 = 2 - (+7) = 2 + (-7)$$
$$(-5) - 8 = (-5) - (+8) = (-5) + (-8)$$

Complete the following conversions to addition.

   a) $5 - 20 = 5 + (-20)$ c) $1.8 - 7.7 = 1.8 + (-7.7)$

   b) $(-30) - 10 = (-30) + (-10)$ d) $(-9.6) - 2.2 = (-9.6) + (-2.2)$

a) 5 + (-10)
b) 5 + (+10)
c) (-9) + (-3)
d) (-9) + (+3)

30. Any subtraction can be performed by converting to the equivalent addition and then performing the addition. For example:

$$4 - 7 = 4 + (-7) = -3$$
$$8 - (-3) = 8 + 3 = 11$$

Following the examples, complete these:

a) $(-5) - 6 = (-5) + (-6) =$ _____   b) $(-7) - (-5) = (-7) + 5 =$ _____

a) $5 + (-20)$

b) $(-30) + (-10)$

c) $1.8 + (-7.7)$

d) $(-9.6) + (-2.2)$

---

31. Complete each subtraction by converting to addition and then performing the addition.

ADDITION

a) $(-4) - 9 =$ _____ = _____

b) $(-1) - (-3) =$ _____ = _____

c) $3 - (-7) =$ _____ = _____

d) $2 - 6 =$ _____ = _____

a) -11 b) -2

---

32. Use the same method to complete these:

a) $8 - 9 =$ _____   c) $(-3) - 1 =$ _____

b) $1 - (-1) =$ _____   d) $(-5) - (-5) =$ _____

a) $(-4) + (-9) = -13$

b) $(-1) + 3 = 2$

c) $3 + 7 = 10$

d) $2 + (-6) = -4$

---

33. Complete.

a) $1.9 - 3.7 =$ _____   b) $(-2.6) - (-9.3) =$ _____

a) -1 c) -4

b) 2 d) 0

---

34. We "add the opposite" to subtract a signed number from "0".

$$0 - 7 = 0 + (-7) = -7$$
$$0 - (-3) = 0 + (+3) = 3$$

Complete:

a) $0 - 25 =$ _____   b) $0 - (-5.6) =$ _____

a) -1.8 b) 6.7

---

35. When "0" is subtracted from a signed number, the answer is identical to the signed number. For example:

$$4 - 0 = 0$$
$$(-8) - 0 = -8$$

a) $75 - 0 =$ _____

b) $(-10.8) - 0 =$ _____

a) -25 b) 5.6

---

a) 75 b) -10.8

36.   When an ordinary subtraction is performed correctly, the sum of the
second term and the answer equals the first term.   For example:

$$8 - 3 = 5 \quad \text{and} \quad 3 + 5 = 8$$

$$7 - 1 = 6 \quad \text{and} \quad 1 + 6 = 7$$

If the method we are using to subtract signed numbers makes sense, the
same fact should be true.   Two completed subtractions are given below.

$$4 - 9 = -5 \qquad [\text{from } 4 + (-9) = -5]$$

$$(-8) - (-2) = -6 \qquad [\text{from } (-8) + 2 = -6]$$

Let's add the second term and the answer for each subtraction to see
whether their sum equals the first term.

a) Does $9 + (-5) = 4$? _____   b) Does $(-2) + (-6) = -8$? _____

a) Yes   b) Yes

SELF-TEST 1 (pages 1-10)

In each blank, write ">" or "<".

1. -5 $\leq$ 2   2. $-\frac{1}{4}$ ⟋ $-\frac{3}{4}$

Find these "absolute values".

3. $|-1.84|$ = 1.84   4. $|+\frac{9}{5}|$ = 9/5

Do these additions.

5. $(+50) + (-20)$ = +30

6. $(-6.2) + (-1.3)$ = -7.5

7. $(-9) + 0$ = -9

8. $(-4) + 7 + (-6)$ = -3

9. $8 + (-5) + 6 + (-9)$ = 0

10. $(-12) + (-1) + 10 + (-3) + 9$ = 3

Name the opposite of each number.   11. 5.18 -5.18   12. -14 14

Do these subtractions.

13. $(-5) - (+5)$ = -10

14. $12 - (+8)$ = +20

15. $(-3.7) - (+1.2)$ = -2.5

16. $4 - 11$ = -7

17. $0 - (+1)$ = 1

18. $(-36) - 0$ = -36

ANSWERS: 1. <   5. +30   9. 0   13. -10   17. +1

2. >   6. -7.5   10. +3   14. +20   18. -36

3. 1.84   7. -9   11. -5.18   15. -2.5

4. $\frac{9}{5}$ 8. -3   12. +14   16. -7

1-6   MULTIPLYING SIGNED NUMBERS

In this section, we will discuss the "rules" for multiplying signed numbers.

37.   Various symbols can be used to indicate a multiplication.   For example, all six expressions below mean: <u>multiply 3 and 7</u>.
 	
     3 times 7   3(7)
     3 x 7   (3)7
     3 · 7   (3)(7)
 	
In a multiplication, the two numbers multiplied are called <u>factors</u>; the answer is called the <u>product</u>.   For example:
 	
  In (6)(9) = 54:   the two <u>factors</u> are 6 and 9.
                    the <u>product</u> is 54.
    In (7)(-5) = -35: a) the two <u>factors</u> are __7__ and __-5__ .
                     b) the <u>product</u> is __-35__.	

38.   Any multiplication stands for an addition of identical terms.   <u>The first factor tells how many times the second factor should be used as a term</u>.   For example:
 	
    4(2) = 2 + 2 + 2 + 2   3(-8) = (-8) + (-8) + (-8)
 	
Write each multiplication as an addition.
 	
  a) 2(10) = __10+10__   b) 5(-4) = 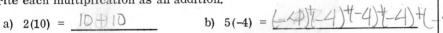
	a) 7 and -5
	
	b) -35

39.   <u>When both factors are positive, the product is positive</u> (because each term in the corresponding addition is positive).   For example:
 	
    3(6) or 3(+6) = (+6) + (+6) + (+6) = +18
 	
Using the above rule, find each product.
 	
  a) 5(+30) = __+150__   b) 7(1. 2) = __+8.4__
	a) 10 + 10
	
	b) (-4) + (-4) + (-4)
	    + (-4) + (-4)

40.   <u>When one factor is positive and one is negative, the product is negative</u> (because each term in the corresponding addition is negative).   For example:
 	
    4(-7) = (-7) + (-7) + (-7) + (-7) = -28
 	
The above rule applies whether the one negative factor is written first or second.   Complete these:
 	
  a) 6(-10) = __-60__   b) (-9)(20) = __-180__
	a) +150 b) +8. 4

	a) -60 b) -180

41.   <u>When</u> <u>both</u> <u>factors</u> <u>are</u> <u>negative</u>, <u>the</u> <u>product</u> <u>is</u> <u>positive</u> (see the patterns below).

$$(3)(-5) = -15 \qquad\qquad (3)(-9) = -27$$
$$(2)(-5) = -10 \qquad\qquad (2)(-9) = -18$$
$$(1)(-5) = -5 \qquad\qquad (1)(-9) = -9$$
$$(0)(-5) = 0 \qquad\qquad (0)(-9) = 0$$
$$(-1)(-5) = +5 \qquad\qquad (-1)(-9) = +9$$
$$(-2)(-5) = +10 \qquad\qquad (-2)(-9) = +18$$
$$(-3)(-5) = +15 \qquad\qquad (-3)(-9) = +27$$

Using the above rule, complete these:

a) (-2)(-7) = __14__   b) (-5)(-1.3) = __6.5__

---

42.   Multiply:

a) (-9)(7) = __-63__   c) 4(-6.3) = __-25.2__

b) (-7)(-8) = __+56__   d) (-1.2)(-3.3) = __+3.96__

a) +14   b) +6.5

---

43.   Three special properties of multiplication are listed below.

1) When one factor is "1", the product is <u>identical</u> to the other factor.

$$(1)(+5) = +5 \qquad\qquad (-3)(1) = -3$$

2) When one factor is "0", the product is "0".

$$(7)(0) = 0 \qquad\qquad (0)(-9) = 0$$

3) When one factor is "-1", the product is the <u>opposite</u> of the other factor.

$$(-1)(6) = -6 \qquad\qquad (-8)(-1) = +8$$

Using the above properties, complete these:

a) (1)(-17) = __-17__   c) (-40)(-1) = __40__

b) (-1)(6.5) = __-6.5__   d) (0)(-8.4) = __0__

a) -63   c) -25.2
b) +56   d) +3.96

---

44.   Complete: a) (1)(-1) = __-1__   c) (-1)(0) = __0__

b) (0)(1) = __0__   d) (-1)(-1) = __1__

a) -17   c) +40
b) -6.5   d) 0

---

a) -1   c) 0
b) 0   d) +1

1-7   ORDER PROPERTIES OF MULTIPLICATION

In this section, we will discuss the two "order" properties of multiplication.   They are the <u>commutative</u> property and the <u>associative</u> property.

45. If we interchange the two factors in a multiplication, we get the same product. For example: Both (3)(-4) and (-4)(3) equal -12. Both (-7)(-8) and (-8)(-7) equal +56. The property above is called the "<u>commutative</u>" property of multiplication. Using the commutative property, complete these: a) (-5)(6) = (6)(-5) b) (-9)(-4) = (-4)(-9)	
46. When a multiplication contains three or more factors, we can multiply "two at a time" as we have done below. $$(-4)(3)(-1)(2)$$ $$\downarrow$$ $$(-12)(-1)(2)$$ $$\downarrow$$ $$(+12)(2) = 24$$ Using the same method, complete these: a) (-8)(-5)(-2) = -80 b) (-1)(5)(-3)(2) = 30 $-5 \cdot -3 \cdot 2$ $15 \cdot 2 = 30$	a) (6)(-5) b) (-4)(-9)

47. When one of the factors is "0", the product is always "0".   For example:
 	
a) (7)(-9)(0)   b) (-4)(0)(8)(-10)
   $\downarrow$ $\downarrow$
  (-63)(0) = 0 (0)(8)(-10)
                            $\downarrow$
                         (0)(-10) = 0
	a) -80 b) 30

48. To complete (4)(-5)(7), we can multiply 4 and -5 first or -5 and 7 first and get the same product.   That is:
 	
(4)[(-5)(7)]   [(4)(-5)](7)
    $\downarrow$ $\downarrow$
(4)(-35) = -140   (-20)(7) = -140
 	
The property above is called the "<u>associative</u>" property of multiplication. The multiplications below also illustrate the associative property. Complete them.
 	
a) [(-1)(-2)](-3)   b) (-1)[(-2)(-3)]
     $\downarrow$ $\downarrow$
  (2)(-3) = -6 (-1)(6) = -6
	a) 0 b) 0

49. The important point about the commutative and associative properties of multiplication is this: the <u>factors</u> <u>can</u> <u>be</u> <u>written</u> <u>in</u> <u>any</u> <u>order</u> <u>or</u> <u>multi-plied</u> <u>in</u> <u>any</u> <u>order</u> <u>without</u> <u>changing</u> <u>the</u> <u>product.</u>   Using that fact, complete these:

    a)   (-2)(4)(8) = (4)(8)(-2)   b)   (7)(-1)(-9) = (-1)(-9)(7)

> a) (2)(-3) = -6
> b) (-1)(6) = -6

> a) (4)(8)(-2)   b) (-1)(-9)(7)

1-8   DIVIDING SIGNED NUMBERS

In this section, we will discuss the "rules" for dividing signed numbers.

50. In algebra, a division is usually written as a fraction.   That is:

$$32 \div 4 \text{ is written } \frac{32}{4} \qquad -16 \div 2 \text{ is written } -16/2$$

51. In $\frac{15}{5} = 3$:   15 is called the <u>numerator</u>.
                  5 is called the <u>denominator</u>.
                  3 is called the <u>quotient</u>.

    In $\frac{10}{-2} = -5$: a)   the <u>denominator</u> is -2 .

                    b)   the <u>quotient</u> is -5 .

                    c)   the <u>numerator</u> is 10 .

> $\frac{-16}{2}$

52. The rules for dividing signed numbers are similar to the rules for multiplying signed numbers.   The three division rules are stated below.

    1) IF <u>BOTH</u> <u>NUMBERS</u> <u>ARE</u> <u>POSITIVE</u>, THE QUOTIENT IS <u>POSITIVE</u>.

$$\frac{30}{5} = +6 \qquad\qquad \frac{27}{3} = +9$$

    2) IF <u>BOTH</u> <u>NUMBERS</u> <u>ARE</u> <u>NEGATIVE</u>, THE QUOTIENT IS <u>POSITIVE</u>.

$$\frac{-20}{-4} = +5 \qquad\qquad \frac{-70}{-10} = +7$$

    3) IF <u>ONE</u> <u>NUMBER</u> <u>IS</u> <u>POSITIVE</u> <u>AND</u> <u>THE</u> <u>OTHER</u> <u>IS</u> <u>NEGATIVE</u>, THE QUOTIENT IS <u>NEGATIVE</u>.

$$\frac{-42}{7} = -6 \qquad\qquad \frac{15}{-3} = -5$$

Using the rules above, complete these:

    a) $\frac{-8}{2} = $ -4   b) $\frac{36}{-12} = $ -3   c) $\frac{-56}{-7} = $ 8

> a) -2
> b) -5
> c) 10

53. When an ordinary division is performed correctly, the product of the
     denominator and the quotient equals the numerator. For example:

     $$\frac{10}{2} = 5, \quad \text{since} \quad (2)(5) = 10$$

     The fact above can be used to justify the rules for dividing signed numbers.
     That is:

     $\frac{-30}{5} = -6$, since $(5)(-6) = -30$ $\frac{-12}{-3} = 4$, since $(-3)(4) = \underline{-12}$

a) −4

b) −3

c) +8

---

54. Four special properties of division are listed below.

     1) When a non-zero number is divided by itself, the quotient is +1.

         $$\frac{7}{7} = +1 \qquad\qquad \frac{-10}{-10} = +1$$

     2) When a number is divided by "+1", the quotient is identical to the
         number.

         $$\frac{4}{1} = 4 \qquad\qquad \frac{-6}{1} = -6$$

     3) When a number is divided by "−1", the quotient is the opposite of
         the number.

         $$\frac{3}{-1} = -3 \qquad\qquad \frac{-8}{-1} = 8$$

     4) When "0" is divided by a non-zero number, the quotient is "0".

         $$\frac{0}{5} = 0 \qquad\qquad \frac{0}{-9} = 0$$

     Using the above properties, complete these:

     a) $\frac{-12}{-12} = \underline{1}$ b) $\frac{20}{-1} = \underline{-20}$ c) $\frac{-37}{1} = \underline{-37}$ d) $\frac{0}{-2} = \underline{0}$

−12

---

55. Complete these:

     a) $\frac{-1}{1} = \underline{-1}$ b) $\frac{-1}{-1} = \underline{1}$ c) $\frac{0}{-1} = \underline{0}$ d) $\frac{1}{-1} = \underline{-1}$

a) 1   c) −37

b) −20   d) 0

---

56. Complete:

     a) $\frac{-4.96}{2} = \underline{-2.48}$ b) $\frac{-31.2}{-2.6} = \underline{12}$

a) −1   c) 0

b) 1   d) −1

 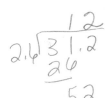

a) −2.48   b) 12

1-9   ORDER OF OPERATIONS

In this section, we will discuss the evaluation of expressions that contain more than one operation. The purpose of the section is to show the proper order of operations.

57.   To evaluate the expressions below, we converted each subtraction to addition and then performed the addition.

$$4 + 8 - (-5) = 4 + 8 + 5 = 17$$

$$9 - (-2) - 7 = 9 + 2 + (-7) = 4$$

Using the same method, evaluate these:

a) $(-1) - 8 + 10 =$ $-1 + -8 + 10 = 1$

b) $6 - (-3) + 5 - 20 =$ $6 + 3 + 5 + -20 = -6$

---

58.   One or both terms in an addition or subtraction can be a multiplication. For example:

In $4(-2) + 10$, the terms are $4(-2)$ and 10.

In $3(8) - (-1)(5)$, the terms are $3(8)$ and $(-1)(5)$.

To evaluate expressions like those above, <u>we perform the multiplications first</u>. That is:

$$\underline{4(-2)} + 10 \qquad \underline{3(8)} - \underline{(-1)(5)}$$
$$\downarrow \qquad\qquad\qquad \downarrow \qquad\quad \downarrow$$
$$(-8) + 10 = 2 \qquad 24 - (-5)$$
$$24 + 5 = 29$$

Using the same method, evaluate these:

a) $3(-10) - (-9)$ b) $(-1)(6) + (-7)(-2)$

   $-30 + 9 = -21$ $-6 + 14 = 8$

a) 1   b) -6

---

59.   One term in an addition or subtraction can be a grouping.   For example:

In $10 + (7 - 11)$,   the terms are 10 and $(7 - 11)$,

In $6 - [(-3) + 5]$, the terms are 6 and $[(-3) + 5]$.

To evaluate expressions like those above, <u>we perform the operation within the grouping first</u>.   That is:

$$10 + (7 - 11) \qquad\qquad 6 - \underline{[(-7) + 5]}$$
$$10 + \underline{[7 + (-11)]} \qquad\qquad\qquad \downarrow$$
$$\downarrow \qquad\qquad\qquad 6 - (-2)$$
$$10 + (-4) = 6 \qquad\qquad 6 + 2 = 8$$

Using the same method, evaluate these:

a) $(-4) - (10 + 2)$ b) $20 + [3 + (-9)]$

   $-4 - (8)$ $20 + -6 = 14$
   $-4 + -8 = -12$

a) -21   b) 8

60.   To evaluate the expression below, we simplified each term first.

$$\underline{7(-5)} + \underline{(6 - 10)}$$
$$\downarrow \qquad \downarrow$$
$$(-35) + \quad (-4) \ = -39$$

Using the same method, evaluate these:

a) 2(5) - (12 - 8)   b) 6(-1) + [(-2) + 5]

a) -12	b) 14

---

61.   One of the factors in a multiplication can be a grouping.   For example:

In (5 - 2)6 , the factors are (5 - 2) and 6.

In 4[(-1) + (-7)] , the factors are 4 and [(-1) + (-7)]

To evaluate expressions like those above, <u>we can perform the operation within the grouping first</u>.   That is:

$$\underline{(5 - 2)}6 \qquad\qquad \underline{4[(-1) + (-7)]}$$
$$\downarrow \qquad\qquad\qquad\qquad \downarrow$$
$$(3) \quad 6 = 18 \qquad\qquad 4 \quad (-8) \quad = -32$$

Using the same method, evaluate these:

a) [(-9) + 6]7   b) 2(4 + 10)

$$-3 \cdot 7 = 21 \qquad\qquad 2 \cdot -6 = -12$$

a) 6	b) -3

---

62.   In the expression below, the terms are 5(3 + 4) and 8.   To evaluate it, we simplified 5(3 + 4) first.

$$5(3 + 4) + 8$$
$$5(7) + 8$$
$$35 + 8 = 43$$

Using the same method, evaluate these:

a) 30 - 2(4 + 5)   b) 4[(-2) + (-3)] + 10

$$30 + 18 = 12$$

a) -21	b) -12

---

63.   To evaluate the expression below, we simplified each term first.

$$5(-2) + 3[(-2) + 4]$$
$$(-10) + 3(2)$$
$$(-10) + 6 \quad = -4$$

Using the same method, evaluate these:

a) 6(5 + 2) - 3(4)   b) 5(-8) - 2[(-5) + (-15)]

$$6(7) + 12 \qquad\qquad -40 + 40 = 0$$
$$42 + 12 = 30$$

a) 12	b) -10

64. One or both terms in an addition or subtraction can be a division. For example:

In $\frac{-12}{4} + 7$, the terms are $\frac{-12}{4}$ and 7.

In $\frac{10}{2} - \frac{20}{-5}$, the terms are $\frac{10}{2}$ and $\frac{20}{-5}$.

To simplify expressions like those above, <u>we perform the divisions first</u>. That is:

$$\frac{-12}{4} + 7$$
$$\downarrow$$
$$(-3) + 7 = 4$$

$$\frac{10}{2} - \frac{20}{-5}$$
$$\downarrow \qquad \downarrow$$
$$5 - (-4)$$
$$5 + 4 = 9$$

Using the same method, evaluate these:

a) $\frac{6}{-1} + \frac{12}{2}$

b) $2 - \frac{-9}{3}$

a) 30   b) 0

---

65. The numerator and denominator of a fraction can be a multiplication. In such cases, we perform the multiplication before dividing. For example:

$$\frac{3(-4)}{2} = \frac{-12}{2} = -6$$

Evaluate these:

a) $\frac{24}{2(-3)} =$

b) $\frac{2(-20)}{4(-2)} =$

a) 0   5) 5

---

66. The numerator and denominator of a fraction can be an addition or subtraction. In such cases, we perform the addition or subtraction before dividing. For example:

$$\frac{7 - 11}{2} = \frac{-4}{2} = -2$$

Evaluate these:

a) $\frac{10}{(-2) + (-3)}$

b) $\frac{2 - 20}{2 - 8} =$

a) −4   b) 5

---

67. Following the example, evaluate the other expression.

$$\frac{3(7 + 5)}{6} = \frac{3(12)}{6} = \frac{36}{6} = 6$$

$$\frac{5[(-1) - 9]}{10} =$$

a) −2   b) 3

68.   Evaluate:

    a) $10 - \left(\dfrac{14}{2} - 5\right)$ b) $7\left[\dfrac{(-9) + 1}{2}\right]$

-5

---

a) 8   b) -28

SELF-TEST 2 (pages 11-19)

Do these multiplications.

1. $(4)(-7) =$ _-28_   3. $(-2.5)(1.2) =$ _-3_   5. $(-7)(-3)(-4) =$ _-84_

2. $(-6)(-9) =$ _54_   4. $(-1)(-16.7) =$ _16.7_   6. $(5)(-1)(6)(-2) =$ _60_

Do these divisions.

7. $\dfrac{42}{-7} =$ _-6_   9. $\dfrac{-25.2}{4} =$ _-6.3_   11. $\dfrac{-8}{-8} =$ _1_

8. $\dfrac{-63}{-9} =$ _7_   10. $\dfrac{0}{-5} =$ _0_   12. $\dfrac{1.37}{-1} =$ _-1.37_

Do these problems involving order of operations.

    36+12

13. $4(9) - 6(-2) =$ _48_   15. $\dfrac{(-6)(4)}{(8)(-1)} =$ _3_   -24   17. $(-1)(9) - 3[(-10) + 7] =$ _0_

     5+35+18   -8   -9 +30+21=

14. $5(1 - 7) + 18 =$ _-12_   16. $\dfrac{5 + 2}{2 + 5} =$ _-1_   3/-3   18. $8 - \dfrac{7(15 + 9)}{3} =$ _-6_

                                                         $8 - \dfrac{42}{3}$ 8 +14

ANSWERS:

1. -28	4. +16.7	7. -6	10. 0	13. 48	16. -1
2. +54	5. -84	8. +7	11. +1	14. -12	17. 0
3. -3	6. +60	9. -6.3	12. -1.37	15. 3	18. -6

1-10   MULTIPLYING SIGNED FRACTIONS

In this section, we will discuss the procedure for multiplications involving signed fractions.

> Note: In this book, we will always use a horizontal fraction line instead of a slanted fraction line.   That is:
>
> $$\text{We will use } \frac{7}{8} \text{ instead of } 7/8.$$
>
> We expect you to also use horizontal fraction lines.

---

69. To multiply fractions, we simply <u>multiply</u> <u>their</u> <u>numerators</u> <u>and</u> <u>their</u> <u>denominators</u>. That is: $$\left(\frac{2}{3}\right)\left(\frac{5}{7}\right) = \frac{(2)(5)}{(3)(7)} = \frac{10}{21} \qquad \left(\frac{1}{2}\right)\left(\frac{1}{4}\right) = \frac{(1)(1)}{(2)(4)} = \underline{\frac{1}{8}}$$	
70. Any whole number is equal to a fraction whose denominator is "1". For example: $$1 = \frac{1}{1} \qquad\qquad 5 = \frac{5}{1} \qquad\qquad 13 = \frac{13}{1}$$ Therefore, any multiplication of a whole number and a fraction can be converted into a multiplication of two fractions. That is: $$3\left(\frac{4}{7}\right) = \frac{3}{1}\left(\frac{4}{7}\right) = \frac{3(4)}{1(7)} = \frac{12}{7} \qquad \frac{3}{4}(1) = \frac{3}{4}\left(\frac{1}{1}\right) = \frac{3(1)}{4(1)} = \frac{3}{4}$$ Using the same method, complete these: a) $\frac{1}{6}(5) = \frac{1}{6}\left(\frac{5}{1}\right) = \underline{5/6}$ b) $1\left(\frac{9}{2}\right) = \frac{1}{1}\left(\frac{9}{2}\right) = \underline{9/2}$	$\dfrac{1}{8}$
71. The shorter method below is ordinarily used to multiply a whole number and a fraction. That is, we get the numerator of the product <u>by</u> <u>simply</u> <u>multiplying</u> <u>the</u> <u>whole</u> <u>number</u> <u>and</u> <u>the</u> <u>original</u> <u>numerator</u>. $$5\left(\frac{2}{3}\right) = \frac{5(2)}{3} = \frac{10}{3} \qquad \frac{8}{7}(1) = \frac{8(1)}{7} = \underline{4/7}$$	a) $\dfrac{5}{6}$ b) $\dfrac{9}{2}$
72. <u>When</u> <u>one</u> <u>factor</u> <u>is</u> <u>positive</u> <u>and</u> <u>the</u> <u>other</u> <u>is</u> <u>negative</u>, <u>the</u> <u>product</u> <u>is</u> <u>negative</u>. For example: $$\left(\frac{1}{3}\right)\left(-\frac{4}{5}\right) = -\frac{4}{15} \qquad (-7)\left(\frac{1}{5}\right) = -\frac{7}{5}$$ Using the above rule, complete these: a) $\left(-\frac{5}{2}\right)\left(\frac{3}{7}\right) = \underline{-15/14}$ b) $2\left(-\frac{3}{5}\right) = \underline{-6/5}$	$\dfrac{8}{7}$
	a) $-\dfrac{15}{14}$ b) $-\dfrac{6}{5}$

73.　　When <u>both</u> <u>factors</u> <u>are</u> <u>negative</u>, <u>the</u> <u>product</u> <u>is</u> <u>positive</u>.　For example:

$$\left(-\frac{7}{8}\right)\left(-\frac{1}{2}\right) = \frac{7}{16} \qquad\qquad \left(-\frac{3}{7}\right)(-4) = \frac{12}{7}$$

Using the above rule, complete these:

a) $\left(-\frac{3}{2}\right)\left(-\frac{5}{4}\right) = \underline{\frac{15}{8}}$　　　　　b) $(-2)\left(-\frac{1}{7}\right) = \underline{\frac{2}{7}}$

74.　　Three special properties of multiplication are listed below.

　　1) When a fraction is multiplied by "1", the product is <u>identical</u> to the fraction.

$$1\left(\frac{3}{5}\right) = \frac{3}{5} \qquad\qquad \left(-\frac{7}{2}\right)(1) = -\frac{7}{2}$$

　　2) When a fraction is multiplied by "-1", the product is the <u>opposite</u> of the fraction.

$$(-1)\left(\frac{9}{4}\right) = -\frac{9}{4} \qquad\qquad \left(-\frac{1}{6}\right)(-1) = \frac{1}{6}$$

　　3) When a fraction is multiplied by "0", the product is "0".

$$0\left(\frac{5}{8}\right) = 0 \qquad\qquad \left(-\frac{10}{3}\right)(0) = 0$$

Using the properties above, complete these:

a) $(-1)\left(\frac{1}{4}\right) = \underline{-\frac{1}{4}}$　　　　c) $0\left(-\frac{7}{5}\right) = \underline{0}$

b) $\left(-\frac{3}{2}\right)(1) = \underline{-\frac{3}{2}}$　　　　d) $\left(-\frac{4}{3}\right)(-1) = \underline{\frac{4}{3}}$

---

a) $\frac{15}{8}$　　　b) $\frac{2}{7}$

a) $-\frac{1}{4}$　　b) $-\frac{3}{2}$　　c) 0　　d) $\frac{4}{3}$

1-11　EQUIVALENT FRACTIONS

In this section, we will define what is meant by "equivalent" fractions.　Then we will discuss the procedures for raising fractions to higher terms and reducing fractions to lowest terms.

75.　　The numerator and denominator of a fraction are called its "terms". Fractions that are equal even though they have different terms are called "equivalent" fractions.　For example:

$$\frac{1}{2} \quad \text{and} \quad \frac{3}{6} \quad \text{are "equivalent" fractions}$$

When two fractions are equivalent, we say that the one with larger terms is "<u>in higher terms</u>".　That is:

$$\frac{3}{6} \text{ is "in higher terms" than } \frac{1}{2},$$

　　　　because 3 and 6 are larger than 1 and 2.

Continued on following page.

75.   Continued

In each pair of equivalent fractions, identify the one that is <u>in higher terms</u>.

a) $\frac{5}{4}$ and $\frac{10}{8}$ b) $\frac{10}{20}$ and $\frac{2}{4}$ c) $-\frac{3}{2}$ and $-\frac{6}{4}$

---

76.   All of the fractions below are equal because each equals "1". Which fraction is "in highest terms"? _____

$$\frac{2}{2} = \frac{3}{3} = \frac{4}{4} = \frac{5}{5} = \frac{7}{7} = \frac{10}{10}$$

a) $\frac{10}{8}$ c) $-\frac{6}{4}$

b) $\frac{10}{20}$

---

77.   To raise a fraction to higher terms, we can multiply it by any fraction that equals "1" That is:

$$\frac{2}{3} = \frac{2}{3}(1) = \frac{2}{3}\left(\frac{4}{4}\right) = \frac{8}{12}$$

a) By substituting $\frac{3}{3}$ for the "1", raise $\frac{1}{4}$ to higher terms. $\frac{1}{4} = \frac{1}{4}(1) = \frac{1}{4}\left(\quad\right) = $ _____

b) By substituting $\frac{5}{5}$ for the "1", raise $-\frac{3}{2}$ to higher terms. $-\frac{3}{2} = -\frac{3}{2}(1) = -\frac{3}{2}\left(\quad\right) = $ _____

$\frac{10}{10}$

---

78.   In the problem below, we are asked to raise $\frac{3}{4}$ to an equivalent fraction whose denominator is "8". Since the "4" has to be multiplied by 2 to get the new denominator "8", we multiplied $\frac{3}{4}$ by $\frac{2}{2}$ to get the equivalent fraction.

$$\frac{3}{4} = \frac{\boxed{}}{8} \qquad \frac{3}{4} = \frac{3}{4}\left(\frac{2}{2}\right) = \frac{\boxed{6}}{8}$$

To decide what fraction to multiply by, we divide the new denominator by the original denominator. For example, since $12 \div 3 = 4$, we must multiply by $-\frac{5}{3}$ by $\frac{4}{4}$ below. Do so.

$$-\frac{5}{3} = -\frac{\boxed{}}{12} \qquad -\frac{5}{3} = \left(-\frac{5}{3}\right)\left(\quad\right) = -\frac{\boxed{}}{12}$$

a) $\frac{1}{4}\left(\frac{3}{3}\right) = \frac{3}{12}$

b) $-\frac{3}{2}\left(\frac{5}{5}\right) = -\frac{15}{10}$

---

79.   A two-step shortcut can be used for the same type of problems. Two examples are discussed.

1) Divide 18 by 6 and get 3.
2) Multiply 5 by 3 and get 15. $\frac{5}{6} = \frac{\boxed{15}}{18}$

1) Divide 16 by 2 and get 8.
2) Multiply 3 by 8 and get 24. $-\frac{3}{2} = -\frac{\boxed{24}}{16}$

$\left(-\frac{5}{3}\right)\left(\frac{4}{4}\right) = -\frac{\boxed{20}}{12}$

Continued on following page.

79.   Continued

Using the shortcut, complete these:

a) $\dfrac{3}{10} = \dfrac{\boxed{}}{20}$

b) $-\dfrac{7}{3} = -\dfrac{\boxed{}}{15}$

c) $-\dfrac{1}{2} = -\dfrac{\boxed{}}{100}$

---

80.   To convert a whole number to a fraction whose denominator is other than "1", we can also multiply it by any fraction that equals "1". For example:

$$3 = 3(1) = 3\left(\dfrac{5}{5}\right) = \dfrac{15}{5}$$

a) By substituting $\dfrac{2}{2}$ for the "1", convert 5 to a fraction.

$5 = 5(1) = 5\left(\right) = \underline{}$

b) By substituting $\dfrac{4}{4}$ for the "1", convert –6 to a fraction.

$-6 = -6(1) = -6\left(\right) = \underline{}$

a) $\dfrac{\boxed{6}}{20}$

b) $-\dfrac{\boxed{35}}{15}$

c) $-\dfrac{\boxed{50}}{100}$

---

81.   In the problem below, we are asked to raise "3" to an equivalent fraction whose denominator is "4". To do so, we multiplied 3 by $\dfrac{4}{4}$.

$$3 = \dfrac{\boxed{}}{4} \qquad 3 = 3\left(\dfrac{4}{4}\right) = \dfrac{\boxed{12}}{4}$$

To get a fraction whose denominator is "10" below, we must multiply "–1" by $\dfrac{10}{10}$. Do so.

$$-1 = \dfrac{\boxed{}}{10} \qquad -1 = -1\left(\right) = -\dfrac{\boxed{}}{10}$$

a) $5\left(\dfrac{2}{2}\right) = \dfrac{10}{2}$

b) $-6\left(\dfrac{4}{4}\right) = -\dfrac{24}{4}$

---

82.   The same type of problems can be done <u>by</u> <u>simply</u> <u>multiplying</u> <u>the</u> <u>absolute</u> <u>value</u> <u>of</u> <u>the</u> <u>whole</u> <u>number</u> <u>and</u> <u>the</u> <u>desired</u> <u>denominator</u>.

$$4 = \dfrac{\boxed{24}}{6} \qquad \text{(Since } 4 \times 6 = 24)$$

$$-2 = -\dfrac{\boxed{16}}{8} \qquad \text{(Since } 2 \times 8 = 16)$$

Using the shortcut above, complete these:

a) $3 = \dfrac{\boxed{}}{5}$ b) $1 = \dfrac{\boxed{}}{16}$ c) $-5 = -\dfrac{\boxed{}}{4}$

$-1\left(\dfrac{10}{10}\right) = -\dfrac{10}{10}$

---

83.   To factor a number, we write it as a multiplication. For example:

15 can be factored into (3)(5)

42 can be factored into (7)(6)

To reduce a fraction to lower terms, we can factor the original terms to get a fraction that equals "1". That is:

$$\dfrac{6}{10} = \dfrac{(3)(2)}{(5)(2)} = \left(\dfrac{3}{5}\right)\left(\dfrac{2}{2}\right) = \dfrac{3}{5}(1) = \dfrac{3}{5}$$

a) $\dfrac{\boxed{15}}{5}$

b) $\dfrac{\boxed{16}}{16}$

c) $-\dfrac{\boxed{20}}{4}$

Continued on following page.

83.   Continued

Using the same method, complete each reduction below.

a) $\dfrac{6}{9} = \dfrac{(2)(3)}{(3)(3)} = \left(\dfrac{2}{3}\right)\left(\dfrac{3}{3}\right) = \dfrac{2}{3}(\quad) = $ _____

b) $\dfrac{24}{20} = \dfrac{(6)(4)}{(5)(4)} = \left(\dfrac{6}{5}\right)\left(\dfrac{4}{4}\right) = \dfrac{6}{5}(\quad) = $ _____

---

84.   A fraction can be reduced to lower terms <u>only</u> <u>if</u> <u>both</u> <u>terms</u> <u>contain</u> <u>a</u> <u>common</u> <u>factor</u>.   For example:

$\dfrac{10}{14}$ <u>can</u> be reduced because of the common factor "2".

$\dfrac{6}{11}$ <u>cannot</u> be reduced because there is no common factor.

For each fraction, identify the factor common to both terms <u>if</u> <u>there</u> <u>is</u> <u>one</u>.

a) $\dfrac{10}{25}$ _____   b) $\dfrac{7}{9}$ _____   c) $\dfrac{21}{14}$ _____

> a) $\dfrac{2}{3}(1) = \dfrac{2}{3}$
>
> b) $\dfrac{6}{5}(1) = \dfrac{6}{5}$

---

85.   If the terms of a fraction do not contain a common factor, the fraction cannot be reduced to lower terms.   We say that a fraction of that type is "<u>in</u> <u>lowest</u> <u>terms</u>".   Which fractions below are "<u>in</u> <u>lowest</u> <u>terms</u>"?

a) $\dfrac{2}{3}$ b) $\dfrac{14}{16}$ c) $\dfrac{3}{9}$ d) $\dfrac{5}{8}$

> a) 5
>
> b) None
>
> c) 7

---

86.   The terms of $\dfrac{8}{24}$ have three common factors: 2, 4, and 8.   We used each to reduce the fraction to lower terms below.

$$\dfrac{8}{24} = \left(\dfrac{4}{12}\right)\left(\dfrac{2}{2}\right) = \left(\dfrac{4}{12}\right)(1) = \dfrac{4}{12}$$

$$\dfrac{8}{24} = \left(\dfrac{2}{6}\right)\left(\dfrac{4}{4}\right) = \left(\dfrac{2}{6}\right)(1) = \dfrac{2}{6}$$

$$\dfrac{8}{24} = \left(\dfrac{1}{3}\right)\left(\dfrac{8}{8}\right) = \left(\dfrac{1}{3}\right)(1) = \dfrac{1}{3}$$

All three fractions on the right are in lower terms.   However, only one of them is <u>in</u> <u>lowest</u> <u>terms</u>.   Which one? _____

> Only (a) and (d)

---

87.   When the terms of a fraction have <u>more</u> <u>than</u> <u>one</u> <u>common</u> <u>factor</u>:

1) We can reduce <u>to</u> <u>lower</u> <u>terms</u> by factoring out any of them.

2) We can reduce <u>to</u> <u>lowest</u> <u>terms</u> in one step only by factoring out the <u>largest</u> one.

Identify the <u>largest</u> common factor for the terms of each fraction.

a) $\dfrac{12}{18}$ _____   b) $\dfrac{36}{24}$ _____   c) $\dfrac{44}{66}$ _____

> $\dfrac{1}{3}$

88. Since it is sometimes difficult to identify the largest common factor, we sometimes need <u>more</u> <u>than</u> <u>one</u> <u>step</u> to reduce a fraction to lowest terms. For example, we used 8 as the common factor to reduce the fraction below.

$$\frac{32}{48} = \left(\frac{4}{6}\right)\left(\frac{8}{8}\right) = \frac{4}{6}(1) = \frac{4}{6}$$

Since $\frac{4}{6}$ is not in lowest terms, we must reduce further. We get:

$$\frac{32}{48} = \frac{4}{6} = \underline{}$$

a) 6

b) 12

c) 22

---

89. Reduce each fraction to lowest terms.

a) $\frac{4}{8} = \underline{}$ b) $\frac{25}{10} = \underline{}$ c) $\frac{48}{18} = \underline{}$

$\frac{2}{3}$

---

90. Negative fractions can also be reduced to lowest terms. For example:

$$-\frac{8}{12} = -\frac{2}{3} \qquad\qquad -\frac{20}{16} = -\frac{5}{4}$$

Reduce each fraction to lowest terms.

a) $-\frac{16}{12} = \underline{}$ b) $-\frac{14}{21} = \underline{}$ c) $-\frac{27}{81} = \underline{}$

a) $\frac{1}{2}$

b) $\frac{5}{2}$

c) $\frac{8}{3}$

---

91. A fraction that equals a whole number can be converted to a whole number by reducing to lowest terms. For example:

$$\frac{24}{8} = \left(\frac{3}{1}\right)\left(\frac{8}{8}\right) = \left(\frac{3}{1}\right)(1) = \frac{3}{1} \text{ or } 3$$

However, we usually make such conversions by simply performing a division. That is:

$$\frac{24}{8} = 24 \div 8 = 3 \qquad \text{a) } \frac{30}{6} = \underline{} \qquad \text{b) } -\frac{14}{2} = \underline{}$$

a) $-\frac{4}{3}$

b) $-\frac{2}{3}$

c) $-\frac{1}{3}$

---

a) 5   b) −7

1-12   "CANCELLING" IN MULTIPLICATIONS INVOLVING FRACTIONS

In this section, we will discuss multiplications with products that are not in lowest terms.   We will show how the "cancelling" process can be used to simplify multiplications of that type.

---

92.   When multiplying fractions, the product should always be reduced to lowest terms.   For example:

$$\left(\frac{1}{2}\right)\left(\frac{4}{5}\right) = \frac{4}{10} = \frac{2}{5} \qquad\qquad 8\left(\frac{3}{4}\right) = \frac{24}{4} = 6$$

Complete.   Reduce each product to lowest terms.

   a) $\left(\frac{4}{3}\right)\left(-\frac{6}{7}\right) =$ _____   b) $\left(-\frac{1}{5}\right)(-5) =$ _____

---

93.   We had to reduce the product at the left below to lowest terms.   To avoid that reduction, we can use a process called "cancelling" in which we take out common factors before multiplying.   We did so at the right below.

$$\left(\frac{4}{5}\right)\left(\frac{3}{8}\right) = \frac{12}{40} = \frac{3}{10} \qquad\qquad \left(\frac{\overset{1}{\cancel{4}}}{5}\right)\left(\frac{3}{\underset{2}{\cancel{8}}}\right) = \frac{3}{10}$$

   Notice these points about the "cancelling" process.

   1) We divided the "4" and the "8" by 4.

   2) We got the terms of the product by the multiplications (1)(3) and (5)(2).

Use "cancelling" to complete these:

   a) $\left(\frac{3}{4}\right)\left(\frac{5}{6}\right) =$ _____   b) $\left(-\frac{7}{6}\right)\left(\frac{4}{5}\right) =$ _____

Answer: a) $-\frac{8}{7}$ b) $+1$

---

94.   Two "cancellings" were possible below.   We divided both 9 and 6 by 3. We divided both 5 and 10 by 5.   Use "cancelling" to complete the other multiplication.

$$\left(\frac{\overset{3}{\cancel{9}}}{\underset{2}{\cancel{10}}}\right)\left(\frac{\overset{1}{\cancel{5}}}{\underset{2}{\cancel{6}}}\right) = \frac{3}{4} \qquad\qquad \left(-\frac{9}{14}\right)\left(-\frac{7}{12}\right) =$$ _____

Answer:
a) $\left(\frac{\overset{1}{\cancel{3}}}{4}\right)\left(\frac{5}{\underset{2}{\cancel{6}}}\right) = \frac{5}{8}$

b) $\left(-\frac{7}{\underset{3}{\cancel{6}}}\right)\left(\frac{\overset{2}{\cancel{4}}}{5}\right) = -\frac{14}{15}$

---

95.   We got the numerator of the product below by multiplying "1" and "1". Complete the other multiplication.

$$\left(\frac{\overset{1}{\cancel{4}}}{\underset{3}{\cancel{21}}}\right)\left(\frac{\overset{1}{\cancel{7}}}{\cancel{8}}\right) = \frac{1}{6} \qquad\qquad \left(\frac{3}{10}\right)\left(-\frac{5}{6}\right) =$$ _____

Answer: $\left(-\frac{\overset{3}{\cancel{9}}}{\underset{2}{\cancel{14}}}\right)\left(-\frac{7}{\underset{4}{\cancel{12}}}\right) = \frac{3}{8}$

---

96.   We got the denominator of the product below by multiplying "1" and "1". Complete the other multiplication.

$$\left(\frac{\overset{3}{\cancel{9}}}{\underset{1}{\cancel{3}}}\right)\left(\frac{\overset{1}{\cancel{2}}}{\underset{1}{\cancel{2}}}\right) = \frac{3}{1} = 3 \qquad\qquad \left(-\frac{11}{15}\right)\left(-\frac{15}{11}\right) =$$ _____

Answer: $\left(\frac{\overset{1}{\cancel{3}}}{\underset{2}{\cancel{10}}}\right)\left(-\frac{\overset{1}{\cancel{5}}}{\underset{2}{\cancel{6}}}\right) = -\frac{1}{4}$

97. We can also cancel when multiplying a whole number and a fraction. For example:

$$\cancel{3}^{1}\left(\frac{5}{\cancel{9}_{3}}\right) = \frac{5}{3} \qquad\qquad \cancel{10}^{2}\left(\frac{2}{\cancel{5}_{1}}\right) = \frac{4}{1} = 4$$

Use "cancelling" to complete these:

a) $8\left(-\dfrac{5}{6}\right) =$ _____

b) $\left(-\dfrac{4}{7}\right)(-7) =$ _____

$\left(-\dfrac{\cancel{11}^{1}}{\cancel{15}_{1}}\right)\left(-\dfrac{\cancel{15}^{1}}{\cancel{11}_{1}}\right) = \dfrac{1}{1} = 1$

---

98. Cancelling is not possible with all multiplications. Some examples are shown.

$$\left(\frac{3}{7}\right)\left(\frac{4}{5}\right) \qquad 24\left(-\frac{2}{5}\right) \qquad \left(-\frac{8}{9}\right)(-13)$$

Note: Even if you do cancel, always check to see that the product is in lowest terms.

a) $\cancel{8}^{4}\left(-\dfrac{5}{\cancel{6}_{3}}\right) = -\dfrac{20}{3}$

b) $\left(-\dfrac{4}{7}\right)\left(-\cancel{7}^{1}\right) = \dfrac{4}{1} = 4$

1-13 RECIPROCALS AND DIVISION

In this section, we will define "reciprocals" (also called "multiplicative inverses"). We will show how any division can be converted to an equivalent multiplication. We will also show that division by "0" is an impossible operation.

99. Two numbers whose product is "+1" are called "reciprocals" of each other. For example:

Since $\left(\dfrac{1}{3}\right)(3) = +1$:   The reciprocal of $\dfrac{1}{3}$ is 3.
The reciprocal of 3 is $\dfrac{1}{3}$.

Since $\left(-\dfrac{1}{5}\right)(-5) = +1$:   The reciprocal of $-\dfrac{1}{5}$ is -5.
The reciprocal of -5 is $-\dfrac{1}{5}$.

Write the reciprocals of these:

a) $\dfrac{1}{7}$ ____   b) $-\dfrac{1}{6}$ ____   c) 10 ____   d) -4 ____

---

100. Some more examples of "reciprocals" are shown below.

Since $\left(\dfrac{5}{3}\right)\left(\dfrac{3}{5}\right) = +1$:   The reciprocal of $\dfrac{5}{3}$ is $\dfrac{3}{5}$.
The reciprocal of $\dfrac{3}{5}$ is $\dfrac{5}{3}$.

Since $\left(-\dfrac{2}{7}\right)\left(-\dfrac{7}{2}\right) = +1$:   The reciprocal of $-\dfrac{2}{7}$ is $-\dfrac{7}{2}$.
The reciprocal of $-\dfrac{7}{2}$ is $-\dfrac{2}{7}$.

a) 7   c) $\dfrac{1}{10}$

b) -6   d) $-\dfrac{1}{4}$

Continued on following page.

100.   Continued

Write the reciprocals of these:

a) $\dfrac{3}{4}$ ____   b) $-\dfrac{9}{8}$ ____   c) $\dfrac{6}{11}$ ____   d) $-\dfrac{25}{33}$ ____

---

101. There are two numbers that are their own reciprocals.

a) Since $(+1)(+1) = +1$, the reciprocal of $+1$ is ____.

b) Since $(-1)(-1) = +1$, the reciprocal of -1 is ____.

a) $\dfrac{4}{3}$ c) $\dfrac{11}{6}$

b) $-\dfrac{8}{9}$ d) $-\dfrac{33}{25}$

---

102. To see that "0" has no reciprocal, answer these:

a) Whenever one factor is "0", the product is always ____

b) Can we get $+1$ as a product when one factor is "0"? ____

c) Therefore, does "0" have a reciprocal? ____

a) $+1$ b) -1

---

103. Any division (or fraction) can be converted to an equivalent multiplication <u>by multiplying the numerator by the reciprocal of the denominator.</u> For example:

$$\frac{12}{3} = 12\,(\text{the reciprocal of 3}) = 12\left(\frac{1}{3}\right)$$

$$\frac{7}{8} = 7\,(\text{the reciprocal of 8}) = 7\left(\frac{1}{8}\right)$$

Divisions involving signed numbers can be converted to multiplications in the same way. For example:

$$\frac{10}{-5} = 10\,(\text{the reciprocal of } -5) = 10\left(-\frac{1}{5}\right)$$

$$\frac{-3}{-4} = -3\,(\text{the reciprocal of } -4) = -3\left(-\frac{1}{4}\right)$$

Convert each division to an equivalent multiplication.

a) $\dfrac{-16}{8} =$ _____   b) $\dfrac{20}{-2} =$ _____   c) $\dfrac{-35}{-7} =$ _____

a) 0

b) No

c) No

---

104. When a division by "0" is converted to multiplication, one of the factors is "the reciprocal of 0". For example:

$$\frac{6}{0} = 6\,(\text{the reciprocal of 0})$$

$$\frac{-4}{0} = -4\,(\text{the reciprocal of 0})$$

$$\frac{0}{0} = 0\,(\text{the reciprocal of 0})$$

a) $-16\left(\dfrac{1}{8}\right)$

b) $20\left(-\dfrac{1}{2}\right)$

c) $-35\left(-\dfrac{1}{7}\right)$

Continued on following page.

104. Continued

Since "0" has no reciprocal, <u>DIVISION BY "0" IS IMPOSSIBLE</u>.
Perform each possible division below.

a) $\frac{0}{1} = $ _____   b) $\frac{1}{0} = $ _____   c) $\frac{0}{-3} = $ _____   d) $\frac{-3}{0} = $ _____

---

105. Any fraction with a <u>negative numerator alone</u> or a <u>negative denominator alone</u> is equal to a negative fraction.   For example:

$\frac{-8}{2}$ and $\frac{8}{-2}$ equal $-\frac{8}{2}$, because each equals -4

The same fact applies to fractions that do not equal whole numbers.
That is:

$\frac{-3}{4}$ and $\frac{3}{-4}$ equal $-\frac{3}{4}$

Write each fraction in two equivalent forms.

a) $\frac{-15}{3} = $ _____ and _____   b) $-\frac{1}{5} = $ _____ and _____

a) 0

b) impossible

c) 0

d) impossible

a) $\frac{15}{-3}$ and $-\frac{15}{3}$ b) $\frac{-1}{5}$ and $\frac{1}{-5}$

1-14   DIVIDING SIGNED FRACTIONS

In this section, we will discuss the procedure for dividing signed fractions.

106. Any division involving a fraction is written as a "complex" fraction in algebra.   For example:

$\frac{2}{3} \div \frac{4}{7}$ is written $\frac{\frac{2}{3}}{\frac{4}{7}}$

$\frac{5}{8} \div 6$ is written $\frac{\frac{5}{8}}{6}$

$10 \div \frac{1}{2}$ is written ——

$\frac{10}{\frac{1}{2}}$

107.   In a "complex" fraction, the longer fraction line is called the "major" fraction line.   The numerator or denominator or both are fractions.
For example:

In $\dfrac{\frac{1}{4}}{\frac{3}{8}}$: the numerator is $\frac{1}{4}$; the denominator is $\frac{3}{8}$

In $\dfrac{\frac{3}{10}}{9}$: the numerator is $\frac{3}{10}$; the denominator is 9

In $\dfrac{2}{\frac{5}{6}}$:   a) the numerator is _____; b) the denominator is _____

108.   To perform a division in complex-fraction form, <u>we multiply the numerator by the reciprocal of the denominator.</u>   For example:

$$\dfrac{\frac{3}{7}}{\frac{4}{9}} = \frac{3}{7}\left(\text{the reciprocal of } \frac{4}{9}\right) = \frac{3}{7}\left(\frac{9}{4}\right) = \frac{27}{28}$$

$$\dfrac{6}{\frac{5}{3}} = 6\left(\text{the reciprocal of } \frac{5}{3}\right) = 6\left(\frac{3}{5}\right) = \frac{18}{5}$$

$$\dfrac{\frac{1}{4}}{10} = \frac{1}{4}\,(\text{the reciprocal of } 10) = \frac{1}{4}\left(\quad\right) = \underline{\quad}$$

a) 2   b) $\frac{5}{6}$

109.   The rules for dividing signed numbers also apply to divisions involving signed fractions.

<u>When one term is positive and the other negative, the quotient is negative.</u>

$$\dfrac{-\frac{2}{3}}{\frac{7}{11}} = \left(-\frac{2}{3}\right)\left(\frac{11}{7}\right) = -\frac{22}{21}$$

<u>When both terms are negative, the quotient is positive.</u>

$$\dfrac{-\frac{1}{2}}{-3} = \left(-\frac{1}{2}\right)\left(-\frac{1}{3}\right) = \frac{1}{6}$$

Complete each division.

a) $\dfrac{\frac{2}{3}}{-\frac{3}{7}} = $ _____

b) $\dfrac{-\frac{1}{5}}{-\frac{3}{2}} = $ _____

$\dfrac{1}{4}\left(\dfrac{1}{10}\right) = \dfrac{1}{40}$

110. When dividing fractions, we always reduce the quotient to lowest terms. For example:

$$\frac{\frac{9}{8}}{\frac{3}{4}} = \left(\frac{\overset{3}{\cancel{9}}}{\cancel{8}_2}\right)\left(\frac{\overset{1}{\cancel{4}}}{\cancel{3}_1}\right) = \frac{3}{2}$$

Complete. Reduce each quotient to lowest terms.

a) $\dfrac{-9}{\frac{6}{5}}$ = _____

b) $\dfrac{-\frac{2}{3}}{-8}$ = _____

a) $-\dfrac{14}{3}$, from $2\left(-\dfrac{7}{3}\right)$

b) $\dfrac{2}{15}$, from $\left(-\dfrac{1}{5}\right)\left(-\dfrac{2}{3}\right)$

---

111. Three properties of division are extended to divisions involving fractions below.

   1) When a fraction is divided by itself, the quotient is "1".

$$\frac{\frac{3}{5}}{\frac{3}{5}} = \left(\frac{3}{5}\right)\left(\frac{5}{3}\right) = 1 \qquad \frac{-\frac{7}{2}}{-\frac{7}{2}} = \left(-\frac{7}{2}\right)\left(-\frac{2}{7}\right) = 1$$

   2) When a fraction is divided by "1", the quotient is identical to the fraction.

$$\frac{\frac{9}{8}}{1} = \frac{9}{8}(1) = \frac{9}{8} \qquad \frac{-\frac{1}{4}}{1} = \left(-\frac{1}{4}\right)(1) = -\frac{1}{4}$$

   3) When a fraction is divided by "-1", the quotient is the opposite of the fraction.

$$\frac{\frac{4}{5}}{-1} = \left(\frac{4}{5}\right)(-1) = -\frac{4}{5} \qquad \frac{-\frac{8}{3}}{-1} = \left(-\frac{8}{3}\right)(-1) = \frac{8}{3}$$

Using the properties above, complete these:

a) $\dfrac{-\frac{5}{8}}{\frac{5}{8}}$ = _____

b) $\dfrac{-\frac{7}{16}}{1}$ = _____

c) $\dfrac{-\frac{1}{2}}{-1}$ = _____

a) $-\dfrac{15}{2}$ b) $\dfrac{1}{12}$

---

112. Two more properties of division are extended to divisions involving fractions below.

   1) When "0" is divided by a fraction, the quotient is "0".

$$\frac{0}{\frac{3}{4}} = 0\left(\frac{4}{3}\right) = 0 \qquad \frac{0}{-\frac{6}{5}} = 0\left(-\frac{5}{6}\right) = 0$$

   2) Dividing a fraction by "0" is IMPOSSIBLE because one factor in the equivalent multiplication is "the reciprocal of 0".

$$\frac{\frac{5}{2}}{0} = \frac{5}{2}\text{(the reciprocal of 0)} \qquad \frac{-\frac{6}{7}}{0} = -\frac{6}{7}\text{(the reciprocal of 0)}$$

a) 1

b) $-\dfrac{7}{16}$

c) $\dfrac{1}{2}$

Continued on following page.

112.   Continued

Using the properties, complete these:

a) $\dfrac{0}{\frac{1}{4}}$ = ___   b) $\dfrac{\frac{1}{4}}{0}$ = ___   c) $\dfrac{0}{-\frac{7}{2}}$ = ___   d) $\dfrac{-\frac{7}{2}}{0}$ = ___

a) 0   b) impossible   c) 0   d) impossible

SELF-TEST 3 (pages 20-32)

Multiply.   Report each product in lowest terms.

1. $\left(\dfrac{5}{8}\right)\left(\dfrac{1}{7}\right)$ = 5/56

   $\dfrac{5}{56}$ =

2. $\left(-\dfrac{2}{3}\right)\left(\dfrac{3}{8}\right)$ = 2/8 = ⁻1/4

3. $\left(-\dfrac{6}{5}\right)\left(-\dfrac{15}{16}\right)$ = 9/8

   $\dfrac{18}{16} = \dfrac{9}{8}$

4. $\dfrac{(10)}{1}\left(-\dfrac{3}{4}\right)$ = ⁻15/2

Complete:

5. $\dfrac{4}{5} = \dfrac{\boxed{12}}{15}$

6. $-\dfrac{1}{8} = -\dfrac{\boxed{5}}{40}$

Reduce each fraction to lowest terms.

7. $-\dfrac{27}{45}$ = 9/15 ⁻3/5

8. $\dfrac{60}{12}$ = 5

Divide.   Report each quotient in lowest terms.

9. $\dfrac{\frac{1}{2}}{\frac{3}{4}}$ =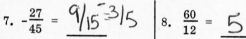

10. $\dfrac{-\frac{5}{2}}{-\frac{1}{6}}$ = 15

11. $\dfrac{2}{-\frac{8}{5}}$ = ⁻5/4

12. $\dfrac{-\frac{2}{3}}{12}$ =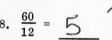

ANSWERS:

1. $\dfrac{5}{56}$ 3. $\dfrac{9}{8}$ 5. $\dfrac{\boxed{12}}{15}$ 7. $-\dfrac{3}{5}$ 9. $\dfrac{2}{3}$ 11. $-\dfrac{5}{4}$

2. $-\dfrac{1}{4}$ 4. $-\dfrac{15}{2}$ 6. $-\dfrac{\boxed{5}}{40}$ 8. 5   10. 15   12. $-\dfrac{1}{18}$

1-15   ADDING SIGNED FRACTIONS WITH LIKE DENOMINATORS

In this section, we will discuss the procedure for adding signed fractions with "like" or "common" denominators.

---

113.   To add fractions with "like" or "common" denominators, we add their numerators and keep the same denomonator.   For example:

$$\frac{4}{7} + \frac{1}{7} = \frac{4 + 1}{7} = \frac{5}{7}$$

$$\frac{(-1)}{5} + \frac{(-3)}{5} = \frac{(-1) + (-3)}{5} = \frac{-4}{5} \text{ or } -\frac{4}{5}$$

Complete these additions:

   a) $\frac{1}{3} + \frac{4}{3} = $ _____

   b) $\frac{(-6)}{7} + \frac{2}{7} = $ _____

---

114.   When adding signed fractions, it is helpful to <u>put all negative signs in the numerator</u>.   For example:

$$\frac{7}{8} + \left(-\frac{2}{8}\right) = \frac{7}{8} + \frac{(-2)}{8} = \frac{5}{8}$$

$$\left(-\frac{7}{5}\right) + \frac{3}{5} = \frac{(-7)}{5} + \frac{3}{5} = \frac{-4}{5} \text{ or } -\frac{4}{5}$$

Complete these additions:

   a) $\frac{2}{3} + \left(-\frac{1}{3}\right) = $ _____

   b) $\left(-\frac{5}{9}\right) + \left(-\frac{5}{9}\right) = $ _____

a) $\frac{5}{3}$

b) $-\frac{4}{7}$

---

115.   When adding fractions, <u>we always reduce the sum to lowest terms</u>.
For example:

$$\frac{1}{10} + \frac{3}{10} = \frac{4}{10} = \frac{2}{5} \qquad \frac{2}{3} + \frac{7}{3} = \frac{9}{3} = 3$$

Complete. Reduce each sum to lowest terms.

   a) $\frac{3}{4} + \frac{3}{4} = $ _____

   b) $\frac{8}{9} + \frac{1}{9} = $ _____

a) $\frac{1}{3}$

b) $-\frac{10}{9}$

---

116.   In each addition below, we reduced the sum to lowest terms.

$$\frac{3}{8} + \left(-\frac{7}{8}\right) = \frac{3}{8} + \frac{(-7)}{8} = \frac{-4}{8} \text{ or } -\frac{4}{8} = -\frac{1}{2}$$

$$\left(-\frac{5}{2}\right) + \left(-\frac{3}{2}\right) = \frac{(-5)}{2} + \frac{(-3)}{2} = \frac{-8}{2} = -4$$

Complete. Reduce each sum to lowest terms.

   a) $\left(-\frac{11}{16}\right) + \left(-\frac{5}{16}\right) = $ _____

   b) $\left(-\frac{5}{8}\right) + \frac{3}{8} = $ _____

a) $\frac{3}{2}$

b) 1

117. When two fractions <u>with</u> <u>the</u> <u>same</u> <u>terms</u> <u>but</u> <u>different</u> <u>signs</u> are added, their sum is "0". For example:

$$\frac{3}{4} + \left(-\frac{3}{4}\right) = \frac{3}{4} + \frac{(-3)}{4} = \frac{0}{4} = 0 \qquad \frac{5}{2} + \left(-\frac{5}{2}\right) = \underline{\hspace{2cm}}$$

a) -1   b) $-\frac{1}{4}$

118. Two fractions whose sum is "0" are called "opposites". Therefore:

Since $\frac{3}{4} + \left(-\frac{3}{4}\right) = 0$:   The opposite of $\frac{3}{4}$ is $-\frac{3}{4}$.

The opposite of $-\frac{3}{4}$ is $\frac{3}{4}$.

Write the opposite of each fraction.

a) $\frac{3}{8}$ _____   b) $-\frac{7}{4}$ _____   c) $-\frac{1}{2}$ _____   d) $\frac{11}{16}$ _____

0

a) $-\frac{3}{8}$ b) $\frac{7}{4}$ c) $\frac{1}{2}$ d) $-\frac{11}{16}$

1-16   ADDING SIGNED FRACTIONS WITH UNLIKE DENOMINATORS

In this section, we will discuss the procedure for adding fractions with "unlike" denominators. Before doing so, we will review the meaning of "multiples".

119. One number is a "<u>multiple</u>" of a second only if it can be exactly divided by the second. That is:

24 <u>is</u> a multiple of 4, since $24 \div 4 = 6$

23 <u>is</u> <u>not</u> a multiple of 4, since $23 \div 4 = 5$ r3

a) Is 32 a multiple of 8? _____   b) Is 47 a multiple of 5? _____

120. The multiples of a number can be obtained by counting by that number. For example:

<u>Multiples of 3</u>:   3, 6, 9, 12, 15, 18, 21, 24, 27...

<u>Multiples of 8</u>:   8, 16, 24, 32, 40, 48, 56, 64, 72...

Which of the following are multiples of 7? _____

a) 42   b) 61   c) 77   d) 7   e) 19

a) Yes

b) No

(a), (c), and (d)

121. We can add fractions by simply adding their numerators only when their denominators are "like" or "common". To get such an addition when the larger denominator is a multiple of the smaller, we substitute an equivalent fraction for the fraction with the smaller denominator. Two examples are shown:

We substituted $\frac{6}{8}$ for $\frac{3}{4}$ to get "like" or "common" denominators.

$$\frac{3}{4} + \frac{1}{8} = \frac{6}{8} + \frac{1}{8} = \frac{7}{8}$$

We substituted $-\frac{4}{10}$ for $-\frac{2}{5}$ to get "like" or "common" denominators.

$$\frac{3}{10} + \left(-\frac{2}{5}\right) = \frac{3}{10} + \left(-\frac{4}{10}\right) = -\frac{1}{10}$$

Using the same steps, complete these:

a) $\frac{13}{16} + \frac{1}{4} = $ _____

b) $\left(-\frac{1}{2}\right) + \left(-\frac{3}{8}\right) = $ _____

---

122. Complete. Reduce each sum to lowest terms.

a) $\frac{5}{6} + \left(-\frac{1}{3}\right) = $ _____

b) $\left(-\frac{4}{5}\right) + \left(-\frac{8}{15}\right) = $ _____

Answers:
a) $\frac{17}{16}$

b) $-\frac{7}{8}$

---

123. To add a fraction and a whole number, we convert the whole number to a fraction with the same denominator. For example:

$$3 + \frac{1}{2} = \frac{6}{2} + \frac{1}{2} = \frac{7}{2}$$

$$\frac{5}{8} + (-1) = \frac{5}{8} + \left(-\frac{8}{8}\right) = -\frac{3}{8}$$

Using the same steps, complete these:

a) $1 + \frac{5}{6} = $ _____

b) $\left(-\frac{4}{3}\right) + (-2) = $ _____

Answers:
a) $\frac{1}{2}$, from $\frac{3}{6}$

b) $-\frac{4}{3}$, from $-\frac{20}{15}$

Answers:
a) $\frac{11}{6}$

b) $-\frac{10}{3}$

124. To add fractions when the larger denominator is not a multiple of the smaller, we must substitute for <u>both</u> fractions. Any multiple of both denominators can be used as the common denominator. For example:

For $\frac{1}{2} + \frac{2}{5}$, any multiple of both 2 and 5 can be used as the common denominator. For example, we can use 10 or 20.

Using 10, we get: $\frac{1}{2} + \frac{2}{5} = \frac{5}{10} + \frac{4}{10} = \frac{9}{10}$

Using 20, we get: $\frac{1}{2} + \frac{2}{5} = \frac{10}{20} + \frac{8}{20} = \frac{18}{20} = \frac{9}{10}$

Though more than one common denominator can be used in the addition above, it is simpler to use the smallest possible one. It is called the "<u>lowest</u> <u>common</u> <u>denominator</u>".

For $\frac{1}{2}$ and $\frac{2}{5}$, the <u>lowest</u> <u>common</u> <u>denominator</u> is _____.

---

125. To find the lowest common denominator, we can <u>test</u> <u>each</u> <u>multiple</u> <u>of</u> <u>the</u> <u>larger</u> <u>denominator</u> until we find the smallest one that is also a multiple of the smaller denominator. Two examples are discussed.

For $\frac{5}{6} + \frac{3}{8}$, we tested the multiples of 8 below.

Is 8 a multiple of 6?   No.
Is 16 a multiple of 6?   No.
Is 24 a multiple of 6?   Yes.

Therefore, 24 is the lowest common denominator.

For $\frac{4}{5} + \left(-\frac{1}{6}\right)$, we tested the multiples of 6 below.

Is 6 a multiple of 5?   No.
Is 12 a multiple of 5?   No.
Is 18 a multiple of 5?   No.
Is 24 a multiple of 5?   No.
Is 30 a multiple of 5?   Yes.

Therefore, 30 is the lowest common denominator.

Using the method above, find the lowest common denominator for each addition.

a) $\frac{1}{5} + \frac{1}{3}$ _____

b) $\left(-\frac{11}{12}\right) + \frac{9}{8}$ _____

c) $\frac{7}{8} + \left(-\frac{1}{3}\right)$ _____

d) $\left(-\frac{5}{6}\right) + \left(-\frac{1}{10}\right)$ _____

10

a) 15   c) 24
b) 24   d) 30

126. To perform $\frac{3}{7} + \frac{2}{5}$, we use the following steps:

  1) Find the lowest common denominator.   It is 35.

  2) Substitute $\frac{15}{35}$ for $\frac{3}{7}$ and $\frac{14}{35}$ for $\frac{2}{5}$, $\frac{3}{7} + \frac{2}{5} = \frac{15}{35} + \frac{14}{35} = \frac{29}{35}$

  and then add in the usual way.

  Using the steps above, complete these:

  a) $\frac{3}{4} + \left(-\frac{1}{10}\right) =$ _____

  b) $\left(-\frac{2}{3}\right) + \left(-\frac{1}{2}\right) =$ _____

---

127. Complete. Reduce each sum to lowest terms.

  a) $\left(-\frac{1}{4}\right) + \frac{7}{12} =$ _____

  b) $\left(-\frac{2}{7}\right) + \left(-\frac{1}{6}\right) =$ _____

a) $\frac{13}{20}$	b) $-\frac{7}{6}$

---

a) $\frac{1}{3}$	b) $-\frac{19}{42}$

1-17   SUBTRACTING SIGNED FRACTIONS

In this section, we will discuss the procedure for subtracting signed fractions.

---

128. To subtract fractions with "like" denominators, we simply subtract their numerators. The answer is always reduced to lowest terms. For example:

$$\frac{4}{5} - \frac{2}{5} = \frac{4-2}{5} = \frac{2}{5}$$

$$\frac{11}{4} - \frac{5}{4} = \frac{11-5}{4} = \frac{6}{4} = \frac{3}{2}$$

$$\frac{9}{8} - \frac{1}{8} = \frac{9-1}{8} = \frac{8}{8} = 1$$

Complete. Reduce each answer to lowest terms.

  a) $\frac{9}{10} - \frac{7}{10} =$ _____   b) $\frac{9}{2} - \frac{3}{2} =$ _____

---

129. When a fraction is subtracted from itself, the answer is "0". That is:

$$\frac{11}{4} - \frac{11}{4} = \frac{0}{4} = 0 \qquad \frac{1}{16} - \frac{1}{16} =$$ _____

a) $\frac{1}{5}$	b) 3

38   Signed Numbers

130. To subtract fractions with "unlike" denominators, we must get common denominators first. For example:

$$\frac{13}{12} - \frac{1}{4} = \frac{13}{12} - \frac{3}{12} = \frac{10}{12} = \frac{5}{6}$$

$$\frac{1}{2} - \frac{2}{5} = \frac{5}{10} - \frac{4}{10} = \frac{1}{10}$$

Complete each subtraction.

a) $\frac{7}{3} - \frac{5}{6} = $ _____

b) $\frac{11}{6} - \frac{3}{8} = $ _____

---

131. To subtract a fraction from a larger whole number, we convert the whole number to a fraction with the same denominator. That is:

$$1 - \frac{6}{7} = \frac{7}{7} - \frac{6}{7} = \frac{1}{7} \qquad 2 - \frac{5}{8} = \underline{\quad} - \frac{5}{8} = \underline{\quad}$$

---

132. Any subtraction of fractions can be converted to an equivalent addition by "adding the opposite of the second fraction". That is:

$$\frac{2}{5} - \frac{4}{5} = \frac{2}{5} + \left(\text{the opposite of } \frac{4}{5}\right) = \frac{2}{5} + \left(-\frac{4}{5}\right)$$

$$\left(-\frac{7}{9}\right) - \left(-\frac{1}{9}\right) = \left(-\frac{7}{9}\right) + \left(\text{the opposite of } -\frac{1}{9}\right) = \left(-\frac{7}{9}\right) + \frac{1}{9}$$

Convert each subtraction to an equivalent addition.

a) $\left(-\frac{5}{4}\right) - \frac{7}{4} = $ _____ + _____

b) $\frac{7}{8} - \left(-\frac{3}{8}\right) = $ _____ + _____

---

133. To perform each subtraction below, we converted to addition and then found the sum.

$$\frac{3}{8} - \frac{7}{8} = \frac{3}{8} + \left(-\frac{7}{8}\right) = -\frac{4}{8} = -\frac{1}{2}$$

$$\left(-\frac{5}{4}\right) - \left(-\frac{1}{4}\right) = \left(-\frac{5}{4}\right) + \frac{1}{4} = -\frac{4}{4} = -1$$

Using the same method, complete these:

a) $\left(-\frac{1}{5}\right) - \frac{9}{5} = $ _____

b) $\frac{11}{16} - \left(-\frac{3}{16}\right) = $ _____

Answers column:

0

a) $\frac{3}{2}$ b) $\frac{35}{24}$

$\frac{16}{8} - \frac{5}{8} = \frac{11}{8}$

a) $\left(-\frac{5}{4}\right) + \left(-\frac{7}{4}\right)$

b) $\frac{7}{8} + \frac{3}{8}$

a) -2

b) $\frac{7}{8}$

134. To perform each subtraction below, we had to get common denominators before performing the equivalent addition.

$$\frac{1}{4} - \frac{3}{2} = \frac{1}{4} + \left(-\frac{3}{2}\right) = \frac{1}{4} + \left(-\frac{6}{4}\right) = -\frac{5}{4}$$

$$\left(-\frac{1}{6}\right) - \left(-\frac{3}{8}\right) = \left(-\frac{1}{6}\right) + \frac{3}{8} = \left(-\frac{4}{24}\right) + \frac{9}{24} = \frac{5}{24}$$

Using the same method, complete these:

a) $\frac{5}{12} - \left(-\frac{1}{4}\right) = $ _____

b) $\left(-\frac{1}{3}\right) - \frac{1}{2} = $ _____

135. When the subtraction involves a whole number, we convert the whole number to a fraction before performing the equivalent addition. For example:

$$1 - \frac{7}{4} = 1 + \left(-\frac{7}{4}\right) = \frac{4}{4} + \left(-\frac{7}{4}\right) = -\frac{3}{4}$$

$$\left(-\frac{2}{3}\right) - (-2) = \left(-\frac{2}{3}\right) + 2 = \left(-\frac{2}{3}\right) + \frac{6}{3} = \frac{4}{3}$$

Using the same method, complete these:

a) $\left(-\frac{4}{5}\right) - 1 = $ _____

b) $\left(-\frac{5}{2}\right) - (-3) = $ _____

a) $\frac{2}{3}$ b) $-\frac{5}{6}$

a) $-\frac{9}{5}$ b) $\frac{1}{2}$

1-18   ORDER OF OPERATIONS WITH FRACTIONS

In this section, we will discuss some combined operations with fractions. The purpose of the section is to show the proper order of operations.

136. To evaluate the expression below, we performed the multiplication first. Complete the other evaluation.

$$3\left(\frac{5}{8}\right) - 1 \qquad\qquad 2 - 3\left(\frac{1}{4}\right)$$
$$\downarrow$$
$$\frac{15}{8} - 1$$
$$\frac{15}{8} - \frac{8}{8} = \frac{7}{8}$$

137. To evaluate the expression below, we performed the operation within the grouping first. Complete the other evaluation.

$$3\left(\frac{4}{5} + 1\right) \qquad\qquad 2\left(\frac{5}{3} - 1\right)$$

$$3\,\left(\frac{9}{5}\right) \;=\; \frac{27}{5}$$

$\dfrac{5}{4}$

---

138. To evaluate the expression below, we performed the multiplication in the numerator first. Evaluate the other expression.

$$\frac{3\left(\frac{2}{5}\right)}{7} \;=\; \frac{\frac{6}{5}}{7} \;=\; \left(\frac{6}{5}\right)\left(\frac{1}{7}\right) \;=\; \frac{6}{35}$$

$$\frac{1}{3\left(\frac{7}{8}\right)} \;=$$

$\dfrac{4}{3}$

---

139. To evaluate the expression below, we simplified the denominator first. Evaluate the other expression.

$$\frac{3}{2\left(\frac{4}{3}\right) - 1} \;=\; \frac{3}{\frac{8}{3} - 1} \;=\; \frac{3}{\frac{5}{3}} \;=\; 3\left(\frac{3}{5}\right) \;=\; \frac{9}{5}$$

$$\frac{4\left(-\frac{3}{5}\right) + 1}{5} \;=$$

$\dfrac{8}{21}$

---

140. To evaluate the expression below, we simplified the numerator first. Evaluate the other expression.

$$\frac{2\left(\frac{3}{8} + 1\right)}{5} \;=\; \frac{2\left(\frac{11}{8}\right)}{5} \;=\; \frac{\frac{11}{4}}{5} \;=\; \frac{11}{4}\left(\frac{1}{5}\right) \;=\; \frac{11}{20}$$

$$\frac{1}{3\left(\frac{5}{4} - 1\right)} \;=$$

$-\dfrac{7}{25}$

---

141. To evaluate the expression below, we performed the multiplication and the operation within the grouping first. Evaluate the other expression.

$$3\left(\frac{5}{4}\right) - \left(1 - \frac{5}{4}\right)$$

$$\frac{15}{4} \;-\; \left(-\frac{1}{4}\right) \;=\; \frac{15}{4} + \frac{1}{4} \;=\; \frac{16}{4} \;=\; 4$$

$$5\left(\frac{1}{4} + 1\right) - 3\left(\frac{1}{4}\right)$$

$\dfrac{4}{3}$

$\dfrac{11}{2}$

1-19   IMPROPER FRACTIONS AND MIXED NUMBERS

In this section, we will define "proper" and "improper" fractions and "mixed numbers".   Then we will show the relationship between improper fractions and mixed numbers.

---

142.   A fraction is a "proper" fraction if its numerator is smaller than its denominator.

$$\frac{2}{3}, \quad \frac{5}{11}, \quad \text{and} \quad \frac{1}{4} \quad \text{are "proper" fractions.}$$

A fraction is an "improper" fraction if its numerator is equal to or larger than its denominator.

$$\frac{7}{4}, \quad \frac{15}{5}, \quad \text{and} \quad \frac{8}{8} \quad \text{are "improper" fractions.}$$

Which fractions below are "improper" fractions?   _____

a) $\frac{12}{6}$ b) $\frac{4}{5}$ c) $\frac{4}{4}$ d) $\frac{1}{3}$ e) $\frac{4}{3}$

---

143.   Any addition of a whole number and a fraction can be written as a "mixed number".   For example:

$$1 + \frac{3}{4} \quad \text{is written} \quad 1\frac{3}{4} \qquad 5 + \frac{1}{3} \quad \text{is written} \quad \text{_____}$$

(a), (c), and (e)

---

144.   Any improper fraction whose numerator cannot be exactly divided by its denominator can be converted to a mixed number.   For example:

$$\frac{9}{2} = \frac{8}{2} + \frac{1}{2} \qquad\qquad \frac{11}{7} = \frac{7}{7} + \frac{4}{7}$$
$$\quad\; = 4 + \frac{1}{2} = 4\frac{1}{2} \qquad\qquad = 1 + \frac{4}{7} = 1\frac{4}{7}$$

A shorter method for the same conversions is shown below.   In the shorter method, we divide the numerator by the denominator and write the remainder as a fraction.

$$\frac{9}{2} = 2\overline{)9}^{\;4\;r1} = 4\frac{1}{2} \qquad\qquad \frac{11}{7} = 7\overline{)11}^{\;1\;r4} = 1\frac{4}{7}$$

Convert each fraction to a mixed number.

a) $\frac{8}{3} =$ _____   b) $\frac{28}{5} =$ _____   c) $\frac{79}{10} =$ _____

$5\frac{1}{3}$

---

145.   When converting a fraction to a mixed number, the fraction part is always reduced to lowest terms.   That is:

$$\frac{15}{6} = 2\frac{3}{6} = 2\frac{1}{2} \qquad\qquad \frac{30}{8} = \text{_____}$$

a) $2\frac{2}{3}$ c) $7\frac{9}{10}$

b) $5\frac{3}{5}$

146.   Any negative improper fraction is equal to a negative mixed number. That is:

$$-\frac{4}{3} = -1\frac{1}{3} \qquad -\frac{18}{4} = -4\frac{1}{2} \qquad -\frac{10}{8} = \underline{\qquad}$$

> **Note:** In this course, we will usually report answers as improper fractions rather than mixed numbers.

$3\frac{3}{4}$, from $3\frac{6}{8}$

$-1\frac{1}{4}$

SELF-TEST 4 (pages 33-42)

Add. Report each sum in lowest terms.

1. $\frac{5}{4} + \left(-\frac{3}{4}\right) =$

2. $\left(-\frac{3}{2}\right) + \left(-\frac{1}{6}\right) =$

3. $\left(-\frac{2}{5}\right) + \frac{7}{3} =$

Subtract. Report each difference in lowest terms.

4. $\frac{2}{9} - \frac{8}{9} =$

5. $\left(-\frac{3}{8}\right) - \left(-\frac{5}{12}\right) =$

6. $\frac{8}{5} - 4 =$

7. Evaluate.

$3\left(2 - \frac{7}{3}\right) =$

8. Evaluate.

$\dfrac{5\left(\frac{3}{4}\right) + 3}{6} =$

9. Convert $-\frac{20}{12}$ to a mixed number. $-\frac{20}{12} = \underline{\qquad}$

ANSWERS:

1. $\frac{1}{2}$ 3. $\frac{29}{15}$ 5. $\frac{1}{24}$ 7. -1 9. $-1\frac{2}{3}$

2. $-\frac{5}{3}$ 4. $-\frac{2}{3}$ 6. $-\frac{12}{5}$ 8. $\frac{9}{8}$

SUPPLEMENTARY PROBLEMS - CHAPTER 1

<u>Note</u>: Answers to all supplementary problems are in the back of the book.

Assignment 1

Write either ">" or "<" in each blank.

1. -1 ____ -3

2. -2.7 ____ 2.7

3. -6 ____ 0

4. $-\dfrac{3}{2}$ ____ $-\dfrac{7}{2}$

Find these "absolute values".

5. $\left|+\dfrac{3}{5}\right|$ = _____

6. $|-16|$ = _____

7. $\left|-\dfrac{8}{3}\right|$ = _____

8. $|0|$ = _____

Do these additions.

9. (-7) + (-1)

10. 9 + (-5)

11. -3.8 + 1.2

12. (-5.4) + (-2.5)

13. 28 + (-15)

14. (-32) + 23

15. 0 + (-1.29)

16. (-40) + 0

17. (-8.72) + 8.72

18. 17 + (-17)

19. 52 + (-46)

20. 9 + (-8) + 6

21. (-7) + 4 + (-11)

22. 3 + (-10) + 12 + (-6)

23. (-5) + 6 + (-2) + (-8) + 9

Do these subtractions.

24. 3 - (-5)

25. (-7) - 4

26. 6 - 13

27. (-8) - 0

28. (-18) - 12

29. (-2) - (-9)

30. (-20) - (-12)

31. 0 - 35

32. 0 - (-1.75)

33. 2.3 - 7.8

34. (-5.3) - 4.9

35. (-14.1) - (-22.7)

Assignment 2

Do these multiplications.

1. (-6)(8)

2. (2)(-7)

3. (-5)(-7)

4. (-9)(-4)

5. (-12)(0)

6. (0)(-1)

7. (-4.8)(1)

8. (-1)(-24.6)

9. (-3)(5)(-4)

10. (-8)(-1)(-7)

11. (4)(-2)(0)(-5)

12. (1)(-6)(3)(-7)

Do these divisions.

13. $\dfrac{-42}{7}$

14. $\dfrac{63}{-9}$

15. $\dfrac{-13}{-13}$

16. $\dfrac{-5}{-1}$

17. $\dfrac{0}{-2}$

18. $\dfrac{-20}{1}$

19. $\dfrac{12}{-1}$

20. $\dfrac{-200}{-4}$

21. $\dfrac{-17.2}{2}$

22. $\dfrac{-3.82}{-3.82}$

23. $\dfrac{8.4}{-1.2}$

24. $\dfrac{0}{-31.6}$

Do the following problems involving order of operations.

25. 7 - (-4) + 9

26. (-8) - 4 + 6

27. 13 - (8)(-5)

28. (-3)(6) + 2(4 - 7)

29. $\dfrac{24}{-8} - \dfrac{-6}{2}$

30. $\dfrac{8 - 7}{7 - 8}$

31. $\dfrac{3[(-2) - 6]}{4}$

32. $10 - \dfrac{18}{3(-1)}$

33. $7\left(\dfrac{3 - 9}{2}\right)$

Assignment 3

Report each product in lowest terms.

1. $\left(-\dfrac{1}{3}\right)\left(\dfrac{1}{5}\right)$
2. $\left(\dfrac{3}{7}\right)\left(-\dfrac{7}{4}\right)$
3. $\left(-\dfrac{4}{3}\right)\left(-\dfrac{5}{2}\right)$
4. $\left(\dfrac{9}{16}\right)\left(\dfrac{10}{3}\right)$

5. $5\left(-\dfrac{7}{20}\right)$
6. $(-2)\left(\dfrac{3}{8}\right)$
7. $\left(-\dfrac{9}{5}\right)(-1)$
8. $\left(-\dfrac{1}{6}\right)(-12)$

9. $\left(-\dfrac{3}{2}\right)(0)$
10. $\left(-\dfrac{5}{6}\right)\left(-\dfrac{3}{10}\right)$
11. $\left(-\dfrac{21}{16}\right)\left(\dfrac{8}{9}\right)$
12. $\left(\dfrac{36}{5}\right)\left(-\dfrac{15}{4}\right)$

Complete the following:

13. $-\dfrac{4}{9} = -\dfrac{\boxed{}}{27}$
14. $-\dfrac{7}{4} = -\dfrac{\boxed{}}{40}$
15. $5 = \dfrac{\boxed{}}{20}$
16. $-3 = -\dfrac{\boxed{}}{6}$

Reduce each fraction to lowest terms.

17. $\dfrac{18}{60}$
18. $-\dfrac{32}{24}$
19. $-\dfrac{14}{42}$
20. $-\dfrac{55}{11}$

Report each quotient in lowest terms.

21. $\dfrac{\frac{1}{8}}{-\frac{3}{4}}$
22. $\dfrac{-\frac{5}{3}}{-\frac{3}{10}}$
23. $\dfrac{-\frac{7}{16}}{\frac{1}{4}}$
24. $\dfrac{-\frac{45}{4}}{-\frac{15}{8}}$

25. $\dfrac{-\frac{8}{7}}{2}$
26. $\dfrac{-15}{-\frac{3}{5}}$
27. $\dfrac{-\frac{2}{3}}{-1}$
28. $\dfrac{-\frac{9}{5}}{\frac{9}{5}}$

Assignment 4

Report each sum in lowest terms.

1. $\dfrac{11}{8} + \left(-\dfrac{3}{8}\right)$
2. $\left(-\dfrac{5}{7}\right) + \dfrac{5}{7}$
3. $\left(-\dfrac{3}{4}\right) + \left(-\dfrac{11}{4}\right)$
4. $\left(-\dfrac{7}{3}\right) + \left(-\dfrac{2}{3}\right)$

5. $\left(-\dfrac{1}{5}\right) + \left(-\dfrac{9}{20}\right)$
6. $\dfrac{7}{2} + \left(-\dfrac{5}{6}\right)$
7. $\left(-\dfrac{9}{8}\right) + \dfrac{11}{12}$
8. $\left(-\dfrac{1}{6}\right) + \left(-\dfrac{3}{10}\right)$

9. $(-3) + \dfrac{5}{4}$
10. $\left(-\dfrac{7}{3}\right) + 5$
11. $\left(-\dfrac{9}{5}\right) + (-1)$
12. $(-4) + \left(-\dfrac{5}{8}\right)$

Report each answer in lowest terms.

13. $\dfrac{4}{9} - \left(-\dfrac{7}{9}\right)$
14. $\left(-\dfrac{5}{6}\right) - \dfrac{1}{6}$
15. $\left(-\dfrac{5}{2}\right) - \dfrac{5}{2}$
16. $\left(-\dfrac{3}{4}\right) - \left(-\dfrac{3}{4}\right)$

17. $\left(-\dfrac{9}{2}\right) - \dfrac{3}{8}$
18. $\dfrac{13}{10} - \left(-\dfrac{5}{2}\right)$
19. $\dfrac{7}{10} - \dfrac{11}{15}$
20. $\left(-\dfrac{1}{12}\right) - \dfrac{1}{15}$

21. $\dfrac{4}{7} - (-2)$
22. $(-1) - \left(-\dfrac{2}{3}\right)$
23. $\left(-\dfrac{3}{8}\right) - 5$
24. $1 - \dfrac{23}{16}$

Evaluate the following expressions.

25. $\dfrac{3\left(\frac{1}{5} - 1\right)}{8}$
26. $\dfrac{1}{5\left(-\frac{3}{2}\right) + 7}$
27. $6\left(\dfrac{1}{4}\right) - \left(3 - \dfrac{5}{3}\right)$

Convert each fraction to a mixed number.

28. $-\dfrac{16}{5}$
29. $\dfrac{18}{12}$
30. $-\dfrac{30}{8}$

Chapter 2 NON-FRACTIONAL EQUATIONS

In this chapter, we will discuss the algebraic principles and processes that are used to solve non-fractional equations. Both the addition axiom and the multiplication axiom are introduced. The meaning of formulas and formula evaluation are discussed. Some evaluations that require equation-solving are included.

2-1   AN INTRODUCTION TO EQUATIONS

In this section, we will discuss equations and the meaning of "solving an equation".

1. An equation is a mathematical sentence with an "=" in it. An equation may be either true or false. For example: a) The equation $4 + 8 = 12$ is ___T___ (true/false). b) The equation $9 - 4 = 6$ is ___F___ (true/false).	
2. The equation below is neither true nor false because we do not know what number "x" stands for. $$x + 2 = 5$$ However, as soon as we substitute a number for "x", the equation becomes either true or false. a) If we substitute -1 for "x", is the equation true? ___F no___ b) If we substitute 3 for "x", is the equation true? ___T yes___	a) true b) false
3. When substituting a number for a letter, the number that makes the equation true is said to <u>satisfy</u> the equation. What number satisfies each equation below? a) $y + 1 = 5$ ___4___ b) $3 + d = 9$ ___6___	a) No b) Yes
4. The number that satisfies an equation is called its <u>solution</u> or <u>root</u>. Find the solution or root of each equation. a) $10 + t = 19$ ___9___ b) $10 - p = 2$ ___8___	a) 4 b) 6
5. Finding the solution or root of an equation is called "<u>solving the equation</u>". Solve each equation. a) $9 - m = 2$ ___7___ b) $F - 2 = 6$ ___8___	a) 9 b) 8

6.   The solution or root of an equation can be a negative number.
      For example:

         The solution of   x + (-2) = -5 is -3.

         The solution of   (-1) + y = -9 is _____.

| a) 7 | b) 8 |

---

7.   The solution of   x + 4 = 10 is 6. Instead of saying "the solution is 6",
      we simply say x = 6. Write the solution of each equation below.

         a)   4 + Q = 9   b)   9 - y = 6

              Q = _____   y = _____

-8

---

a) Q = 5   b) y = 3

2-2   THE ADDITION AXIOM FOR EQUATIONS

To solve equations like x + 9 = 6 and 10 = y - 3 , we use "the addition axiom for equations". We will
discuss the addition axiom in this section.

---

8.   An equation contains a left side and a right side connected by an equal
      sign. That is:

         | Left Side | = | Right Side |

      In the equation below, each number and each letter is a term. We have
      drawn a box around each term.

         7 + x = 15   9 = y - 2

         a) In 7 + x = 15 , there are _____ terms on the left side.

         b) In 9 = y - 2 ,   there are _____ terms on the right side.

---

9.   The ADDITION AXIOM FOR EQUATIONS says this:

         IF WE ADD THE SAME QUANTITY TO BOTH SIDES
         OF AN EQUATION THAT IS TRUE, THE NEW
         EQUATION IS ALSO TRUE.

      When an equation contains a letter and a number on the same side, we use
      the addition axiom to isolate the letter and solve the equation. To do so,
      we "get rid of" the number on that side by adding its opposite to both
      sides. Two examples are shown.

a) two

b) two

ued on following page.

9.   Continued

To isolate "x" below, we "got rid of" the 2 by adding its opposite (-2) to both sides.

$$x + 2 = 7$$
$$x + \underline{2 + (-2)} = 7 + (-2)$$
$$x + \quad 0 \quad = 5$$
$$x = 5$$

To isolate "y" below, we "got rid of" the -3 by adding its opposite (3) to both sides.

$$-9 = (-3) + t$$
$$3 + (-9) = \underline{3 + (-3)} + t$$
$$-6 = \quad 0 \quad + t$$
$$t = -6$$

To check a solution, we substitute it in the original equation to see that it satisfies the equation. We checked the solution of x + 2 = 7 below. Check the other solution.

Check

$$x + 2 = 7$$
$$5 + 2 = 7$$
$$7 = 7$$

Check

$$-9 = (-3) + t$$

---

10.   What number would we add to both sides to solve each equation?

a) 12 + x = 4 _____

b) 20 = p + 30 _____

c) V + (-4) = -5 _____

d) 2 = (-9) + y _____

$$-9 = (-3) + t$$
$$-9 = (-3) + (-6)$$
$$-9 = -9$$

---

11.   Use the addition axiom to solve each equation.

a) y + (-1) = 8

b) 7 = 10 + d

y = _____   d = _____

a) -12   c) 4

b) -30   d) 9

---

12.   To solve the equation below, we can convert the subtraction to addition before using the addition axiom. For example:

$$x - 7 = 5$$
$$x + (-7) = 5$$
$$x + \underline{(-7) + 7} = 5 + 7$$
$$x + \quad 0 \quad = 12$$
$$x = 12$$

a) y = 9   b) d = -3

Continued on following page.

12.   Continued

However, <u>it</u> <u>is</u> <u>simpler</u> <u>to</u> <u>solve</u> <u>it</u> <u>without</u> <u>converting</u> <u>the</u> <u>subtraction</u> <u>to</u>
<u>addition.</u> To do so, we treat the subtraction sign as if it were the sign
of the number. That is:

$$x - 7 = 5$$
$$x - 7 + 7 = 5 + 7$$
$$x + 0 = 12$$
$$x = 12$$

Using the second method, solve these:

a) $y - 2 = -3$ b) $20 = Q - 15$

y = _____   Q = _____

---

13.   To solve the equation below, we added (-5) to both sides. Solve the
other equation.

$$m + 5 = 0 \qquad\qquad (-7) + F = 0$$
$$m + 5 + (-5) = 0 + (-5)$$
$$m + 0 = -5$$
$$m = -5 \qquad\qquad\qquad F = \text{\_\_\_\_\_}$$

a) y = -1   b) Q = 35

---

14.   To solve the equation below, we added 2 to both sides. Solve the
other equation.

$$x - 2 = 0 \qquad\qquad 0 = y - 10$$
$$x - 2 + 2 = 0 + 2$$
$$x + 0 = 2$$
$$x = 2 \qquad\qquad\qquad y = \text{\_\_\_\_\_}$$

F = 7

y = 10

2-3   THE MULTIPLICATION AXIOM FOR EQUATIONS

To solve equations like $3x = 15$ and $-9 = 7y$, we use "<u>the</u> <u>multiplication</u> <u>axiom</u> <u>for</u> <u>equations</u>". We will
discuss the multiplication axiom in this section.

15.   To signify a multiplication of a number and a letter, we could write parentheses around one or both factors.   For example:

"5 times x" could be written: 5(x),   (5)x,   or   (5)(x)

"-3 times y" could be written: -3(y),   (-3)y,   or   (-3)(y)

However, we ordinarily write the number and letter side by side with no parentheses.   That is:

"5 times x" is ordinarily written: 5x

"-3 times y" is ordinarily written: -3y

Therefore: When x = 4,   5x = 5(4) = 20

When y = 2,   -3y = -3(2) = _____

---

16.   Expressions like "9t" or "-4d" are called <u>letter-terms</u>.   In such terms, the number is called the <u>coefficient</u> of the letter.   That is:

In 9t,   9 is called the <u>coefficient</u> of "t".

In -4d, _____ is called the <u>coefficient</u> of "d".

-6

---

17.   In the equation below, the letter-term stands for a multiplication.

$$4m = 28 \qquad\qquad 2x = -6$$

Therefore, solving the equations above is the same as "<u>finding the missing factor</u>".

a) Is +7 or -7 the solution of 4m = 28?   _____

b) Is +3 or -3 the solution of 2x = -6?   _____

-4

---

18.   The <u>MULTIPLICATION AXIOM FOR EQUATIONS</u> says this:

> IF WE MULTIPLY BOTH SIDES OF AN EQUATION THAT IS TRUE BY THE SAME QUANTITY, THE NEW EQUATION IS ALSO TRUE.

The multiplication axiom was used to solve the equations below.   Notice that we multiplied both sides <u>by the reciprocal of the coefficient of the letter</u>.

$$5x = 20 \qquad\qquad\qquad -3y = 15$$

$$\frac{1}{5}(5x) = \frac{1}{5}(20) \qquad\qquad -\frac{1}{3}(-3y) = -\frac{1}{3}(15)$$

$$1x = 4 \qquad\qquad\qquad 1\ \ y = -5$$

$$x = 4 \qquad\qquad\qquad y = -5$$

a) +7, since 4(7) = 28

b) -3, since 2(-3) = -6

Continued on following page.

18. Continued

You can see that using the multiplication axiom is the same as <u>dividing</u> <u>the</u> <u>number-term</u> <u>by</u> <u>the</u> <u>coefficient</u> <u>of</u> <u>the</u> <u>letter.</u> Therefore, we can use a shortcut to solve the equations on the last page.

$$\text{For } 5x = 20, \quad x = \frac{20}{5} = 4$$

$$\text{For } -3y = 15, \quad y = \frac{15}{-3} = -5$$

Use the shortcut to solve the equations below.

a) $7m = 56$ b) $-2y = 20$ c) $-10F = -40$

    m = _____   y = _____   F = _____

19. When the letter-term is on the right side, <u>be</u> <u>sure</u> <u>to</u> <u>divide</u> <u>the</u> <u>number-</u> <u>term</u> <u>by</u> <u>the</u> <u>coefficient</u> <u>of</u> <u>the</u> <u>letter.</u>   That is:

$$\text{For } -16 = 2t, \quad t = \frac{-16}{2} = -8$$

$$\text{For } -30 = -5V, \quad V = \frac{-30}{-5} = 6$$

Use the shortcut to solve these:

a) $4 = 4x$ b) $-30 = 3t$ c) $-48 = -6m$

    x = _____   t = _____   m = _____

a) m = 8

b) y = –10

c) F = 4

20. When the number-term is "0", the solution is "0". That is:

$$\text{For } 5x = 0, \quad x = \frac{0}{5} = 0$$

$$\text{For } -9y = 0, \quad y = \frac{0}{-9} = 0$$

Solve these:

a) $-7 = 7d$ b) $0 = -4R$ c) $-2p = -2$

    d = _____   R = _____   p = _____

a) x = 1

b) t = –10

c) m = 8

21. Two solutions are shown below. We checked the first solution. Check the other solution.

$$\text{For } -3x = -18, \quad x = \frac{-18}{-3} = 6 \qquad \text{For } 0 = -7y, \quad y = \frac{0}{-7} = 0$$

    <u>Check</u>   <u>Check</u>

    $-3x = -18$ $0 = -7y$

    $-3(6) = -18$

    $-18 = -18$

a) d = –1

b) R = 0

c) p = 1

22. The solution of an equation can be a fraction. Fractional solutions are always reduced to lowest terms. For example:

$$\text{For } 5x = 4, \quad x = \frac{4}{5}$$

$$\text{For } 15 = 9m, \quad m = \frac{15}{9} = \frac{5}{3}$$

Solve. Reduce each root to lowest terms.

a) 11R = 13   b) 2 = 6y   c) 30p = 45

R = _____   y = _____   p = _____

$0 = -7y$
$0 = -7(0)$
$0 = 0$

---

23. When a fractional solution is negative, we write the negative sign in front of the fraction. That is:

$$\text{For } 7y = -3, \quad y = \frac{-3}{7} = -\frac{3}{7}$$

$$\text{For } 9 = -2d, \quad d = \frac{9}{-2} = -\frac{9}{2}$$

Solve. Reduce each root to lowest terms.

a) -6x = 1   b) -2 = 4R   c) -8h = 10

x = _____   R = _____   h = _____

a) $R = \dfrac{13}{11}$

b) $y = \dfrac{1}{3}$

c) $p = \dfrac{3}{2}$

---

24. When both terms in the division are negative, the solution is positive. That is:

$$\text{For } -3y = -1, \quad y = \frac{-1}{-3} = \frac{1}{3}$$

$$\text{For } -7 = -5m, \quad m = \frac{-7}{-5} = \frac{7}{5}$$

Solve. Reduce each root to lowest terms.

a) -9x = -2   b) -1 = -5p   c) -12d = -15

x = _____   p = _____   d = _____

a) $x = -\dfrac{1}{6}$

b) $R = -\dfrac{1}{2}$

c) $h = -\dfrac{5}{4}$

---

a) $x = \dfrac{2}{9}$

b) $p = \dfrac{1}{5}$

c) $d = \dfrac{5}{4}$

25.   Two solutions are shown below. We checked the first solution. Check the other solution.

For $8x = 2$, $x = \dfrac{2}{8} = \dfrac{1}{4}$ For $-8 = 6y$, $y = \dfrac{-8}{6} = -\dfrac{4}{3}$

<u>Check</u>   <u>Check</u>

$8x = 2$ $-8 = 6y$

$8\left(\dfrac{1}{4}\right) = 2$

$2 = 2$

$-8 = 6y$

$-8 = 6\left(-\dfrac{4}{3}\right)$

$-8 = -8$

2-4   USING BOTH AXIOMS TO SOLVE EQUATIONS

To solve equations like $2x + 3 = 11$ or $3y - 5 = 20$, both the addition axiom and the multiplication axiom are used. We will discuss solutions of that type in this section.

---

26.   Both axioms are used to solve an equation like $3x + 5 = 26$. That is:

        The <u>addition axiom</u> is used to isolate the $3x$ by "getting rid of" the 5 on the left side.

        The <u>multiplication axiom</u> is then used to solve the resulting equation.

The equation is solved and checked below.

$$3x + 5 = 26$$
$$3x + \underline{5 + (-5)} = 26 + (-5)$$
$$3x + 0 = 21$$
$$3x = 21$$
$$x = \dfrac{21}{3} = 7$$

<u>Check</u>

$$3x + 5 = 26$$
$$3(7) + 5 = 26$$
$$21 + 5 = 26$$
$$26 = 26$$

Using the same steps, solve each equation below.

  a) $7y + 24 = 10$ b) $87 = 51 + 9F$

        $y =$ _____ $F =$ _____

---

a) $y = -2$ b) $F = 4$

27. Following the example, solve the other equation.

$$7 = 6y + (-17)$$ $$7m + (-1) = -15$$

$$17 + 7 = 6y + \underline{(-17) + 17}$$
$$\downarrow$$
$$24 = 6y + \quad 0$$

$$24 = 6y$$

$$y = \frac{24}{6} = 4$$ m = _____

---

28. Following the example, solve the other equation.   m = -2

$$4d - 13 = 15$$ $$83 = 10t - 7$$

$$4d \underline{- 13 + 13} = 15 + 13$$
$$\downarrow$$
$$4d + \quad 0 \quad = 28$$

$$4d = 28$$

$$d = \frac{28}{4} = 7$$ t = _____

---

29. Following the example, solve the other equation.   t = 9

$$20 - 5y = 65$$ $$79 = 23 - 8p$$

$$\underline{(-20) + 20} - 5y = 65 + (-20)$$
$$\downarrow$$
$$0 \quad - 5y = 45$$

$$-5y = 45$$

$$y = \frac{45}{-5} = -9$$ p = _____

---

30. Following the example, solve the other equation.   p = -7

$$3x + 18 = 0$$ $$28 + 7y = 0$$

$$3x + \underline{18 + (-18)} = 0 + (-18)$$
$$\downarrow$$
$$3x + \quad 0 \quad = -18$$

$$3x = -18$$

$$x = \frac{-18}{3} = -6$$ y = _____

---

31. Following the example, solve the other equation.   y = -4

$$6p - 30 = 0$$ $$12 - 4d = 0$$

$$6p \underline{- 30 + 30} = 0 + 30$$
$$\downarrow$$
$$6p + \quad 0 \quad = 30$$

$$6p = 30$$

$$p = \frac{30}{6} = 5$$ d = _____

54   Non-Fractional Equations

32.   Solve. Reduce the solution to lowest terms.

   a)   7t + 9 = 8   b)   9 - 6n = 5

   t = _____   n = _____

d = 3

a) t = $-\frac{1}{7}$ b) n = $\frac{2}{3}$

SELF-TEST 5 (pages 45-54)

Solve each equation. Report fractional solutions in lowest terms.

1. 25 = x + 9	2. p + 2 = 0	3. w - 5 = -8
4. 9P = 54	5. -7 = -28t	6. -6y = 15
7. 10s + 19 = 15	8. 8 = 8 + 3x	9. 9r - 21 = 0

ANSWERS: 1. x = 16   4. P = 6   7. s = $-\frac{2}{5}$

2. p = -2   5. t = $\frac{1}{4}$ 8. x = 0

3. w = -3   6. y = $-\frac{5}{2}$ 9. r = $\frac{7}{3}$

2-5   ADDING LETTER TERMS

In this section, we will show how the "distributive principle of multiplication over addition" can be used to perform additions like   3x + 5x   or   7y + (-2y).   Then we will solve equations that contain additions of that type.

---

33.   One of the factors in a multiplication can be an addition.   In such cases, we can use either of two methods to find the product.

We can use the "adding before multiplying" method.

$$(5 + 2)6 \qquad\qquad [4 + (-6)]5$$
$$\downarrow \qquad\qquad\qquad \downarrow$$
$$(7)\ 6 = 42 \qquad\qquad (-2)\ 5 = -10$$

We can use the "multiplying before adding" method.

$$(5 + 2)6 = 5(6) + 2(6) \qquad [4 + (-6)]5 = 4(5) + (-6)(5)$$
$$= 30 + 12 = 42 \qquad\qquad = \underline{\ \ } + \underline{\ \ } = \underline{\ \ }$$

---

34.   In the "multiplying before adding" method, the first step involves the DISTRIBUTIVE PRINCIPLE OF MULTIPLICATION OVER ADDITION. Two examples are diagrammed below.

The arrows show that 7 is multiplied by both 8 and 5.	The arrows show that 9 is multiplied by both 4 and -7.

$$(8 + 5)7 = 8(7) + 5(7) \qquad [4 + (-7)]9 = 4(9) + (-7)(9)$$

The process above is called "multiplying by the distributive principle". We can also multiply by the distributive principle when a letter is involved.   For example:

$$(2 + 3)x = 2(x) + 3(x) = 2x + 3x$$
$$[5 + (-4)]y = 5(y) + (-4)(y) = 5y + (-4y)$$
$$[(-6) + (-8)]t = (-6)(t) + (-8)(t) = \underline{\ \ \ \ } + \underline{\ \ \ \ }$$

20 + (-30) = -10

---

35.   Two multiplications by the distributive principle are shown below.

$$(5 + 4)x = 5x + 4x$$
$$(2 + 8)p = 2p + 8p$$

By simply reversing the process, we can break up each product into the original factors.   Doing so is called "factoring by the distributive principle".   That is:

5x + 4x   can be factored to get   (5 + 4)x

2p + 8p   can be factored to get   _____

(-6t) + (-8t)

(2 + 8)p

36.   "Factoring by the distributive principle" can be used to add letter terms. For example:

$$3x + 9x = \underset{\downarrow}{(3 + 9)}x \qquad 7y + (-5y) = \underset{\downarrow}{[7 + (-5)]}y$$

$$= 12 \; x \qquad\qquad\qquad = 2 \quad y$$

Using the same process, complete these additions:

a) $8t + 7t = (8 + 7)t$

    $=$ _____

b) $(-5R) + (-4R) = [(-5) + (-4)]R$

    $=$ _____

---

37.   The distributive-principle method is the same as adding the two coefficients. That is:

$$9y + 4y = 13y, \quad \text{since} \quad 9 + 4 = 13$$

$$(-8t) + 3t = -5t, \quad \text{since} \quad (-8) + 3 = -5$$

$$(-5R) + (-4R) = -9R, \quad \text{since} \quad (-5) + (-4) = -9$$

By simply adding the coefficients, complete these additions.

a) $11m + 3m =$ _____

b) $(-2V) + 7V =$ _____

c) $6t + (-9t) =$ _____

d) $(-5x) + (-5x) =$ _____

a) 15t   b) −9R

---

38.   To solve the equation below, we added the two letter terms and then used the multiplication axiom.

Solution	Check
$5y + 2y = 21$	$5y + 2y = 21$
$7y = 21$	$5(3) + 2(3) = 21$
$y = \dfrac{21}{7} = 3$	$15 + 6 = 21$
	$21 = 21$

Using the same steps, solve each equation.

a) $9x + (-4x) = 25$

b) $45 = (-6y) + (-3y)$

a) 14m   c) −3t

b) 5V   d) −10x

---

39.   Solve. Reduce each solution to lowest terms.

a) $(-4t) + 8t = 1$

b) $24 = (-10m) + (-6m)$

a) x = 5   b) y = −5

40.   The solution of the equation below is "0". Solve the other equation.

$$3x + 4x = 0 \qquad\qquad (-7d) + 3d = 0$$

$$7x = 0$$

$$x = \frac{0}{7} = 0$$

a) $t = \dfrac{1}{4}$

b) $m = -\dfrac{3}{2}$

---

41.   Since "0x" means "0 times x", "0x" = 0. Therefore, the sum below is "0".

$$5x + (-5x) = 0x = 0$$

Two quantities are called <u>opposites</u> if their sum is "0". Therefore:

The opposite of 5x is –5x.

The opposite of –5x is 5x.

It should be obvious that two letter terms containing the same letter are opposites <u>if</u> <u>their</u> <u>coefficients</u> <u>are</u> <u>opposites</u>. That is:

3y and (–3y) are opposites, since 3 + (–3) = 0

Write the opposite of each letter term.

a) 2x ____   b) –4y ____   c) 12F ____   d) –20V ____

d = 0

a) –2x   b) 4y   c) –12F   d) 20V

2-6   SUBTRACTING LETTER TERMS

In this section, we will show how the "<u>distributive</u> <u>principle</u> <u>of</u> <u>multiplication</u> <u>over</u> <u>subtraction</u>" can be used to perform subtractions like 7x – 5x and 2d – 9d. Then we will solve equations that contain subtractions of that type.

---

42.   One of the factors in a multiplication can be a subtraction. In such cases, we can use either of two methods to find the product.

We can use the "<u>subtracting</u> <u>before</u> <u>multiplying</u>" method.

$$(7 - 5)8 \qquad\qquad (4 - 9)7$$

$$(2)\ \ 8 = 16 \qquad\qquad (-5)\ \ 7 = -35$$

We can use the "<u>multiplying</u> <u>before</u> <u>subtracting</u>" method.

$$(7 - 5)8 = 7(8) - 5(8) \qquad\qquad (4 - 9)7 = 4(7) - 9(7)$$

$$= 56 - 40 = 16 \qquad\qquad = \underline{\ \ \ } - \underline{\ \ \ } = \underline{\ \ \ }$$

$28 - 63 = -35$

43. In the "multiplying before subtracting" method, the first step involves the <u>DISTRIBUTIVE PRINCIPLE OF MULTIPLICATION OVER SUBTRAC-TION</u>. Two examples are diagrammed below.

The arrows show that 3 is multiplied by both 4 and 2.

The arrows show that 5 is multiplied by both 6 and 9.

$$(4 - 2)3 = 4(3) - 2(3) \qquad (6 - 9)5 = 6(5) - 9(5)$$

The process above is called "<u>multiplying by the distributive principle</u>". We can also multiply by the distributive principle when a letter is involved. That is:

$$(8 - 3)x = 8(x) - 3(x) = 8x - 3x$$

$$(2 - 7)y = 2(y) - 7(y) = \underline{} - \underline{}$$

---

44. Two multiplications by the distributive principle are shown below.

$$(8 - 3)x = 8x - 3x$$

$$(2 - 7)y = 2y - 7y$$

By simply reversing the process, we can break up each product into the original factors. Doing so is called "<u>factoring by the distributive principle</u>". That is:

$8x - 3x$ can be factored to get $(8 - 3)x$

$2y - 7y$ can be factored to get $\underline{}$

> 2y – 7y

---

45. "<u>Factoring by the distributive principle</u>" can be used to subtract letter terms. For example:

$$6d - 4d = (6 - 4)d \qquad 3m - 8m = (3 - 8)m$$
$$= 2d \qquad\qquad = -5m$$

By simply subtracting the coefficients, complete these subtractions:

a) $10R - 2R = \underline{}$ c) $(-2x) - 5x = \underline{}$

b) $5v - 15v = \underline{}$ d) $9y - 9y = \underline{}$

> (2 – 7)y

---

46. To solve the equation below, we subtracted the two letter terms and then used the multiplication axiom.

$$2x - 5x = 12$$
$$-3x = 12$$
$$x = \frac{12}{-3} = -4$$

Using the same steps, solve these:

a) $8y - 3y = 20$ b) $40 = 3t - 7t$

> a) 8R   c) –7x
>
> b) –10v   d) 0,
>   from 0y

47.   Solve. Reduce each solution to lowest terms.

    a) $10x - 7x = 1$ b) $-12 = 10V - 20V$

a) $y = 4$ b) $t = -10$

48.   The solution of the equation below is "0". Solve the other equation.

    $8t - 5t = 0$ $2p - 8p = 0$

    $3t = 0$

    $t = \dfrac{0}{3} = 0$

a) $x = \dfrac{1}{3}$ b) $V = \dfrac{6}{5}$

$p = 0$

2-7   SUBSTITUTING "1x" FOR "x" AND "-1x" FOR "-x"

In this section, we will show that "x" equals "1x" and "-x" equals "-1x". Then we will use those facts as a help in solving equations.

49.   Any letter without a coefficient really has a coefficient of "1". That is:

    $x = 1x$ $y = 1y$

When a letter without a coefficient appears in an addition, it is helpful to write the "1" explicitly. For example:

    $7x + x = 7x + 1x = 8x$

    $y + (-4y) = 1y + (-4y) = -3y$

    $t + t = 1t + 1t = 2t$

Write the coefficient "1" explicitly and then complete these:

    a) $x + 3x =$ _____   b) $(-9t) + t =$ _____   c) $m + m =$ _____

50.   To solve the equation below, we began by writing the coefficient "1" explicitly. Solve the other equation.

    $m + 6m = 28$ $1 = (-5d) + d$

    $1m + 6m = 28$

    $7m = 28$

    $m = \dfrac{28}{7} = 4$

a) $4x$

b) $-8t$

c) $2m$

$d = -\dfrac{1}{4}$

51. When a letter without a coefficient appears in a subtraction, it is also helpful to write the "1" explicitly. For example:

$$x - 7x = 1x - 7x = -6x$$

$$3y - y = 3y - 1y = 2y$$

$$t - t = 1t - 1t = 0t = 0$$

Write the coefficient "1" explicitly and then complete these:

   a) 10d - d = _____   b) V - 4V = _____   c) p - p = _____

---

52. To solve the equation below, we began by writing the coefficient "1" explicitly. Solve the other equation.

$$9x - x = 40 \qquad\qquad 10 = d - 4d$$

$$9x - 1x = 40$$

$$8x = 40$$

$$x = \frac{40}{8} = 5$$

a) 9d

b) −3V

c) 0

---

53. Following the example, solve the other equation.

$$6x - 5x = 7 \qquad\qquad 2v - v = 0$$

$$1x = 7$$

$$x = \frac{7}{1} = 7$$

$$d = -\frac{10}{3}$$

---

54. Any letter with a "−" in front of it really has a coefficient of "−1". That is:

$$-x = -1x \qquad\qquad -y = -1y$$

When a letter with only a "−" in front of it appears in an addition, it is helpful to write the "−1" explicitly. For example:

$$5x + (-x) = 5x + (-1x) = 4x$$

$$(-t) + (-8t) = (-1t) + (-8t) = -9t$$

$$(-y) + (-y) = (-1y) + (-1y) = -2y$$

Write the coefficient "−1" explicitly and then complete these:

   a) (−p) + 10p = _____   b) (−3d) + (−d) = _____   c) (−F) + (−F) = _____

v = 0, from 1v = 0

---

55. To solve the equation below, we began by writing the coefficient "−1" explicitly. Solve the other equation.

$$7x + (-x) = 3 \qquad\qquad -36 = (-q) + (-3q)$$

$$7x + (-1x) = 3$$

$$6x = 3$$

$$x = \frac{3}{6} = \frac{1}{2}$$

a) 9p

b) −4d

c) −2F

56. To solve each equation below, we began by writing the coefficient "–1" explicitly.

$$-y = 10 \qquad\qquad -4 = -x$$
$$-1y = 10 \qquad\qquad -4 = -1x$$
$$y = \frac{10}{-1} = -10 \qquad\qquad x = \frac{-4}{-1} = 4$$

Using the same method, solve these:

a) $-t = 7$ b) $-p = -9$ c) $10 = -V$

$q = 9$

---

57. Following the example, solve the other equation.

$$y - 2y = 0 \qquad\qquad 4x - 5x = 8$$
$$1y - 2y = 0$$
$$-1y = 0$$
$$y = \frac{0}{-1} = 0$$

a) $t = -7$

b) $p = 9$

c) $V = -10$

---

58. To solve the equation below, we wrote the coefficient "1" explicitly before using the addition axiom. Use the same steps to solve the other equation.

$$7 - x = 3 \qquad\qquad 15 = 9 - y$$
$$7 - 1x = 3$$
$$\underline{(-7) + 7} - 1x = 3 + (-7)$$
$$\downarrow$$
$$0 \quad - 1x = -4$$
$$-1x = -4$$
$$x = \frac{-4}{-1} = 4$$

$x = -8$

---

59. We added "x" and "–x" below. Since their sum is "0", "x" and "–x" are <u>opposites</u>.

$$x + (-x) = 1x + (-1x) = 0x = 0$$

Therefore: The opposite of "x" is "–x".

The opposite of "–x" is "x".

Write the opposite of each of these:

a) $1y$ _____   b) $-1t$ _____   c) V _____   d) $-p$ _____

$y = -6$ (from $6 = -1y$)

---

a) $-1y$ b) $1t$ c) $-V$ d) p

2-8   EQUATIONS WITH A LETTER TERM ON BOTH SIDES

To solve equations like $7x = 2x + 20$ or $40 - 3m = 7m$, we also use both axioms. We will solve equations of that type in this section.

60.   The equation $7x = 2x + 20$ has a letter term on both sides. To solve it, we use both axioms.

   The <u>addition axiom</u> is used to get both letter terms on one side so that they can be combined.

   The <u>multiplication axiom</u> is then used to solve the resulting equation.

The steps used to solve the equation are described below.

   1) To get both letter terms on the same side so that they can be combined, we add -2x to both sides.

   2) Then we use the multiplication axiom to solve $5x = 20$.

$$7x = 2x + 20$$
$$(-2x) + 7x = \underline{(-2x) + 2x} + 20$$
$$\downarrow$$
$$5x = \quad 0 \quad + 20$$
$$5x = 20$$
$$x = \frac{20}{5} = 4$$

Complete the checking of "4" as the solution of the original equation at the right.

$$7x = 2x + 20$$
$$7(4) = 2(4) + 20$$
$$\underline{\quad} = \underline{\quad} + 20$$
$$\underline{\quad} = \underline{\quad}$$

---

61.   When an equation has a letter term on both sides, we use the addition axiom to "get rid of" the letter term <u>on the same side as the number</u>. To do so, we add its <u>opposite</u> to both sides.

   In $5y + 36 = 9y$, we "get rid of" the 5y by adding -5y to both sides.

   In $2t = 45 + (-7t)$, we "get rid of" the (-7t) by adding 7t to both sides.

Solve each of the equations above.

   a) $5y + 36 = 9y$ b) $2t = 45 + (-7t)$

$28 = 8 + 20$
$28 = 28$

62. Following the example, solve the other equation.

$$9R - 7 = 4R \qquad\qquad 7d = 8d - 60$$

$$\underline{(-9R) + 9R} - 7 = 4R + (-9R)$$
$$\downarrow$$
$$0 \quad - 7 = -5R$$
$$-7 = -5R$$
$$R = \frac{-7}{-5} = \frac{7}{5}$$

a) y = 9

b) t = 5

---

63. Following the example, solve the other equation.

$$7x = 40 - 3x \qquad\qquad 11 - 4y = 4y$$

$$3x + 7x = 40 \underline{- 3x + 3x}$$
$$\downarrow$$
$$10x = 40 + \quad 0$$
$$10x = 40$$
$$x = \frac{40}{10} = 4$$

d = 60

---

64. To solve the equation below, we began by writing the coefficient "1" of the "y" explicitly. Solve the other equation.

$$y = 14 + 3y \qquad\qquad 7t = 10 + t$$

$$1y = 14 + 3y$$
$$(-3y) + 1y = 14 + \underline{3y + (-3y)}$$
$$\downarrow$$
$$-2y = 14 + \quad 0$$
$$-2y = 14$$
$$y = \frac{14}{-2} = -7$$

$y = \dfrac{11}{8}$

---

65. To solve the equation below, we began by writing the coefficient "1" of the "x" explicitly. Solve the other equation.

$$4 - x = 4x \qquad\qquad y = 12 - y$$

$$4 - 1x = 4x$$
$$4 \underline{- 1x + 1x} = 4x + 1x$$
$$\downarrow$$
$$4 + \quad 0 \quad = 5x$$
$$4 = 5x$$
$$x = \frac{4}{5}$$

$t = \dfrac{5}{3}$

---

y = 6 (from 2y = 12)

2-9   EQUATIONS WITH A LETTER TERM AND A NUMBER TERM ON BOTH SIDES

To solve equations like $7x + 9 = 4x + 21$, <u>the</u> <u>addition</u> <u>axiom</u> <u>is</u> <u>used</u> <u>twice</u>. We will discuss the method in this section.

66.   The equation $7x + 9 = 4x + 21$ has a letter term and a number term on both sides. To solve it, we must get both letter terms on one side and both number terms on the other side so that they can be combined. To do so, <u>we must "get rid of" a letter term from one side and a number term from the other side</u>. We have a choice of two pairs:

            1) getting rid of 4x and 9

    or   2) getting rid of 7x and 21

With either choice, <u>we must use the addition axiom twice</u>. We solved below by getting rid of 4x and 9.

    To get rid of 4x, we add -4x to both sides.

$$7x + 9 = 4x + 21$$
$$(-4x) + 7x + 9 = \underline{(-4x) + 4x} + 21$$
$$3x + 9 = \quad 0 \quad + 21$$
$$3x + 9 = 21$$

    To get rid of 9, we now add -9 to both sides.

$$3x + \underline{9 + (-9)} = 21 + (-9)$$
$$3x + \quad 0 \quad = 12$$
$$3x = 12$$
$$x = \frac{12}{3} = 4$$

Complete the check of the above solution at the right.

$$7x + 9 = 4x + 21$$
$$7(4) + 9 = 4(4) + 21$$

---

67.   To solve $5y + 16 = 7y + 10$, we also have to use the addition axiom twice. We have a choice of getting rid of two pairs.

    1) 5y and 10   or   2) 7y and 16

The steps in both solutions are shown at the top of the next page. Notice that we get "3" as the solution with each method.

$$28 + 9 = 16 + 21$$
$$37 = 37$$

Continued on following page.

67.   Continued

1) <u>Getting rid of 5y and 10.</u>

$$5y + 16 = 7y + 10$$

$$(-5y) + 5y + 16 = (-5y) + 7y + 10$$

$$16 = 2y + 10$$

$$(-10) + 16 = 2y + 10 + (-10)$$

$$6 = 2y$$

$$y = 3$$

2) <u>Getting rid of 7y and 16.</u>

$$5y + 16 = 7y + 10$$

$$(-7y) + 5y + 16 = (-7y) + 7y + 10$$

$$-2y + 16 = 10$$

$$-2y + 16 + (-16) = 10 + (-16)$$

$$-2y = -6$$

$$y = 3$$

By getting rid of either possible pair, solve these:

a) $3x + 8 = 9x + 2$ b) $8y + 1 = 3y + 7$

---

68.   Following the example, solve the other equation.

$$9a - 4 = 3a + 8$$ $$5t + 1 = 4t - 9$$

$$(-3a) + 9a - 4 = (-3a) + 3a + 8$$

$$6a - 4 = 8$$

$$6a - 4 + 4 = 8 + 4$$

$$6a = 12$$

$$a = 2$$

a) $x = 1$ b) $y = \dfrac{6}{5}$

---

69.   Following the example, solve the other equation.

$$30 - 5d = 20 - 9d$$ $$15 - 2h = 8 - h$$

$$30 - 5d + 5d = 20 - 9d + 5d$$

$$30 = 20 - 4d$$

$$(-20) + 30 = (-20) + 20 - 4d$$

$$10 = -4d$$

$$d = -\dfrac{10}{4} = -\dfrac{5}{2}$$

t = −10

h = 7

SELF-TEST 6 (pages 55-66)

Solve each equation. Report each solution in lowest terms.

1.   (-2d) + 9d = 0

2. 2 = r - 5r

3. 6 - x = 7

4. 7t + 8 = 3t

5. w = 10 - 5w

6. y + 1 = 2y + 7

7. 11 - 2P = P + 11

8. 6h - 13 = 22 - 9h

ANSWERS: 1. d = 0   3. x = -1   5. w = $\frac{5}{3}$ 7. P = 0

2. r = -$\frac{1}{2}$ 4. t = -2   6. y = -6   8. h = $\frac{7}{3}$

2-10   THE DISTRIBUTIVE PRINCIPLE OVER ADDITION AND EQUATIONS

In this section, we will discuss "multiplying by the distributive principle of multiplication over addition". Then we will use that process to solve equations like 2(x + 7) = 22 and 5(2y + 3) = 15y.

---

70.   We can also multiply by the distributive principle when the second factor is an addition. An example is shown. Notice that both the "7" and the "4" are multiplied by "5".

$$5(7 + 4) = 5(7) + 5(4)$$

When one of the factors in the addition is a letter, the distributive principle works the same way. For example:

$$3(x + 5) = 3(x) + 3(5) \qquad 8(2 + y) = 8(2) + 8(y)$$

Following the examples, fill in the blanks below.

a) 9(t + 6) = (   )(t) + (   )(6)   b) 4(1 + m) = (   )(1) + (   )(m)

71.   After multiplying by the distributive principle, we simplify the terms in the product.   For example:

$$6(x + 4) = \underline{6(x)} + \underline{6(4)}$$
$$= 6x + 24$$

$$4(7 + y) = \underline{4(7)} + \underline{4(y)}$$
$$= 28 + 4y$$

Simplify the terms in each product below.

a) $8(t + 5) = 8(t) + 8(5) = $ _____ + _____

b) $3(9 + d) = 3(9) + 3(d) = $ _____ + _____

a) $\underline{(9)}(t) + \underline{(9)}(6)$

b) $\underline{(4)}(1) + \underline{(4)}(m)$

72.   In each multiplication below, the letter has a coefficient.   We multiply by the distributive principle in the usual way.

$$6(3x + 7) = \underline{6(3x)} + \underline{6(7)}$$
$$= 18x + 42$$

$$2(5 + 8y) = \underline{2(5)} + \underline{2(8y)}$$
$$= 10 + 16y$$

Note: $6(3x)$ and $2(8y)$ are three-factor multiplications.   We simplify those terms by multiplying the two numerical factors.   That is:

$$6(3x) = \underline{6(3)}(x)$$
$$= 18 \; x$$

$$2(8y) = \underline{2(8)}(y)$$
$$= 16 \; y$$

Simplify the terms in each product below.

a) $3(5x + 7) = 3(5x) + 3(7) = $ _____ + _____

b) $7(2 + 9y) = 7(2) + 7(9y) = $ _____ + _____

a) $8t + 40$

b) $27 + 3d$

73.   We ordinarily multiply and simplify the terms of the product in one step.   For example:

$$8(t + 4) = 8t + 32$$
$$9(5 + d) = 45 + 9d$$

$$3(2x + 7) = 6x + 21$$
$$5(6 + 8y) = 30 + 40y$$

Multiply and simplify the terms of the product in one step.

a) $5(x + 1) = $ _____ + _____

b) $6(10 + y) = $ _____ + _____

c) $4(8d + 3) = $ _____ + _____

d) $9(4 + 6p) = $ _____ + _____

a) $15x + 21$

b) $14 + 63y$

74.   There is a <u>common error</u> when multiplying by the distributive principle. The <u>error</u> is <u>forgetting</u> to <u>multiply the second</u> term <u>of</u> the <u>addition by the first factor</u>.   Some examples are shown.

Error	Correct
$3(x + 8) = 3x + 8$	$3(x + 8) = 3x + 24$
$9(4 + y) = 36 + y$	$9(4 + y) = 36 + 9y$
$8(3b + 5) = 24b + 5$	$8(3b + 5) = 24b + 40$
$7(1 + 6m) = 7 + 6m$	$7(1 + 6m) = 7 + 42m$

a) $5x + 5$

b) $60 + 6y$

c) $32d + 12$

d) $36 + 54p$

Continued on following page.

74.   Continued

Avoiding the common error, complete these:

a) $2(x + 7)$ = _____   c) $5(6 + y)$ = _____

b) $7(5b + 1)$ = _____   d) $8(5 + 7m)$ = _____

---

75.   Expressions like those below are called "<u>instances</u> <u>of</u> <u>the</u> <u>distributive</u> <u>principle</u>".

$$3(x + 7) \qquad 5(4 + y) \qquad 8(2m + 1)$$

When an equation contains an "instance of the distributive principle", we solve it by multiplying by the distributive principle first and then proceed in the usual way. An example is shown. Solve the other equation.

$$3(x + 2) = 21 \qquad\qquad 18 = 6(5 + 2y)$$
$$3x + 6 = 21$$
$$3x + 6 + (-6) = 21 + (-6)$$
$$3x = 15$$
$$x = 5$$

a) $2x + 14$

b) $35b + 7$

c) $30 + 5y$

d) $40 + 56m$

---

76.   Following the example, solve the other equation.

$$5(3 + 2y) = 20y \qquad\qquad 7x = 4(1 + x)$$
$$15 + 10y = 20y$$
$$15 + 10y + (-10y) = 20y + (-10y)$$
$$15 = 10y$$
$$y = \frac{15}{10} = \frac{3}{2}$$

$y = -1$

---

77.   To solve the equation below, we began by multiplying by the distributive principle on the right side. Solve the other equation.

$$2x + 9 = 3(x + 2) \qquad\qquad 5(y + 1) = 7y - 2$$
$$2x + 9 = 3x + 6$$
$$(-2x) + 2x + 9 = (-2x) + 3x + 6$$
$$9 = 1x + 6$$
$$(-6) + 9 = 1x + 6 + (-6)$$
$$3 = 1x$$
$$x = 3$$

$x = \dfrac{4}{3}$

$y = \dfrac{7}{2}$

78. To solve the equation below, we began by multiplying by the distributive principle on both sides.   Solve the other equation.

$$7(d + 2) = 2(3 + d)$$

$$7d + 14 = 6 + 2d$$

$$(-7d) + 7d + 14 = 6 + 2d + (-7d)$$

$$14 = 6 + (-5d)$$

$$(-6) + 14 = (-6) + 6 + (-5d)$$

$$8 = -5d$$

$$d = -\frac{8}{5}$$

$$3(2m + 5) = 5(1 + m)$$

m = -10

2-11   THE DISTRIBUTIVE PRINCIPLE OVER SUBTRACTION AND EQUATIONS

In this section, we will discuss "multiplying by the distributive principle over subtraction". Then we will use that process to solve equations like $4(t - 1) = 6$ and $7(3m - 4) = 14m$.

79. We can multiply by the distributive principle when the second factor is a subtraction. An example is shown. The arrows show that both "8" and "5" are multiplied by "4".

$$4(8 - 5) = 4(8) - 4(5)$$

When the subtraction contains a letter (with or without a coefficient), the distributive principle works the same way. For example:

$$7(2x - 3) = 7(2x) - 7(3) \qquad 9(6 - y) = 9(6) - 9(y)$$

Following the examples, fill in the blanks below.

a) 9(t - 1) = (   )(t) - (   )(1)   b) 4(6 - 3p) = (   )(6) - (   )(3p)

80. When multiplying by the distributive principle over subtraction, we ordinarily write the simplified product in one step. Remember that the terms are subtracted, not added. For example:

$$9(x - 3) = 9x - 27 \qquad 7(2y - 6) = 14y - 42$$

$$2(7 - d) = 14 - 2d \qquad 4(6 - 3t) = 24 - 12t$$

Following the examples, complete these:

a) 6(m - 1) = _____   c) 3(5x - 2) = _____

b) 5(4 - h) = _____   d) 10(1 - 4y) = _____

a) (9)(t) - (9)(1)

b) (4)(6) - (4)(3p)

81.   When multiplying by the distributive principle, <u>don't forget to multiply</u> <u>the second term of the subtraction by the first factor</u>.   Complete these:

    a) $7(1 - x) =$ _____

    b) $6(y - 9) =$ _____

    c) $4(2a - 1) =$ _____

    d) $9(7 - 8b) =$ _____

a) $6m - 6$

b) $20 - 5h$

c) $15x - 6$

d) $10 - 40y$

---

82.   The equation below contains an "instance of the distributive principle over subtraction".   To solve it, we began by multiplying by the distributive principle.   Solve the other equation.

$$4(t - 1) = 16$$
$$4t - 4 = 16$$
$$4t - 4 + 4 = 16 + 4$$
$$4t = 20$$
$$t = 5$$

$$9(2b - 4) = 18$$

a) $7 - 7x$

b) $6y - 54$

c) $8a - 4$

d) $63 - 72b$

---

83.   Following the example, solve the other equation.

$$x = 2(9 - 4x)$$
$$1x = 18 - 8x$$
$$8x + 1x = 18 - 8x + 8x$$
$$9x = 18$$
$$x = 2$$

$$3(7 - 4m) = m$$

$b = 3$

---

84.   Solve each equation:

    a) $2(x - 1) = 3x - 7$

    b) $5(y + 3) = 3(y - 5)$

$m = \dfrac{21}{13}$

a) $x = 5$

b) $y = -15$

2-12   COMBINING LIKE TERMS BEFORE USING THE AXIOMS

When solving equations, we always combine "like" terms on each side before using the axioms. We will discuss that process in this section.

---

85.   Two or more terms that can be combined into one term are called "like" terms.   Two types are identified below.

    1) Two or more <u>number terms</u> are "like" terms.   For example:

       $3 + (-4) = -1$ $(-2) + 9 + (-5) = 2$

    2) Two or more <u>letter terms</u> <u>containing</u> <u>the</u> <u>same</u> <u>letter</u> are "like" terms.   For example:

       $7x + x = 8x$ $3y + (-8y) + 2y = -3y$

Combine the "like" terms:

    a) $12 + (-10) + (-5) = $ _____   b) $(-5t) + 10t + (-t) = $ _____

---

86.   To solve each equation below, we combined like terms and then used the multiplication axiom.

        $2x + 5x = -14$ $4d = (-8) + (-12)$

            $7x = -14$ $4d = -20$

             $x = -2$ $d = -5$

Using the same steps, solve these:

    a) $10y - y = 7$ b) $5p = (-4) + 1$

> a) -3   b) 4t

---

87.   The expressions below were simplified by combining like terms.   Notice that we wrote the letter term first in $3x + 2$ and $-5y + 4$.

       $7 + 3x + (-5) = 3x + 2$ $-6y + 4 + y = -5y + 4$

Simplify each expression by combining like terms.

    a) $10 + 7x + 5 = $ _____   c) $(-1) + 8d + (-3) = $ _____

    b) $y + 9 + y = $ _____   d) $m + (-5) + (-9m) = $ _____

> a) $y = \dfrac{7}{9}$
>
> b) $p = -\dfrac{3}{5}$

> a) $7x + 15$
>
> b) $2y + 9$
>
> c) $8d + (-4)$
>
> d) $-8m + (-5)$

88. To simplify the expressions below, we began by converting each subtraction to addition.

$$10 - 2x - 3 = 10 + (-2x) + (-3) = -2x + 7$$

$$5d - 4 - 2d = 5d + (-4) + (-2d) = 3d + (-4)$$

However, we ordinarily simplify expressions like those above directly without converting the subtractions to additions. To do so, we treat the subtraction sign as if it were the sign of the term. We get:

$$10 - 2x - 3 = -2x + 7$$

$$5d - 4 - 2d = 3d - 4$$

Simplify these directly by combining like terms.

a) 4 + 5x - 7 = _____   c) 5 - 6t - 4 = _____

b) 9y + 2 - 12y = _____   d) 10V - 8 - V = _____

---

89. To solve the equation below, we began by combining like terms on the left side. Solve the other equation.

$$7 + 3x + (-9) = 6 \qquad\qquad 4 + (-5y) + (-3) = 2$$

$$3x - 2 = 6$$

$$3x - 2 + 2 = 6 + 2$$

$$3x = 8$$

$$x = \frac{8}{3}$$

a) 5x - 3

b) -3y + 2

c) -6t + 1

d) 9V - 8

---

90. To solve the equation below, we began by combining like terms on the right side. Solve the other equation.

$$10 = x - 5x + 4 \qquad\qquad 8 = 7y - 4y - 1$$

$$10 = -4x + 4$$

$$10 + (-4) = -4x + 4 + (-4)$$

$$6 = -4x$$

$$x = -\frac{6}{4} = -\frac{3}{2}$$

$y = -\dfrac{1}{5}$

---

91. When solving an equation, we always combine like terms on each side before using the axioms. Solve these:

a) t + (-6t) + (-9) = 0   b) 4x - 1 = 7 - 2x - 3

$y = 3$

---

a) $t = -\dfrac{9}{5}$ b) $x = \dfrac{5}{6}$

2-13   ADDITION GROUPINGS AND EQUATIONS

In this section, we will discuss the method for solving equations like   10 + (2x + 1) = 7   and
10 - (2x + 1) = 7.   Before solving the second type, we will define the opposite of an addition grouping.

92.   "Addition" groupings are groupings that contain an addition, like (3 + 5)
or (x + 7). When an addition grouping follows an <u>addition sign</u>, <u>the group-
ing symbols</u> <u>can</u> <u>be</u> <u>dropped</u> <u>without</u> <u>changing</u> <u>the</u> <u>value</u> <u>of</u> <u>the</u> <u>expression.</u>
As an example, we evaluated 10 + (3 + 5) two different ways below.

By performing the operation within the grouping first.

$$10 + (3 + 5)$$
$$10 + \quad 8 \quad = 18$$

By dropping the grouping symbols first.

$$10 + (3 + 5)$$
$$10 + 3 + 5$$
$$13 \quad + 5 = 18$$

After an addition sign, the grouping symbols <u>can also be dropped</u> when
the grouping contains a letter.   That is:

7 + (x + 5)   can be written: 7 + x + 5

3 + (2y + 1)   can be written: _____

---

93.   To solve the equation below, we dropped the grouping symbols, combined
like terms on the left side, and then used the axioms.   Solve the other
equation.

3 + (2y + 1) = 9	8 + (t + 3) = 0

$$3 + 2y + 1 = 9$$
$$2y + 4 = 9$$
$$2y + 4 + (-4) = 9 + (-4)$$
$$2y = 5$$
$$y = \frac{5}{2}$$

[margin: 3 + 2y + 1]

---

94.   When an addition grouping follows a <u>subtraction sign</u>, <u>we</u> <u>cannot</u> <u>simply</u>
<u>drop</u> <u>the</u> <u>grouping</u> <u>symbols</u> <u>without</u> <u>changing</u> <u>the</u> <u>value</u> <u>of</u> <u>the</u> <u>expression.</u>
As an example, we evaluated   10 - (3 + 5)   two different ways at the top
of the next page.

[margin: t = -11]

Continued on following page.

94.   Continued

By performing the operation within the grouping first.

$$10 - (3 + 5)$$
$$10 - 8 = 2 \qquad (correct)$$

By dropping the grouping symbols first.

$$10 - (3 + 5)$$
$$10 - 3 + 5$$
$$7 + 5 = 12 \qquad (incorrect)$$

After a subtraction sign, the grouping symbols <u>also</u> <u>cannot</u> <u>be</u> <u>dropped</u> when the grouping contains a letter. That is:

7 - (x + 5) <u>cannot</u> be written   7 - x + 5

3 - (2y + 1) <u>cannot</u> be written   _____

---

95.   To get rid of the grouping symbols in an expression like   5 - (3y + 2), we must convert the subtraction to addition. To do so, we <u>add</u> <u>the</u> <u>opposite</u> <u>of</u> <u>the grouping</u>.   That is:

5 - (3y + 2) = 5 + [the opposite of (3y + 2)]

To get the opposite of an addition grouping, <u>we</u> <u>replace</u> <u>each</u> <u>term</u> <u>with</u> <u>its</u> <u>opposite</u>.   For example:

The opposite of (4 + 5) is [(-4) + (-5)].

The opposite of (3y + 2) is [(-3y) + (-2)].

The groupings above are opposites because the sum of each pair is "0". That is:

$$(4 + 5) + [(-4) + (-5)] = \underbrace{4 + (-4)}_{} + \underbrace{5 + (-5)}_{}$$
$$= 0 0 = 0$$

$$(3y + 2) + [(-3y) + (-2)] = \underbrace{3y + (-3y)}_{} + \underbrace{2 + (-2)}_{}$$
$$= \underline{} \underline{} = \underline{}$$

3 - 2y + 1

---

96.   The general pattern for the opposite of an addition grouping is:

The opposite of (a + b) is [(-a) + (-b)]

Following the pattern, write the opposite of each grouping.

a) (x + 7) [+   ]   b) (5d + 1) [+   ]

0 + 0 = 0

---

a) [(-x) + (-7)]

b) [(-5d) + (-1)]

97. To convert the subtraction of a grouping to an equivalent addition, <u>we add the opposite of the grouping.</u>   For example:

$$5 - (2y + 6) = 5 + [\text{the opposite of } (2y + 6)]$$
$$= 5 + (-2y) + (-6)$$

You should be able to make conversions like those above directly.   That is:

$$x - (5y + 7) = x + (-5y) + (-7)$$

a) $9 - (p + 1) = 9 + \underline{\hspace{1cm}} + \underline{\hspace{1cm}}$

b) $2d - (4d + 8) = 2d + \underline{\hspace{1cm}} + \underline{\hspace{1cm}}$

---

98. To solve the equation below, we converted the subtraction of a grouping to an equivalent addition first.   Then we combined like terms on the left side and used the axioms.   Solve the other equation.

$$5x - (2x + 1) = 4 \qquad\qquad F = 35 - (6F + 7)$$
$$5x + (-2x) + (-1) = 4$$
$$3x + (-1) = 4$$
$$3x + (-1) + 1 = 4 + 1$$
$$3x = 5$$
$$x = \frac{5}{3}$$

a) $9 + (-p) + (-1)$

b) $2d + (-4d) + (-8)$

---

99. To solve the equation below, we began by converting the subtraction of a grouping to addition.   Notice that we substituted $(-1y)$ for $(-y)$.   Solve the other equation.

$$7y - (y + 3) = 15 \qquad\qquad 15 = 10 - (y + 4)$$
$$7y + (-1y) + (-3) = 15$$
$$6y + (-3) = 15$$
$$6y + (-3) + 3 = 15 + 3$$
$$6y = 18$$
$$y = 3$$

F = 4

---

100. Solve each equation:

a) $y - (3y + 2) = 8 \qquad\qquad$ b) $4t - (t + 2) = 0$

y = -9

---

a) y = -5   b) $t = \dfrac{2}{3}$

2-14   SUBTRACTION GROUPINGS AND EQUATIONS

In this section, we will discuss the method for solving equations like $10 + (2x - 1) = 7$ and $10 - (2x - 1) = 7$. Before solving the second type, we will define the opposite of a subtraction grouping.

101.   "Subtraction" groupings are groupings that contain a subtraction, like $(5 - 3)$ or $(x - 7)$. When a subtraction grouping follows an <u>addition</u> sign, the grouping symbols <u>can</u> <u>be</u> <u>dropped</u> <u>without</u> <u>changing</u> <u>the</u> <u>value of the</u> <u>expression.</u> As an example, we evaluated $10 + (5 - 3)$ two different ways below.

By performing the operation within the grouping first.

$$10 + (5 - 3)$$
$$10 + \quad 2 \quad = 12$$

By dropping the grouping symbols first.

$$10 + (5 - 3)$$
$$10 + 5 - 3$$
$$15 \quad - 3 = 12$$

After an addition sign, the grouping symbols <u>can</u> <u>also</u> <u>be</u> <u>dropped</u> when the grouping contains a letter. That is:

$7 + (x - 5)$ can be written: $7 + x - 5$

$3 + (2y - 1)$ can be written: _____

---

102.   To solve the equation below, we dropped the grouping symbols, combined like terms on the left side, and then used the axioms.   Solve the other equation

$5 + (2x - 1) = 10$	$9 + (y - 7) = 0$

$$5 + (2x - 1) = 10$$
$$5 + 2x - 1 = 10$$
$$2x + 4 = 10$$
$$2x + 4 + (-4) = 10 + (-4)$$
$$2x = 6$$
$$x = 3$$

[margin answer: $3 + 2y - 1$]

---

103.   When a subtraction grouping follows a <u>subtraction</u> <u>sign,</u> <u>we</u> <u>cannot</u> <u>simply</u> <u>drop</u> <u>the</u> <u>grouping</u> <u>symbols</u> <u>without</u> <u>changing</u> <u>the</u> <u>value</u> <u>of the</u> <u>expression.</u> As an example, we evaluated $10 - (5 - 3)$ two different ways.

By performing the operation within the grouping first.

$$10 - (5 - 3)$$
$$10 - \quad 2 \quad = 8 \quad \text{(correct)}$$

[margin answer: $y = -2$]

Continued on following page.

103.   Continued

By dropping the grouping symbols first.

$$10 \ - \ (5 \ - \ 3)$$

$$\underline{10 \ - \ 5} \ - \ 3$$

$$\underset{\downarrow}{} \\ 5 \quad - \ 3 \ = \ 2 \quad \text{(incorrect)}$$

After a subtraction sign, the grouping symbols <u>also</u> <u>cannot</u> <u>be</u> <u>dropped</u> when the grouping contains a letter.   That is:

$$6 \ - \ (x \ - \ 2) \quad \underline{cannot} \text{ be written} \quad 6 \ - \ x \ - \ 2$$

$$4 \ - \ (3y \ - \ 5) \quad \underline{cannot} \text{ be written} \quad \underline{\hphantom{xxxxxxxx}}$$

---

104.   To get rid of the grouping symbols in an expression like $7 \ - \ (2x \ - \ 3)$, we must convert the subtraction of a grouping to addition.   <u>To</u> <u>do</u> <u>so</u>, <u>we</u> <u>add</u> <u>the</u> <u>opposite</u> <u>of</u> <u>the</u> <u>grouping</u>.   That is: | $4 \ - \ 3y \ - \ 5$

$$7 \ - \ (2x \ - \ 3) \ = \ 7 \ + \ [\text{the opposite of } (2x \ - \ 3)]$$

To get the opposite of a subtraction grouping, <u>we</u> <u>replace</u> <u>each</u> <u>term</u> <u>with</u> <u>its</u> <u>opposite</u>.   For example:

The opposite of $(5 \ - \ 3)$ is $[(-5) \ + \ 3]$

The opposite of $(3y \ - \ 4)$ is $[(-3y) \ + \ 4]$

The groupings above are opposites because the sum of each pair is "0". That is:

$$(5 \ - \ 3) \ + \ [(-5) \ + \ 3] \ = \ \underline{5 \ + \ (-5)}_{\downarrow} \ \underline{- \ 3 \ + \ 3}_{\downarrow}$$

$$= \qquad 0 \qquad + \qquad 0 \qquad = \ 0$$

$$(3y \ - \ 4) \ + \ [(-3y) \ + \ 4] \ = \ \underline{3y \ + \ (-3y)}_{\downarrow} \ \underline{- \ 4 \ + \ 4}_{\downarrow}$$

$$= \quad \underline{\hphantom{xxx}} \quad + \quad \underline{\hphantom{xxx}} \quad = \quad \underline{\hphantom{xxx}}$$

---

105.   The general pattern for the opposite of a subtraction grouping is: | $0 \ + \ 0 \ = \ 0$

> The opposite of $(a \ - \ b)$ is $[(-a) \ + \ b]$

Following the pattern, write the opposite of each grouping.

a) $(x \ - \ 6)$ [+   ]   b) $(4t \ - \ 1)$ [+   ]

---

a) $[(-x) \ + \ 6]$

b) $[(-4t) \ + \ 1]$

106.   To convert the subtraction of a grouping to an equivalent addition, we add the opposite of the grouping.   For example:

$$7 - (3x - 9) = 7 + [\text{the opposite of } (3x - 9)]$$
$$= 7 + (-3x) + 9$$

You should be able to make conversions like those above directly.   That is:

$$d - (6d - 2) = d + (-6d) + 2$$

a) $5 - (R - 1) =$ _____

b) $6x - (8x - 5) =$ _____

---

107.   To solve the equation below, we converted the subtraction of a grouping to an equivalent addition first.   Then we combined like terms on the left side and used the axioms.   Solve the other equation.

$$6y - (2y - 3) = 1 \qquad\qquad x = 10 - (5x - 1)$$
$$6y + (-2y) + 3 = 1$$
$$4y + 3 = 1$$
$$4y + 3 + (-3) = 1 + (-3)$$
$$4y = -2$$
$$y = -\frac{2}{4} = -\frac{1}{2}$$

a) $5 + (-R) + 1$

b) $6x + (-8x) + 5$

---

108.   To solve the equation below, we began by converting the subtraction of a grouping to addition.   Notice that we substituted $(-1t)$ for $(-t)$.   Solve the other equation.

$$5t - (t - 4) = 8 \qquad\qquad 25 = 10 - (H - 5)$$
$$5t + (-1t) + 4 = 8$$
$$4t + 4 = 8$$
$$4t + 4 + (-4) = 8 + (-4)$$
$$4t = 4$$
$$t = 1$$

$x = \dfrac{11}{6}$

---

109.   Solve each equation:

a) $m - (4m - 7) = 3$ \qquad b) $8p - (p - 1) = 0$

$H = -10$

a) $m = \dfrac{4}{3}$ \quad b) $p = -\dfrac{1}{7}$

SELF-TEST 7 (pages 66-79)

Solve each equation. Report each solution in lowest terms.

1. $4(3x + 5) = 5$	2. $2w + 7 = 3(5 - 4w)$	3. $3(y - 3) = 2(y + 3)$
4. $2h + (7h + 3) = 0$	5. $5 = 2E - (4E - 1)$	6. $1 - (2 - 3t) = t$

ANSWERS: 1. $x = -\dfrac{5}{4}$ 3. $y = 15$ 5. $E = -2$

2. $w = \dfrac{4}{7}$ 4. $h = -\dfrac{1}{3}$ 6. $t = \dfrac{1}{2}$

2-15   ADDING AND SUBTRACTING INSTANCES OF THE DISTRIBUTIVE PRINCIPLE

In this section, we will discuss additions and subtractions involving instances of the distributive principle. We will also solve equations involving additions and subtractions of that type.

110.   To simplify the addition and subtraction below, we multiplied by the distributive principle and then combined like terms.

$$3(x + 4) + 7 = 3x + 12 + 7 = 3x + 19$$
$$5(2y - 1) - 3y = 10y - 5 - 3y = 7y - 5$$

Following the examples, simplify each expression:

a) $2(4d - 6) + d = $ _____

b) $7(F + 2) - 8 = $ _____

a) $9d - 12$

b) $7F + 6$

111. To solve the equation below, we simplified the expression on the left side and then used the axioms. Solve the other equation.

$$2(m + 4) - 3 = 10 \qquad\qquad 5(2p - 3) + 3p = 1$$

$$2m + 8 - 3 = 10$$

$$2m + 5 = 10$$

$$2m + 5 + (-5) = 10 + (-5)$$

$$2m = 5$$

$$m = \frac{5}{2}$$

---

112. To solve the equation below, we simplified the expression on the right side and then used the axioms. Solve the other equation.

$$4t = 3(2t - 3) + 5 \qquad\qquad d = 7(d + 1) - 5d$$

$$4t = 6t - 9 + 5$$

$$4t = 6t - 4$$

$$4t + (-6t) = (-6t) + 6t - 4$$

$$-2t = -4$$

$$t = \frac{-4}{-2} = 2$$

$p = \dfrac{16}{13}$

---

113. To simplify the addition below, we multiplied by the distributive principle and then combined like terms.

$$10 + 6(y - 4) = 10 + 6y - 24 = 6y - 14$$

Following the example, simplify these expressions:

a) $3t + 7(t + 5) = $ _____

b) $m + 2(5m - 1) = $ _____

$d = -7$

---

114. To solve the equation below, we simplified the expression on the left side and then used the axioms. Solve the other equation.

$$10F + 4(F - 9) = 6 \qquad\qquad x = 15 + 3(x + 1)$$

$$10F + 4F - 36 = 6$$

$$14F - 36 = 6$$

$$14F - 36 + 36 = 6 + 36$$

$$14F = 42$$

$$F = \frac{42}{14} = 3$$

a) $10t + 35$

b) $11m - 2$

$x = -9$

115.   In the expression below,   2(x + 3) is subtracted from 10.

$$10 - 2(x + 3)$$

When subtracting an instance of the distributive principle, <u>it is very</u> <u>helpful</u> to <u>put brackets around it before multiplying</u>.   We get:

$$10 - [2(x + 3)]$$
$$10 - [2x + 6]$$

Now we can convert the subtraction of a grouping to addition in the usual way and simplify.

$$10 + (-2x) + (-6) = 4 + (-2x)$$

Following the example, simplify these.   <u>Begin</u> by <u>drawing brackets</u> <u>around the instance of the distributive principle</u>.

a)   5 - 3(d + 1)   b)   t - 2(4t + 5)

---

116.   To solve the equation below, we began by simplifying the expression on the left side.   Solve the other equation.

$$10 - 3(x + 4) = 16 \qquad\qquad 9y - 7(y + 2) = 4$$
$$10 - [3(x + 4)] = 16$$
$$10 - [3x + 12] = 16$$
$$10 + (-3x) + (-12) = 16$$
$$-3x + (-2) = 16$$
$$-3x + (-2) + 2 = 16 + 2$$
$$-3x = 18$$
$$x = -6$$

a) 2 + (-3d), from:
        5 + (-3d) + (-3)

b) -7t + (-10), from:
        t + (-8t) + (-10)

---

117.   To simplify the expression below, we began by writing brackets around 5(x - 1).

$$6 - 5(x - 1)$$
$$6 - [5(x - 1)]$$
$$6 - [5x - 5]$$
$$6 + (-5x) + 5 = -5x + 11$$

Following the example, simplify these.   Begin by drawing brackets around the instance of the distributive principle.

a)   20 - 4(d - 2)   b)   10y - 3(2y - 9)

y = 9

118. To solve the equation below, we began by simplifying the expression on the right side. Solve the other equation.

$$35 = 47 - 6(4 - x) \qquad\qquad 20 = t - 3(5 - 2t)$$
$$35 = 47 - [6(4 - x)]$$
$$35 = 47 - [24 - 6x]$$
$$35 = 47 + (-24) + 6x$$
$$35 = 23 + 6x$$
$$(-23) + 35 = (-23) + 23 + 6x$$
$$12 = 6x$$
$$x = 2$$

a) $28 + (-4d)$, from: $20 + (-4d) + 8$

b) $4y + 27$, from: $10y + (-6y) + 27$

119. Solve each equation:

a) $10d - 4(5 + 3d) = 0$ b) $1 = 7 - 2(4F - 3)$

t = 5

a) d = -10 b) $F = \frac{3}{2}$

2-16   DECIMAL NUMBERS IN EQUATIONS

When decimal numbers appear in equations, we use the same steps to solve them. We will discuss equations of that type in this section.

120. We used the addition axiom to solve the equation below. Solve the other equation.

$$x + 1.2 = 3.7 \qquad\qquad .03 = .08 + y$$
$$x + 1.2 + (-1.2) = 3.7 + (-1.2)$$
$$x = 2.5$$

121. Following the example, solve the other equation.

$$t - .35 = .41 \qquad\qquad 2.4 = m - 8$$
$$t - .35 + .35 = .41 + .35$$
$$t = .76$$

y = -.05

122. To solve the equation below, we used the multiplication axiom. Solve the other equation.

$$4t = .56 \qquad\qquad 7x = 4.2$$

$$t = \frac{.56}{4} = .14$$

m = 10.4

---

123. Following the example, solve the other equation.

$$.03d = .06 \qquad\qquad .054 = .006p$$

$$d = \frac{.06}{.03} = 2$$

x = .6

---

124. We used both axioms to solve the equation below.

$$.4x + 1.1 = 3.9$$
$$.4x + 1.1 + (-1.1) = 3.9 + (-1.1)$$
$$.4x = 2.8$$
$$x = \frac{2.8}{.4} = 7$$

Following the example, solve these:

a) $.6 - .2x = .8$ 　　　　b) $1.2t - 4.7 = 2.5$

p = 9

---

125. Following the example, solve the other equation.

$$.7 - m = .1 \qquad\qquad 5.49 = 2.18 - d$$
$$(-.7) + .7 - 1m = .1 + (-.7)$$
$$-1m = -.6$$
$$m = \frac{-.6}{-1} = .6$$

a) x = -1

b) t = 6

---

d = -3.31

2-17   FRACTIONS IN EQUATIONS

When fractions appear in equations, we can use the same steps to solve them. We will discuss equations of that type in this section.

126.   We used the addition axiom to solve the equation below. Solve the other equation.

$$x + \frac{2}{5} = \frac{3}{5}$$ $$y + \frac{1}{4} = 1$$

$$x + \frac{2}{5} + \left(-\frac{2}{5}\right) = \frac{3}{5} + \left(-\frac{2}{5}\right)$$

$$x = \frac{1}{5}$$

---

127.   Following the example, solve the other equation.

$$m - \frac{2}{3} = \frac{1}{6}$$ $$d - \frac{1}{4} = \frac{5}{8}$$

$$m - \frac{2}{3} + \frac{2}{3} = \frac{1}{6} + \frac{2}{3}$$

$$m = \frac{1}{6} + \frac{4}{6}$$

$$m = \frac{5}{6}$$

$y = \frac{3}{4}$

---

128.   We used the multiplication axiom to solve the equation below. That is, we multiplied both sides by $\frac{3}{2}$, the reciprocal of $\frac{2}{3}$. Solve the other equation by multiplying both sides by $\frac{2}{7}$.

$$\frac{2}{3}x = 5$$ $$-3 = \frac{7}{2}y$$

$$\frac{3}{2}\left(\frac{2}{3}x\right) = \frac{3}{2}(5)$$

$$1\ x = \frac{15}{2}$$

$$x = \frac{15}{2}$$

$d = \frac{7}{8}$

---

129.   To solve the equation below, we multiplied both sides by the reciprocal of "4". Solve the other equation.

$$4t = \frac{1}{2}$$ $$\frac{7}{8} = 3m$$

$$\frac{1}{4}(4t) = \frac{1}{4}\left(\frac{1}{2}\right)$$

$$1\ t = \frac{1}{8}$$

$$t = \frac{1}{8}$$

$y = -\frac{6}{7}$

$m = \frac{7}{24}$

130. To solve the equation below, we multiplied both sides by the reciprocal of $\frac{1}{3}$. Solve the other equation.

$$\frac{1}{3}p = 4 \qquad\qquad \frac{1}{2}x = 1$$

$$3\left(\frac{1}{3}p\right) = 3(4)$$

$$1 \quad p = 12$$

$$p = 12$$

---

131. Solve each equation.

a) $2x = \frac{8}{9}$ \qquad\qquad b) $\frac{5}{4}F = 0$

$x = 2$

---

132. We used both axioms to solve the equation below. Solve the other equation.

$$2x - \frac{1}{3} = \frac{1}{6} \qquad\qquad 1 - \frac{1}{6}x = 0$$

$$2x - \frac{1}{3} + \frac{1}{3} = \frac{1}{6} + \frac{1}{3}$$

$$2x = \frac{1}{6} + \frac{2}{6}$$

$$2x = \frac{1}{2}$$

$$\frac{1}{2}(2x) = \frac{1}{2}\left(\frac{1}{2}\right)$$

$$x = \frac{1}{4}$$

a) $x = \frac{4}{9}$

b) $F = 0$

$x = 6$

2-18   FORMULA EVALUATION

In this section, we will discuss the meaning of literal expressions and formulas and the use of formula evaluation for problem-solving.

Note: Since we want to show the numerical part of formula evaluation rather than a method for dealing with the "units" in formulas, "units" are generally avoided in this section and the next one.

133.   We can evaluate an expression containing more than one letter by substituting numbers for the letters.   For example:

If $a = 5$ and $b = 7$, $a + b = 5 + 7 = 12$

If $D = 20$ and $M = 15$, $D - M =$ _____

---

134.   When two or more letters are written next to each other, the expression
stands for a multiplication.   For example:

IR   means   (I)(R)

LWH   means   (L)(W)(H)

$\frac{1}{2}$bh   means $\left(\frac{1}{2}\right)$(b)(h)

We can evaluate expressions like those above by substituting numbers for
the letters.   That is:

If $L = 10$, $W = 2$, and $H = 6$, $LWH = (10)(2)(6) = 120$

If $b = 20$ and $h = 8$, $\frac{1}{2}$bh $= \left(\frac{1}{2}\right)(20)(8) =$ _____

5, from: 20 - 15

---

135.   A formula is a shorthand way of stating a relationship that can also be
stated in words.   An example is given below.

There is a relationship between the distance traveled by
a moving object, its average velocity (or speed), and the
time traveled.   That is:

The distance traveled by a moving object can
be found by multiplying its average velocity
by the time traveled.

The relationship above can be written as a formula in
words.   That is:

| Distance Traveled = Average Velocity x Time Traveled |

However, it is usually written with letters that are abbreviations
for the words.   That is:

| $s = vt$ | where: "s" = distance traveled
"v" = average velocity
"t" = _____

80

---

time traveled

136. By substituting numbers for the letters, we can use a formula to solve
applied problems.   For example:

Using $\boxed{s = vt}$, we can find the distance traveled (s) when the
average velocity (v) and the time traveled (t) are known.   That is:

If the average velocity is 50 miles per hour and the time
traveled is 5 hours, the distance traveled is 250 miles,
since:

$$s = vt = 50(5) = 250 \text{ miles}$$

If the average velocity is 300 kilometers per hour and
the time traveled is 10 hours, the distance traveled
is 3,000 kilometers, since:

$$s = vt = 300(10) = \underline{\hspace{2cm}} \text{ kilometers}$$

---

137. a) The formula relating degrees-Celsius (C) and degrees-Kelvin (K)

is $\boxed{C = K - 273°}$.   Find the number of degrees—Celsius cor-

responding to 400° Kelvin.

$$C = K - 273° = \underline{\hspace{2cm}}$$

b) The formula for the area of a triangle is $\boxed{A = \frac{1}{2}bh}$, where

A = area, b = length of the base, and h = height.   Find the area of
a triangle when b = 10 inches and h = 6 inches.

$$A = \frac{1}{2}bh = \underline{\hspace{4cm}}$$

3,000 kilometers

---

138. In the evaluation below, we <u>multiplied before adding</u>.   Complete the
other evaluation.

In $\boxed{d = f + at}$, find "d" when f = 30, a = 4, and t = 5.

$$d = f + at = 30 + \underbrace{(4)(5)}_{}$$
$$= 30 + 20 = 50$$

In $\boxed{P = 2L + 2W}$, find P when L = 25 and W = 10.

$$P = 2L + 2W =$$

a) C = 127°

b) A = 30 square inches

P = 70

139.   In the evaluation below, we simplified the grouping before multiplying. Complete the other evaluation.

In $\boxed{D = P(Q + R)}$, find D when P = 10, Q = 3, and R = 4.

$$D = P(Q + R) = 10\underset{\downarrow}{(3 + 4)}$$
$$= 10 \quad (7) \quad = 70$$

In $\boxed{C = \dfrac{5}{9}(F - 32)}$, find C when F = 50.

$$C = \frac{5}{9}(F - 32) =$$

---

140.   In the evaluation below, we simplified the grouping before subtracting. Complete the other evaluation.

In $\boxed{A = 180 - (B + C)}$, find A when B = 45 and C = 60.

$$A = 180 - (B + C) = 180 - \underset{\downarrow}{(45 + 60)}$$
$$= 180 - \quad 105 \quad = 75$$

In $\boxed{c = d - (a - b)}$, find "c" when d = 40, a = 20, and b = 10.

$$c = d - (a - b) =$$

F = 10

---

141.   a) In $\boxed{K = mt(x - r)}$, find K when m = 5, t = 10, x = 12, and r = 3.

$$K = mt(x - r) =$$

b) In $\boxed{R = a + b(p - q)}$, find R when a = 20, b = 3, p = 15, and q = 10.

$$R = a + b(p - q) =$$

c = 30

---

a) K = 450   b) R = 35

2-19   FORMULA EVALUATIONS REQUIRING EQUATION-SOLVING

In this section, we will discuss some formula evaluations that require solving an equation.

142.   In many equations, there is one letter alone on the left side.   That letter is called the "solved for" letter.   For example:

In $\boxed{A = \dfrac{1}{2}bh}$, the "solved for" letter is A.

When evaluating to find the value of the "solved for" letter, we simply substitute and perform arithmetic operations.   However, when evaluating to find the value of one of the "non-solved-for" letters, we have to solve an equation after substituting.   An example is shown.   Complete the other evaluation.

In the formula below, find V when M = 150.

$$\boxed{M = V + 100}$$
$$150 = V + 100$$
$$150 + (-100) = V + 100 + (-100)$$
$$V = 50$$

In the formula below, find K when C = 25.

$$\boxed{C = K - 273}$$

143.   Following the example, complete the other evaluation.

In the formula below find "s" when P = 100.

$$\boxed{P = 4s}$$
$$100 = 4s$$
$$s = \frac{100}{4} = 25$$

In the formula below, find R when E = 150 and I = 15.

$$\boxed{E = IR}$$

144.   Following the example, complete the other evaluation.

In the formula below, find L when V = 80, W = 2, and H = 4.

$$\boxed{V = LWH}$$
$$40 = L(2)(4)$$
$$40 = 8L$$
$$L = \frac{40}{8} = 5$$

In the formula below, find "b" when A = 28 and h = 8.

$$\boxed{A = \dfrac{1}{2}bh}$$

K = 298

R = 10

b = 7

145.   a) In the formula below,
            find "r" when e = 20,
            E = 50, and I = 2.

$$e = E - Ir$$

b) In the formula below,
    find Q when D = 100,
    P = 5, and **R** = 10.

$$D = P(Q + R)$$

146.   a) In the formula below,
            find B when A = 85
            and C = 35.

$$A = 180 - (B + C)$$

b) In the formula below,
    find "d" when a = 40,
    b = 20, c = 2, and e = 3.

$$a = b + c(d - e)$$

a) r = 15, from:
    20 = 50 - 2r

b) Q = 10, from:
    100 = 5(Q + 10)

a) B = 60, from: 85 = 180 - (B + 35)   b) d = 13, from: 40 = 20 + 2(d - 3)

SELF-TEST 8 (pages 79-91)

Solve each equation. Report each solution in lowest terms.

1. $2t = 9 - 5(2t + 3)$

2. $5w - 2(3w - 1) = 1$

3. $.042 = .007y$

4. $1.7 - 1.1d = 3.9$

5. $5x - \dfrac{3}{2} = \dfrac{9}{4}$

6. In $\boxed{F = ma}$, find F when m = 12 and a = 20.

7. In $\boxed{A = \dfrac{1}{2}h(b + c)}$, find A when h = 8, b = 11, and c = 14.

8. In $\boxed{K = C + 273}$, find C when K = 150.

9. In $\boxed{E = I(R - r)}$, find "r" when E = 40, I = 10, and R = 15.

ANSWERS:
1. $t = -\dfrac{1}{2}$

2. w = 1

3. y = 6

4. d = -2

5. $x = \dfrac{3}{4}$

6. F = 240

7. A = 100

8. C = -123

9. r = 11

SUPPLEMENTARY PROBLEMS - CHAPTER 2

Assignment 5

Solve each equation.   Report each solution in lowest terms.

1. $y + 9 = 5$ 2. $t - 3 = 8$ 3. $29 = x - 12$ 4. $54 = n + 1$

5. $G - 1 = 0$ 6. $5 + d = 0$ 7. $9p = 4$ 8. $-8R = -15$

9. $7F = -7$ 10. $2 = 14y$ 11. $-35 = 21w$ 12. $-3h = 0$

13. $3x + 5 = 17$ 14. $2r + 15 = 9$ 15. $11 - 15m = 21$

16. $7 - 12s = 4$ 17. $3 = 9 + 4H$ 18. $-1 = 1 + 2A$

19. $4y + 7 = 0$ 20. $16 - 20b = 0$ 21. $4x - 1 = 3$

22. $3c - 5 = -6$ 23. $2 - 11p = 2$ 24. $5v + 1 = 1$

Assignment 6

Solve each equation.   Report each solution in lowest terms.

1. $2t + 5t = 21$ 2. $4R + (-6R) = 0$ 3. $13 = (-3y) + (-2y)$

4. $9P - 3P = 2$ 5. $25 = 2x - 12x$ 6. $5w - 2w = 0$

7. $7h - h = 20$ 8. $6 = F - 9F$ 9. $3 - s = 5$

10. $3m + 8 = 9m$ 11. $b = 3b + 4$ 12. $5x - 4 = 11x$

13. $6r - 5 = 2r$ 14. $a = 1 - a$ 15. $10 - V = 9V$

16. $4d + 10 = 2d + 7$ 17. $5N + 1 = 2N + 2$ 18. $2c + 7 = 7 + 9c$

19. $E + 1 = 3 - E$ 20. $11t + 9 = 3t - 5$ 21. $2h - 7 = 3h + 5$

22. $2 + 3y = 10 - y$ 23. $15 - 6w = 25 - 2w$ 24. $5 - k = 1 - 5k$

Assignment 7

Solve each equation.   Report each solution in lowest terms.

1. $4(x + 3) = 5$ 2. $3(2w + 4) = 2w$ 3. $7y = 5(4 + y)$

4. $29 = 7(3 + 2t)$ 5. $2(d + 4) = 3(d + 2)$ 6. $E + 11 = 4(5 + 4E)$

7. $4(P - 1) = 2$ 8. $r = 2(1 - r)$ 9. $5(3F - 4) = 7F$

10. $7y + 15 = 3(2y - 5)$ 11. $2(x - 4) = 5x - 2$ 12. $3(V + 1) = 2(V - 1)$

13. $a + (3a + 2) = 1$ 14. $4R = 3 + (5 + 2R)$ 15. $18 - (2s + 7) = 12$

16. $x = 1 - (5x + 2)$ 17. $6 - (2 + 5w) = 4$ 18. $3b - (b + 5) = 0$

19. $4B + (2B - 1) = 3$ 20. $5 + (1 - 3K) = 0$ 21. $G - (2G - 3) = 5$

22. $9 - (2 - 5h) = 3h$ 23. $2r - (7r - 3) = 1$ 24. $4t - (1 - t) = 0$

Assignment <u>8</u>

Solve each equation. Report each answer in lowest terms.

1. $3w + 5(w + 2) = 50$

2. $7 + 4(3d - 1) = 0$

3. $4V = 13 + 3(2 - 5V)$

4. $25 - 2(4x + 5) = 45$

5. $8y - 3(2y + 7) = 9$

6. $P - 4(1 - P) = 0$

7. $3 = 9 - 2(1 - 4E)$

8. $4 - 5(2r - 7) = 3r$

9. $15 = 7h - 3(2h - 6)$

10. $3x + 2.4 = 5.7$

11. $10.2 - r = 13.9$

12. $.02m = .80$

13. $.7A - 1.4 = 5.6$

14. $\frac{1}{4}y = 5$

15. $3t = \frac{9}{2}$

16. $s - 1 = \frac{1}{5}$

17. $\frac{1}{2} - E = \frac{1}{3}$

Complete each evaluation.

18. In $\boxed{v = at}$, find "v" when $t = 12$ and $a = 8$.

19. In $\boxed{E = e + ir}$, find E when $e = 24$, $i = 4$, and $r = 12$.

20. In $\boxed{B = 180 - (A + C)}$, find B when $A = 25$ and $C = 85$.

21. In $\boxed{V = \frac{1}{3}Bh}$, find V when $B = 40$ and $h = 30$.

22. In $\boxed{F = \frac{9}{5}C + 32}$, find F when $C = 20$.

23. In $\boxed{P = EI}$, find I when $E = 120$ and $P = 600$.

24. In $\boxed{A = \frac{1}{2}bh}$, find "h" when $A = 60$ and $b = 15$.

25. In $\boxed{Z = A - N}$, find N when $Z = 82$ and $A = 93$.

26. In $\boxed{V = LWH}$, find W when $V = 400$, $L = 10$, and $H = 5$.

27. In $\boxed{v = V - at}$, find "t" when $v = 30$, $V = 90$, and $a = 15$.

28. In $\boxed{F = P(A - a)}$, find A when $F = 200$, $P = 50$, and $a = 16$.

Chapter 3 FRACTIONAL EXPRESSIONS AND EQUATIONS

A fractional expression is a fraction that contains one or more letters. A fractional equation is an equation that contains one or more fractional expressions. In this chapter, we will discuss multiplications involving fractional expressions and a method for solving fractional equations. Some formula evaluations, including the type that requires equation-solving, are also discussed.

3-1   FRACTIONAL EXPRESSIONS

In this section, we will discuss the meaning of fractional expressions containing a letter.

1. A fractional expression containing a letter also stands for a division. For example: $\dfrac{10}{y}$ means: "10" divided by "y" $\dfrac{2x}{5}$ means: "2x" divided by "5" $\dfrac{t+3}{t}$ means: <u>t + 3</u> divided by <u>t</u>	
2. When writing a fractional expression containing a letter, the fraction line should be extended below the entire numerator and above the entire denominator. For example: "5" divided by "7w" is written $\dfrac{5}{7w}$. "m + 1" divided by "m" is written $\dfrac{m+1}{m}$. "t" divided by "2(t − 1)" is written $\dfrac{t}{2(t-1)}$	$t + 3 \ \ldots\ldots\ t$
3. In the evaluation below, we substituted and divided. Complete the other evaluation. If x = 12, $\dfrac{x}{3} = \dfrac{12}{3} = 4$ \qquad If y = 2 , $\dfrac{10}{y} = 5$	$\dfrac{t}{2(t-1)}$
	5, from $\dfrac{10}{2}$

4. In the evaluation below, we multiplied before dividing. Complete the other evaluation.

If $d = 10$, $\dfrac{4d}{5} = \dfrac{4(10)}{5} = \dfrac{40}{5} = 8$

If $p = 3$, $\dfrac{36}{3p} = \dfrac{36}{3(3)} = \dfrac{36}{9} = 4$

4, from $\dfrac{36}{9}$

5. In the evaluation below, we added before dividing. Complete the other evaluation.

If $m = 5$, $\dfrac{m + 1}{2} = \dfrac{5 + 1}{2} = \dfrac{6}{2} = 3$

If $t = 10$, $\dfrac{t}{t - 5} = \dfrac{10}{10-5} = \dfrac{10}{5} = 2$

2, from $\dfrac{10}{5}$

6. In the evaluation below, we simplified the numerator before dividing. Complete the other evaluation.

If $x = 2$, $\dfrac{3x + 10}{4} = \dfrac{3(2) + 10}{4} = \dfrac{6 + 10}{4} = \dfrac{16}{4} = 4$

If $y = 8$, $\dfrac{y}{2y - 12} = \dfrac{8}{2(8)-12} = \dfrac{8}{16-12} = \dfrac{8}{4} = 2$

2, from $\dfrac{8}{4}$

7. In the evaluation below, we simplified the denominator before dividing. Complete the other evaluation.

If $x = 6$, $\dfrac{12}{3(x - 4)} = \dfrac{12}{3(6 - 4)} = \dfrac{12}{3(2)} = \dfrac{12}{6} = 2$

If $p = 7$, $\dfrac{5(p + 1)}{8} = \dfrac{5(7+1)}{8} = \dfrac{5(8)}{8} = \dfrac{40}{8} = 5$

5, from $\dfrac{40}{8}$

8. In the fractional expression below, the numerator and denominator are identical. Notice that we got "1" when evaluating. Complete the other evaluation.

If $x = 9$, $\dfrac{2x}{2x} = \dfrac{2(9)}{2(9)} = \dfrac{18}{18} = 1$

If $y = 7$, $\dfrac{y - 1}{y - 1} = \dfrac{7-1}{7-1} = \dfrac{6}{6} = 1$

1, from $\dfrac{6}{6}$

9. Following the example, complete the other evaluation.

If $t = 0$, $\dfrac{2t}{3} = \dfrac{2(0)}{3} = \dfrac{0}{3} = 0$

If $a = 2$, $\dfrac{a - 2}{5} = \dfrac{2-2}{5} = \dfrac{0}{5} = 0$

0, from $\dfrac{0}{5}$

96   Fractional Expressions and Equations

10. When evaluating below, we got "0" as the denominator. <u>Since division by "0" is impossible, the evaluation cannot be done.</u>

$$\text{If } x = 5, \quad \frac{x}{x-5} = \frac{5}{5-5} = \frac{5}{0} \quad \text{(impossible)}$$

Complete each possible evaluation.

a) If $m = 3$, $\frac{2m-6}{9} = $ $\frac{2(3)-6}{9} = \frac{6-6}{9} = \frac{0}{9}$

b) If $p = 0$, $\frac{p+1}{3p} = $ $\frac{0+1}{3(0)} = \frac{1}{0} = \text{imp}$

---

a) 0, from $\frac{0}{9}$ b) Impossible, from $\frac{1}{0}$

3-2   MULTIPLICATIONS INVOLVING FRACTIONAL EXPRESSIONS

In this section, we will discuss the procedure for multiplications involving fractional expressions containing a letter.

11. We use the same procedure to multiply fractions when a letter is involved. That is, we multiply their numerators and their denominators. For example:

$$\left(\frac{2}{3}\right)\left(\frac{x}{7}\right) = \frac{(2)(x)}{(3)(7)} = \frac{2x}{21} \qquad \left(\frac{4}{5}\right)\left(\frac{3}{y}\right) = \frac{(4)(3)}{(5)(y)} = \frac{12}{5y}$$

12. When writing a product containing a letter, <u>the numerical coefficient is always written in front of the letter</u>. For example:

$$\left(\frac{p}{5}\right)\left(\frac{7}{8}\right) = \frac{(p)(7)}{(5)(8)} = \frac{7p}{40} \qquad \left(\frac{3}{x}\right)\left(\frac{1}{4}\right) = \frac{(3)(1)}{(x)(4)} = \frac{3}{4x}$$

$\frac{12}{5y}$

13. Following the example, complete the other multiplication.

$$\left(\frac{5}{4}\right)\left(\frac{3m}{2}\right) = \frac{(5)(3m)}{(4)(2)} = \frac{15m}{8} \qquad \left(\frac{2b}{3}\right)\left(\frac{4}{7}\right) = \frac{(2b)(4)}{(3)(7)} = \frac{8b}{21}$$

$\frac{3}{4x}$

14. Complete the following:

a) $\left(\frac{x}{7}\right)\left(\frac{1}{2}\right) = \frac{1x}{14}$ b) $\left(\frac{3}{4}\right)\left(\frac{7d}{5}\right) = \frac{21d}{20}$ c) $\left(\frac{1}{5y}\right)\left(\frac{3}{2}\right) = \frac{3}{10y}$

$\frac{8b}{21}$

---

a) $\frac{x}{14}\left(\text{from } \frac{1x}{14}\right)$ b) $\frac{21d}{20}$ c) $\frac{3}{10y}$

15. By substituting $\frac{4}{1}$ for 4 and $\frac{2}{1}$ for 2, we converted each multiplication below to a multiplication of two fractions.

$$4\left(\frac{x}{3}\right) = \left(\frac{4}{1}\right)\left(\frac{x}{3}\right) = \frac{4x}{3} \qquad\qquad \left(\frac{3}{7y}\right)(2) = \left(\frac{3}{7y}\right)\left(\frac{2}{1}\right) = \frac{6}{7y}$$

However, we ordinarily use the shorter method below for multiplications of that type.

$$4\left(\frac{x}{3}\right) = \frac{(4)(x)}{3} = \frac{4x}{3} \qquad\qquad \left(\frac{3}{7y}\right)(2) = \frac{(3)(2)}{7y} = \underline{\frac{6}{7y}}$$

16. By substituting $\frac{1}{1}$ for "1", we converted each multiplication below to a multiplication of two fractions.

$\dfrac{6}{7y}$

$$1\left(\frac{x}{4}\right) = \left(\frac{1}{1}\right)\left(\frac{x}{4}\right) = \frac{x}{4} \qquad\qquad \left(\frac{7}{3d}\right)(1) = \left(\frac{7}{3d}\right)\left(\frac{1}{1}\right) = \frac{7}{3d}$$

However, we ordinarily use the shorter method below for multiplications of that type.

$$1\left(\frac{x}{4}\right) = \frac{(1)(x)}{4} = \frac{x}{4} \qquad\qquad \left(\frac{7}{3d}\right)(1) = \frac{(7)(1)}{3d} = \underline{\frac{7}{3d}}$$

17. Any letter or letter with a coefficient can also be written as a fraction whose denominator is "1". For example:

$\dfrac{7}{3d}$

$$x = \frac{x}{1} \qquad\qquad\qquad 3y = \frac{3y}{1}$$

Using the facts above, we converted each multiplication below to a multiplication of two fractions.

$$\frac{7}{8}(x) = \left(\frac{7}{8}\right)\left(\frac{x}{1}\right) = \frac{7x}{8} \qquad\qquad 3y\left(\frac{1}{5}\right) = \left(\frac{3y}{1}\right)\left(\frac{1}{5}\right) = \frac{3y}{5}$$

However, we ordinarily use the shorter method below for multiplications of that type.

$$\frac{7}{8}(x) = \frac{(7)(x)}{8} = \frac{7x}{8} \qquad\qquad 3y\left(\frac{1}{5}\right) = \frac{(3y)(1)}{5} = \underline{\frac{3y}{5}}$$

18. Complete these:

$\dfrac{3y}{5}$

a) $\left(\dfrac{1}{3x}\right)(7) = \underline{\dfrac{7}{3x}}$ b) $1\left(\dfrac{2p}{3}\right) = \underline{\dfrac{2p}{3}}$ c) $\dfrac{1}{2}(y) = \underline{\dfrac{y}{2}}$

19. Complete these:

a) $\dfrac{7}{3x}$ b) $\dfrac{2p}{3}$ c) $\dfrac{y}{2}$

a) $x\left(\dfrac{4}{7}\right) = \underline{\dfrac{4x}{7}}$ b) $5y\left(\dfrac{1}{8}\right) = \underline{\dfrac{5y}{8}}$ c) $\left(\dfrac{4}{3}\right)(2D) = \underline{\dfrac{8D}{3}}$

a) $\dfrac{4x}{7}$ b) $\dfrac{5y}{8}$ c) $\dfrac{8D}{3}$

3-3   REDUCING FRACTIONAL EXPRESSIONS TO LOWEST TERMS

In this section, we will discuss the procedure for reducing fractional expressions to lowest terms.

20.   When any letter is divided by itself, the quotient is "1". For example:

$$\frac{x}{x} = 1 \qquad \frac{y}{y} = 1 \qquad \frac{F}{F} = 1$$

Using the fact above, we reduced the fractional expression below to lowest terms.

$$\frac{2x}{3x} = \left(\frac{2}{3}\right)\left(\frac{x}{x}\right) = \left(\frac{2}{3}\right)(1) = \frac{2}{3}$$

Reduce each fractional expression to lowest terms.

a) $\frac{5y}{4y}$ = $\frac{5}{4}$ b) $\frac{6R}{7R}$ = $\frac{6}{7}$ c) $\frac{13d}{11d}$ = $\frac{13}{11}$

---

21.   In each reduction below, we could also reduce the numerical fraction to lowest terms.

$$\frac{6x}{8x} = \left(\frac{6}{8}\right)\left(\frac{x}{x}\right) = \left(\frac{3}{4}\right)(1) = \frac{3}{4}$$

$$\frac{10y}{5y} = \left(\frac{10}{5}\right)\left(\frac{y}{y}\right) = (2)(1) = 2$$

Reduce each fractional expression to lowest terms.

a) $\frac{4t}{12t}$ = $\frac{1}{3}$ c) $\frac{12m}{2m}$ = 6

b) $\frac{15F}{6F}$ = $\frac{5}{2}$ d) $\frac{35V}{7V}$ = 5

Answers:

a) $\frac{5}{4}$

b) $\frac{6}{7}$

c) $\frac{13}{11}$

---

22.   When a letter with a coefficient is divided by itself, the quotient is "1". That is:

$$\frac{7x}{7x} = 1 \qquad \frac{10D}{10D} = 1 \qquad \frac{3m}{3m} = 1$$

Answers:

a) $\frac{1}{3}$ c) 6

b) $\frac{5}{2}$ d) 5

---

23.   To reduce the fractional expressions below to lowest terms, we began by substituting "1x" for "x" and "1R" for "R".

$$\frac{5x}{x} = \frac{5x}{1x} = \left(\frac{5}{1}\right)\left(\frac{x}{x}\right) = (5)(1) = 5$$

$$\frac{R}{7R} = \frac{1R}{7R} = \left(\frac{1}{7}\right)\left(\frac{R}{R}\right) = \left(\frac{1}{7}\right)(1) = \frac{1}{7}$$

Reduce each fractional expression to lowest terms.

a) $\frac{3m}{m}$ = 3 c) $\frac{10F}{F}$ = 10

b) $\frac{y}{4y}$ = $\frac{1}{4}$ d) $\frac{V}{2V}$ = $\frac{1}{2}$

Answer:

1

24. To reduce the fractional expression below to lowest terms, we divided the coefficient by the denominator.

$$\frac{6x}{2} = \left(\frac{6}{2}\right)(x) = (3)(x) = 3x$$

Following the example, reduce these to lowest terms.

a) $\frac{10y}{5} = \underline{2y}$ b) $\frac{20m}{4} = \underline{5m}$ c) $\frac{12F}{3} = \underline{4F}$

a) 3   c) 10

b) $\frac{1}{4}$ d) $\frac{1}{2}$

25. Reduce each fractional expression to lowest terms.

a) $\frac{4p}{6p} = \underline{\frac{2}{3}}$ c) $\frac{13V}{V} = \underline{13}$

b) $\frac{9d}{9d} = \underline{1}$ d) $\frac{36R}{12} = \underline{3R}$

a) 2y

b) 5m

c) 4F

a) $\frac{2}{3}$ b) 1   c) 13   d) 3R

3-4   REDUCING PRODUCTS TO LOWEST TERMS

In multiplications involving fractional expressions, the product should always be reduced to lowest terms. We will discuss multiplications of that type in this section.

26. In multiplications involving fractional expressions, the product should always be reduced to lowest terms. For example:

$$\left(\frac{3x}{8}\right)\left(\frac{4}{x}\right) = \frac{12x}{8x} = \frac{3}{2}$$

Do these. Reduce each product to lowest terms.

a) $\left(\frac{y}{2}\right)\left(\frac{6}{5y}\right) = \underline{\frac{6y}{10y} = \frac{3}{5}}$ b) $\left(\frac{1}{2m}\right)\left(\frac{8m}{3}\right) = \underline{\frac{8m}{6m} = \frac{4}{3}}$

27. We were able to reduce each product below to a non-fraction.

$$\left(\frac{2R}{5}\right)\left(\frac{10}{R}\right) = \frac{20R}{5R} = 4$$

$$\left(\frac{3}{2}\right)\left(\frac{4x}{3}\right) = \frac{12x}{6} = 2x$$

a) $\frac{3}{5}$ b) $\frac{4}{3}$

Do these. Reduce each product to lowest terms.

a) $\left(\frac{4}{d}\right)\left(\frac{3d}{2}\right) = \underline{\frac{12d}{2d} = 6}$ b) $\left(\frac{20x}{5}\right)\left(\frac{5}{2}\right) = \underline{\frac{100x}{10} = 10x}$

a) 6, from $\frac{12d}{2d}$ b) 10x, from $\frac{100x}{10}$

28. Following the example, complete the other multiplication.

$$\left(\frac{y}{2}\right)\left(\frac{2}{y}\right) = \frac{2y}{2y} = 1 \qquad \left(\frac{3F}{4}\right)\left(\frac{4}{3F}\right) = \underline{\hspace{2cm}}$$

29. We were able to reduce each product below to lowest terms.

$$x\left(\frac{3}{2x}\right) = \frac{3x}{2x} = \frac{3}{2}$$

$$6y\left(\frac{1}{8y}\right) = \frac{6y}{8y} = \frac{3}{4}$$

Do these. Reduce each product to lowest terms.

a) $F\left(\frac{9}{7F}\right) = \underline{\hspace{2cm}}$ b) $10d\left(\frac{1}{4d}\right) = \underline{\hspace{2cm}}$

30. We were able to reduce each product below to a whole number.

$$x\left(\frac{3}{x}\right) = \frac{3x}{x} = 3$$

$$2y\left(\frac{5}{2y}\right) = \frac{10y}{2y} = 5$$

Do these. Reduce each product to lowest terms.

a) $V\left(\frac{7}{V}\right) = \underline{\hspace{2cm}}$ b) $3m\left(\frac{2}{3m}\right) = \underline{\hspace{2cm}}$

31. We were able to reduce each product below to a non-fraction.

$$7\left(\frac{3y}{7}\right) = \frac{21y}{7} = 3y$$

$$8\left(\frac{5x}{4}\right) = \frac{40x}{4} = 10x$$

Do these. Reduce each product to lowest terms.

a) $9\left(\frac{5F}{9}\right) = \underline{\hspace{2cm}}$ b) $6\left(\frac{2d}{3}\right) = \underline{\hspace{2cm}}$

32. We were able to reduce each product below to a non-fraction.

$$4x\left(\frac{7}{4}\right) = \frac{28x}{4} = 7x$$

$$5y\left(\frac{4}{y}\right) = \frac{20y}{y} = \frac{20y}{1y} = 20$$

Do these. Reduce each product to lowest terms.

a) $8m\left(\frac{3}{2}\right) = \underline{\hspace{2cm}}$ b) $10d\left(\frac{1}{5d}\right) = \underline{\hspace{2cm}}$

Answers (right column):

1, from $\frac{12F}{12F}$

a) $\frac{9}{7}$ b) $\frac{5}{2}$

a) 7   b) 2

a) 5F   b) 4d

a) 12m   b) 2

3-5   "CANCELLING" IN MULTIPLICATIONS INVOLVING FRACTIONAL EXPRESSIONS

Any multiplication in which the product can be reduced to lowest terms can be simplified by the "cancelling" process. We will discuss the "cancelling" process in this section.

33.   The need to reduce a product to lowest terms can be avoided by a process called "cancelling". An example is discussed.

The product at the right had to be reduced to lowest terms.

$$\left(\frac{x}{3}\right)\left(\frac{4}{x}\right) = \frac{4x}{3x} = \frac{4}{3}$$

We used "cancelling" for the same multiplication at the right. Notice the steps:

$$\left(\frac{\overset{1}{\cancel{x}}}{3}\right)\left(\frac{4}{\underset{1}{\cancel{x}}}\right) = \frac{4}{3}$$

1) We divided each "x" by "x".

2) We got the terms of the product by the multiplications (1)(4) and (3)(1).

Use "cancelling" to complete these:

a) $\left(\frac{y}{7}\right)\left(\frac{2}{y}\right) = $ _____   b) $\left(\frac{10}{R}\right)\left(\frac{R}{9}\right) = $ _____

34.   We cancelled below by dividing the "4m" and "6m" by "2m".

$$\left(\frac{1}{\underset{2}{\cancel{4m}}}\right)\left(\frac{\overset{3}{\cancel{6m}}}{5}\right) = \frac{3}{10}$$

Following the example, cancel to complete these.

a) $\left(\frac{4x}{7}\right)\left(\frac{1}{8x}\right) = $ _____   b) $\left(\frac{1}{3m}\right)\left(\frac{9m}{7}\right) = $ _____

35.   Two cancellings were possible below. Notice the steps:

1) We divided "5p" and "p" by "p".
2) We divided 6 and 3 by 3.
3) We got the terms of the product by the multiplications (5)(1) and (2)(1).

$$\left(\frac{\overset{5}{\cancel{5p}}}{\underset{2}{\cancel{6}}}\right)\left(\frac{\overset{1}{\cancel{3}}}{\underset{1}{\cancel{p}}}\right) = \frac{5}{2}$$

Use cancelling to complete these.

a) $\left(\frac{x}{2}\right)\left(\frac{8}{3x}\right) = $ _____   b) $\left(\frac{7}{F}\right)\left(\frac{9F}{14}\right) = $ _____

a) $\left(\frac{\overset{1}{\cancel{y}}}{7}\right)\left(\frac{2}{\underset{1}{\cancel{y}}}\right) = \frac{2}{7}$

b) $\left(\frac{10}{\underset{1}{\cancel{R}}}\right)\left(\frac{\overset{1}{\cancel{R}}}{9}\right) = \frac{10}{9}$

a) $\left(\frac{\overset{1}{\cancel{4x}}}{7}\right)\left(\frac{1}{\underset{2}{\cancel{8x}}}\right) = \frac{1}{14}$

b) $\left(\frac{1}{\underset{1}{\cancel{3m}}}\right)\left(\frac{\overset{3}{\cancel{9m}}}{7}\right) = \frac{3}{7}$

a) $\left(\frac{\overset{1}{\cancel{x}}}{\underset{1}{\cancel{2}}}\right)\left(\frac{\overset{4}{\cancel{8}}}{\underset{3}{\cancel{3x}}}\right) = \frac{4}{3}$ b) $\left(\frac{\overset{1}{\cancel{7}}}{\underset{1}{\cancel{F}}}\right)\left(\frac{\overset{9}{\cancel{9F}}}{\underset{2}{\cancel{14}}}\right) = \frac{9}{2}$

36. After cancelling below, we got the denominator of the product by multiplying "1" and "1".

$$\left(\frac{\overset{3}{\cancel{3x}}}{\underset{1}{\cancel{4}}}\right)\left(\frac{\overset{4}{\cancel{16}}}{\underset{1}{\cancel{x}}}\right) = \frac{12}{1} = 12$$

Use cancelling to complete these.

a) $\left(\dfrac{15}{y}\right)\left(\dfrac{2y}{5}\right) =$ _____

b) $\left(\dfrac{7D}{2}\right)\left(\dfrac{10}{D}\right) =$ _____

---

37. When cancelling below, we divided 2 and 4x by 2 and divided 3 and 9 by 3.

$$\left(\frac{\overset{1}{\cancel{3}}}{\underset{1}{\cancel{2}}}\right)\left(\frac{\overset{2x}{\cancel{4x}}}{\underset{3}{\cancel{9}}}\right) = \frac{2x}{3}$$

Use cancelling to complete these.

a) $\left(\dfrac{1}{3}\right)\left(\dfrac{9y}{7}\right) =$ _____

b) $\left(\dfrac{12t}{5}\right)\left(\dfrac{5}{2}\right) =$ _____

a) $\left(\frac{\overset{3}{\cancel{15}}}{\underset{1}{\cancel{y}}}\right)\left(\frac{\overset{2}{\cancel{2y}}}{\underset{1}{\cancel{5}}}\right) = \frac{6}{1} = 6$

b) $\left(\frac{\overset{7}{\cancel{7D}}}{\underset{1}{\cancel{2}}}\right)\left(\frac{\overset{5}{\cancel{10}}}{\underset{1}{\cancel{D}}}\right) = \frac{35}{1} = 35$

---

38. When cancelling below, we divided the 6 and 2 by 2.

$$\overset{3}{\cancel{6}}\left(\frac{x}{\underset{1}{\cancel{2}}}\right) = \frac{3x}{1} = 3x$$

Use cancelling to complete these.

a) $10\left(\dfrac{y}{5}\right) =$ _____

b) $7\left(\dfrac{5x}{7}\right) =$ _____

a) $\left(\dfrac{1}{\underset{1}{\cancel{3}}}\right)\left(\frac{\overset{3y}{\cancel{9y}}}{7}\right) = \frac{3y}{7}$

b) $\left(\frac{\overset{6t}{\cancel{12t}}}{\underset{1}{\cancel{5}}}\right)\left(\frac{\overset{1}{\cancel{5}}}{\underset{1}{\cancel{2}}}\right) = \frac{6t}{1} = 6t$

---

39. When cancelling below, we divided 6x and 2x by 2x.

$$\overset{3}{\cancel{6x}}\left(\frac{7}{\underset{1}{\cancel{2x}}}\right) = \frac{21}{1} = 21$$

Use cancelling to complete these.

a) $10d\left(\dfrac{1}{2d}\right) =$ _____

b) $5m\left(\dfrac{12}{5m}\right) =$ _____

a) $\overset{2}{\cancel{10}}\left(\frac{y}{\underset{1}{\cancel{5}}}\right) = \frac{2y}{1} = 2y$

b) $\overset{1}{\cancel{7}}\left(\frac{5x}{\underset{1}{\cancel{7}}}\right) = \frac{5x}{1} = 5x$

---

a) $\overset{5}{\cancel{10d}}\left(\frac{1}{\underset{1}{\cancel{2d}}}\right) = \frac{5}{1} = 5$

b) $\overset{1}{\cancel{5m}}\left(\frac{12}{\underset{1}{\cancel{5m}}}\right) = \frac{12}{1} = 12$

40. When cancelling below, we divided 3t and "t" by "t".

$$\overset{3}{\cancel{3t}}\left(\frac{10}{\underset{1}{\cancel{t}}}\right) = \frac{30}{1} = 30$$

Use cancelling to complete these.

a) $2V\left(\dfrac{7}{V}\right) = $ _____   b) $12x\left(\dfrac{1}{x}\right) = $ _____

41. When cancelling below, we divided 8x and 2 by 2.

$$\overset{4x}{\cancel{8x}}\left(\frac{3}{\underset{1}{\cancel{2}}}\right) = \frac{12x}{1} = 12x$$

Use cancelling to complete these.

a) $15y\left(\dfrac{1}{3}\right) = $ _____   b) $9m\left(\dfrac{8}{9}\right) = $ _____

a) $\overset{2}{\cancel{2V}}\left(\dfrac{7}{\underset{1}{\cancel{V}}}\right) = \dfrac{14}{1} = 14$

b) $\overset{12}{\cancel{12x}}\left(\dfrac{1}{\underset{1}{\cancel{x}}}\right) = \dfrac{12}{1} = 12$

a) $\overset{5y}{\cancel{15y}}\left(\dfrac{1}{\underset{1}{\cancel{3}}}\right) = \dfrac{5y}{1} = 5y$ b) $\overset{m}{\cancel{9m}}\left(\dfrac{8}{\underset{1}{\cancel{9}}}\right) = \dfrac{8m}{1} = 8m$

SELF-TEST 9 (pages 94-103)

Reduce each fractional expression to lowest terms.

1. $\dfrac{15y}{10y}$	2. $\dfrac{10w}{2}$	3. $\dfrac{h}{4h}$

Multiply. Reduce each product to lowest terms.

4. $\left(\dfrac{y}{6}\right)\left(\dfrac{2}{y}\right)$	5. $\left(\dfrac{5t}{2}\right)\left(\dfrac{8}{t}\right)$	6. $\left(\dfrac{3}{8}\right)\left(\dfrac{4r}{9}\right)$
7. $6\left(\dfrac{2x}{3}\right)$	8. $4y\left(\dfrac{3}{2y}\right)$	9. $15F\left(\dfrac{4}{5}\right)$

ANSWERS:

1. $\dfrac{3}{2}$ 4. $\dfrac{1}{3}$ 7. 4x

2. 5w   5. 20   8. 6

3. $\dfrac{1}{4}$ 6. $\dfrac{r}{6}$ 9. 12F

3-6   FRACTIONAL EQUATIONS CONTAINING ONE FRACTION

The general method for solving fractional equations has two basic steps:

1) Using the multiplication axiom to clear the fraction(s) and get an equivalent non-fractional equation.

2) Then solving the equivalent non-fractional equation.

In this section, we will use that method to solve fractional equations containing one fraction.

42. We used the multiplication axiom to clear the fraction and solve the equation below. That is, we multiplied both sides by 5, <u>the denominator of the fraction</u>. Solve the other equation by multiplying both sides by 3. $$\frac{x}{5} = 7$$ $$5\left(\frac{x}{5}\right) = 5(7)$$ $$x = 35$$ Check $$\frac{35}{5} = 7$$ $$7 = 7$$ $4 = \dfrac{t}{3}$ $12 = t$	
43. To clear the fraction below, we multiplied both sides by "y", <u>the denominator of the fraction</u>. Then we solved "12 = 3y" to get the solution of the original equation. To solve the other equation, begin by multiplying both sides by "R". $$\frac{12}{y} = 3$$ $$y\left(\frac{12}{y}\right) = y(3)$$ $$12 = 3y$$ $$y = \frac{12}{3} = 4$$ Check $$\frac{12}{4} = 3$$ $$3 = 3$$ $6 = \dfrac{42}{R}$ $6R = 42$ $6R = 42$ $R = 7$	t = 12, from: $$3(4) = 3\left(\frac{t}{3}\right)$$ $$12 = t$$
44. To solve the equation below, we began by multiplying both sides by 3. Solve the other equation. Begin by multiplying both sides by 2. $$\frac{2m}{3} = 6$$ $$3\left(\frac{2m}{3}\right) = 3(6)$$ $$2m = 18$$ $$m = 9$$ Check $$\frac{2(9)}{3} = 6$$ $$\frac{18}{3} = 6$$ $$6 = 6$$ $12 = \dfrac{3x}{2}$ $24 = 3x$ $8 = x$	R = 7, from: $$R(6) = R\left(\frac{42}{R}\right)$$ $$6R = 42$$
	x = 8, from: $$2(12) = 2\left(\frac{3x}{2}\right)$$ $$24 = 3x$$

45.   To clear the fraction in any equation containing only one fraction, <u>we</u>
<u>multiply</u> <u>both</u> <u>sides</u> <u>by</u> <u>its</u> <u>denominator</u>. For example, we multiplied both
sides by "3t" below. Solve the other equation.

$$\frac{24}{3t} = 2$$

Check

$$\frac{24}{3(4)} = 2$$

$$3t\left(\frac{24}{3t}\right) = 3t(2)$$

$$\frac{24}{12} = 2$$

$$24 = 6t$$

$$2 = 2$$

$$t = 4$$

$$6 = \frac{60}{5p}$$

$30p = 60$
$p = 2$

---

46.   Fractional equations can have fractional solutions. An example is shown.
Solve the other equations.

$$\frac{5m}{2} = 3$$

a) $4 = \dfrac{7}{x}$

b) $\dfrac{6}{5p} = 2$

$$2\left(\frac{5m}{2}\right) = 2(3)$$

$4x = 7$

$6 = 10p$

$$5m = 6$$

$x = \dfrac{7}{4}$

$\dfrac{6}{10} = \dfrac{3|5}{p}$

$$m = \frac{6}{5}$$

p = 2, from:

$5p(6) = 5p\left(\dfrac{60}{5p}\right)$

$30p = 60$

---

47.   Following the example, solve the other equations.

$$\frac{3x}{4} = 1$$

a) $\dfrac{10}{y} = 1$

b) $1 = \dfrac{5}{9q}$

$$4\left(\frac{3x}{4}\right) = 4(1)$$

$10 = y$

$9q = 5$

$$3x = 4$$

$q = 5|9$

$$x = \frac{4}{3}$$

a) $x = \dfrac{7}{4}$

b) $p = \dfrac{3}{5}$

---

48.   Following the example, solve the other equations.

$$\frac{2m}{7} = 0$$

a) $\dfrac{x}{12} = 0$

b) $0 = \dfrac{9d}{5}$

$$7\left(\frac{2m}{7}\right) = 7(0)$$

$x = 0$

$0 = 9d$

$$2m = 0$$

$0 = d$

$$m = \frac{0}{2} = 0$$

a) $y = 10$

b) $q = \dfrac{5}{9}$

---

a) x = 0   b) d = 0

3-7   EQUATIONS CONTAINING ONE FRACTION WHOSE NUMERATOR OR DENOMINATOR IS AN ADDITION OR SUBTRACTION

In this section, we will discuss the method for solving fractional equations containing one fraction whose numerator or denominator is an addition or subtraction.

49.   To clear a fraction whose numerator is an addition or subtraction, we must perform multiplications like those below.

$$2\left(\frac{x+3}{2}\right) = \frac{2(x+3)}{2} = \left(\frac{2}{2}\right)(x+3) = (1)(x+3) = x+3$$

$$y\left(\frac{y-6}{y}\right) = \frac{y(y-6)}{y} = \left(\frac{y}{y}\right)(y-6) = (1)(y-6) = y-6$$

We can use cancelling to shorten the multiplications above. We get:

$$\cancel{2}\left(\frac{x+3}{\cancel{2}}\right) = x+3 \qquad\qquad \cancel{y}\left(\frac{y-6}{\cancel{y}}\right) = y-6$$

Use cancelling to complete these.

a) $5\left(\dfrac{t-1}{5}\right) =$ _____

b) $m\left(\dfrac{m+9}{m}\right) =$ _____

c) $2p\left(\dfrac{p+4}{2p}\right) =$ _____

d) $3V\left(\dfrac{4V-15}{3V}\right) =$ _____

50.   To solve the equation below, we began by multiplying both sides by 2. Notice that we had to use the addition axiom. Solve the other equation.

$$\frac{x+3}{2} = 5 \qquad\qquad \frac{y-3}{4} = 1$$

$$\cancel{2}\left(\frac{x+3}{\cancel{2}}\right) = 2(5)$$

$$x+3 = 10$$

$$x+3+(-3) = 10+(-3)$$

$$x = 7$$

51.   To solve the equation below, we began by multiplying both sides by 4. Solve the other equation.

$$k = \frac{2k+7}{4} \qquad\qquad 2V = \frac{8-7V}{3}$$

$$4(k) = \cancel{4}\left(\frac{2k+7}{\cancel{4}}\right)$$

$$4k = 2k+7$$

$$4k+(-2k) = (-2k)+2k+7$$

$$2k = 7$$

$$k = \frac{7}{2}$$

Answers:

a) t - 1   c) p + 4

b) m + 9   d) 4V - 15

y = 7

$V = \dfrac{8}{13}$

52. To solve the equation below, we began by multiplying both sides by "x". Solve the other equation.

$$\frac{x+1}{x} = 3 \qquad\qquad 1 = \frac{3y-5}{2y}$$

$$x\left(\frac{x+1}{x}\right) = x(3)$$

$$x + 1 = 3x$$

$$(-1x) + 1x + 1 = 3x + (-1x)$$

$$1 = 2x$$

$$x = \frac{1}{2}$$

---

53. Following the example, solve the other equation.

$$\frac{y-5}{9} = 0 \qquad\qquad \frac{x+3}{x} = 0$$

$$9\left(\frac{y-5}{9}\right) = 9(0)$$

$$y - 5 = 0$$

$$y - 5 + 5 = 0 + 5$$

$$y = 5$$

y = 5

---

54. When an addition or subtraction is divided by itself, the quotient is "1". For example:

$$\frac{x+2}{x+2} = 1 \qquad\qquad \frac{t-3}{t-3} = 1$$

The fact above is used to perform multiplications like those below which are needed to clear a fraction whose denominator is an addition or subtraction.

$$(x+2)\left(\frac{12}{x+2}\right) = \frac{(x+2)(12)}{x+2} = \left(\frac{x+2}{x+2}\right)(12) = (1)(12) = 12$$

$$(t-3)\left(\frac{2t}{t-3}\right) = \frac{(t-3)(2t)}{t-3} = \left(\frac{t-3}{t-3}\right)(2t) = (1)(2t) = 2t$$

We can use cancelling to shorten the multiplications above. We get:

$$(x+2)\left(\frac{12}{x+2}\right) = 12 \qquad\qquad (t-3)\left(\frac{2t}{t-3}\right) = 2t$$

Use cancelling to complete these.

a) $(m+9)\left(\frac{3}{m+9}\right) =$ _____   c) $(x+1)\left(\frac{x-1}{x+1}\right) =$ _____

b) $(p-7)\left(\frac{4p}{p-7}\right) =$ _____   d) $(2w-3)\left(\frac{2w+3}{2w-3}\right) =$ _____

x = -3

a) 3   c) x - 1

b) 4p   d) 2w + 3

108 Fractional Expressions and Equations

55. To solve the equation below, we began by multiplying both sides by $(x + 3)$. Notice that we had to multiply by the distributive principle on the right side. Solve the other equation.

$$\frac{7}{x + 3} = 5 \qquad\qquad 4 = \frac{y}{y - 1}$$

$$(x + 3)\left(\frac{7}{x + 3}\right) = 5(x + 3)$$

$$7 = 5x + 15$$

$$7 + (-15) = 5x + 15 + (-15)$$

$$-8 = 5x$$

$$x = -\frac{8}{5}$$

---

56. Following the example, solve the other equation.

$$\frac{2x - 1}{2x + 1} = 3 \qquad\qquad 4 = \frac{w + 3}{2w - 3}$$

$$(2x + 1)\left(\frac{2x - 1}{2x + 1}\right) = 3(2x + 1)$$

$$2x - 1 = 6x + 3$$

$$(-2x) + 2x - 1 = (-2x) + 6x + 3$$

$$-1 = 4x + 3$$

$$(-1) + (-3) = 4x + 3 + (-3)$$

$$-4 = 4x$$

$$x = \frac{-4}{4} = -1$$

$y = \frac{4}{3}$

---

57. Following the example, solve the other equation.

$$\frac{t}{t + 2} = 0 \qquad\qquad \frac{3x}{x - 5} = 0$$

$$(t + 2)\left(\frac{t}{t + 2}\right) = 0(t + 2)$$

$$t = 0$$

$w = \frac{15}{7}$

---

58. Following the example, solve the other equation.

$$\frac{x + 3}{x + 5} = 0 \qquad\qquad \frac{3d + 10}{d - 3} = 0$$

$$(x + 5)\left(\frac{x + 3}{x + 5}\right) = 0(x + 5)$$

$$x + 3 = 0$$

$$x + 3 + (-3) = 0 + (-3)$$

$$x = -3$$

$x = 0$

$d = -\frac{10}{3}$

59.   Notice how we wrote the coefficient "1" explicitly when solving the equation
        below. Solve the other equation.

$$\frac{3 - x}{7} = 2 \qquad\qquad 1 = \frac{5}{10 - d}$$

$$7\left(\frac{3 - x}{7}\right) = 7(2)$$

$$3 - 1x = 14$$

$$(-3) + 3 - 1x = 14 + (-3)$$

$$-1x = 11$$

$$x = \frac{11}{-1} = -11$$

d = 5

3-8   EQUATIONS CONTAINING ONE FRACTION WHOSE NUMERATOR
        OR DENOMINATOR IS AN INSTANCE OF THE DISTRIBUTIVE PRINCIPLE

In this section, we will discuss the method for solving fractional equations containing one fraction whose
numerator or denominator is an instance of the distributive principle.

60.   To clear a fraction whose numerator is an instance of the distributive
        principle, we must perform multiplications like those below.

$$3\left[\frac{5(x + 2)}{3}\right] = \frac{(3)(5)(x + 2)}{3} = \left(\frac{3}{3}\right)(5)(x + 2) = (1)(5)(x + 2) = 5(x + 2)$$

$$t\left[\frac{4(t - 1)}{t}\right] = \frac{(t)(4)(t - 1)}{t} = \left(\frac{t}{t}\right)(4)(t - 1) = (1)(4)(t - 1) = 4(t - 1)$$

We can use cancelling to shorten the multiplications above. We get:

$$3\left[\frac{5(x + 2)}{3}\right] = 5(x + 2) \qquad\qquad t\left[\frac{4(t - 1)}{t}\right] = 4(t - 1)$$

Use cancelling to complete these.

a) $9\left[\dfrac{2(R + 7)}{9}\right] =$ _____   c) $2y\left[\dfrac{6(1 - y)}{2y}\right] =$ _____

b) $x\left[\dfrac{5(x - 3)}{x}\right] =$ _____   d) $7d\left[\dfrac{8(2d + 3)}{7d}\right] =$ _____

a) 2(R + 7)   c) 6(1 - y)

b) 5(x - 3)   d) 8(2d + 3)

61.   To solve the equation below, we began by multiplying both sides by 3. Solve the other equation.

$$\frac{2(x + 4)}{3} = 5 \qquad\qquad \frac{3(y - 1)}{5} = 1$$

$$\cancel{3}\left[\frac{2(x + 4)}{\cancel{3}}\right] = 3(5)$$

$$2(x + 4) = 15$$

$$2x + 8 = 15$$

$$2x + 8 + (-8) = 15 + (-8)$$

$$2x = 7$$

$$x = \frac{7}{2}$$

---

62.   To solve the equation below, we began by multiplying both sides by "4t". Solve the other equation.

$$2 = \frac{5(t + 3)}{4t} \qquad\qquad 3 = \frac{6(x - 2)}{x}$$

$$4t(2) = \cancel{4t}\left[\frac{5(t + 3)}{\cancel{4t}}\right]$$

$$8t = 5(t + 3)$$

$$8t = 5t + 15$$

$$8t + (-5t) = (-5t) + 5t + 15$$

$$3t = 15$$

$$t = 5$$

$y = \dfrac{8}{3}$

---

63.   Following the example, solve the other equation.

$$\frac{5(y + 2)}{y} = 0 \qquad\qquad \frac{6(m - 3)}{5} = 0$$

$$\cancel{y}\left[\frac{5(y + 2)}{\cancel{y}}\right] = y(0)$$

$$5(y + 2) = 0$$

$$5y + 10 = 0$$

$$5y + 10 + (-10) = 0 + (-10)$$

$$5y = -10$$

$$y = -2$$

$x = 4$

$m = 3$

64. When an instance of the distributive principle is divided by itself, the quotient is "1". For example:

$$\frac{3(x+5)}{3(x+5)} = 1 \qquad\qquad \frac{5(y-2)}{5(y-2)} = 1$$

The fact above is used to perform multiplications like those below which are needed to clear a fraction whose denominator is an instance of the distributive principle.

$$4(x+3)\left[\frac{7}{4(x+3)}\right] = \frac{(4)(x+3)(7)}{4(x+3)} = \left[\frac{4(x+3)}{4(x+3)}\right](7) = (1)(7) = 7$$

$$8(y-5)\left[\frac{3y}{8(y-5)}\right] = \frac{(8)(y-5)(3y)}{8(y-5)} = \left[\frac{8(y-5)}{8(y-5)}\right](3y) = (1)(3y) = 3y$$

We can use cancelling to shorten the multiplications above. We get:

$$\cancel{4(x+3)}\left[\frac{7}{\cancel{4(x+3)}}\right] = 7 \qquad\qquad \cancel{8(y-5)}\left[\frac{3y}{\cancel{8(y-5)}}\right] = 3y$$

Use cancelling to complete these.

a) $2(q-7)\left[\dfrac{10}{2(q-7)}\right] = $ _____

b) $8(x+8)\left[\dfrac{7x}{8(x+8)}\right] = $ _____

---

65. To clear a fraction when the denominator is an instance of the distributive principle, we also have to perform three-factor multiplications like $(3)(5)(x+2)$ and $(2)(4)(y-3)$. The steps are shown.

$$\cancel{(3)(5)}(x+2) \qquad\qquad\qquad \cancel{(2)(4)}(y-3)$$
$$15\ (x+2) = 15x+30 \qquad\qquad 8\ (y-3) = 8y-24$$

Multiply: a) $(5)(4)(p+4) = $ _____

b) $(3)(2)(m-1) = $ _____

Answer column: a) 10   b) 7x

---

66. To solve the equation below, we began by multiplying both sides by $3(t+2)$. Notice that we had to multiply by the distributive principle on the right side. Solve the other equation.

$$\frac{7}{3(t+2)} = 5 \qquad\qquad 2 = \frac{1}{5(x-2)}$$

$$\cancel{3(t+2)}\left[\frac{7}{\cancel{3(t+2)}}\right] = \cancel{(5)(3)}(t+2)$$

$$7 = 15(t+2)$$

$$7 = 15t+30$$

$$7+(-30) = 15t+30+(-30)$$

$$-23 = 15t$$

$$t = -\frac{23}{15}$$

Answer column: a) $20p+80$ b) $6m-6$

Answer column: $x = \dfrac{21}{10}$

67.   Following the example, solve the other equation.

$$\frac{3y}{2(y-5)} = 1 \qquad\qquad \frac{2m}{4(m+1)} = 3$$

$$\cancel{2(y-5)}\left[\frac{3y}{\cancel{2(y-5)}}\right] = (1)(2)(y-5)$$

$$3y = 2(y-5)$$

$$3y = 2y - 10$$

$$3y + (-2y) = (-2y) + 2y - 10$$

$$1y = -10$$

$$y = -10$$

$$m = -\frac{6}{5}$$

68.   Following the example, solve the other equation.

$$\frac{5x}{3(x+3)} = 0 \qquad\qquad \frac{7R}{8(R-5)} = 0$$

$$\cancel{3(x+3)}\left[\frac{5x}{\cancel{3(x+3)}}\right] = (0)(3)(x+3)$$

$$5x = 0(x+3)$$

$$5x = 0$$

$$x = \frac{0}{5} = 0$$

$$R = 0$$

<hr>

SELF-TEST 10 (pages 104-113)

Solve each equation. Reduce each solution to lowest terms.

1. $\dfrac{2w}{5} = 10$	2. $2 = \dfrac{3}{5x}$	3. $\dfrac{2t - 7}{3t} = 3$
4. $1 = \dfrac{2r + 5}{r - 6}$	5. $\dfrac{4(2d - 3)}{3d} = 0$	6. $4 = \dfrac{P}{2(3P + 1)}$

ANSWERS: 1. $w = 25$ 3. $t = -1$ 5. $d = \dfrac{3}{2}$

　　　　　　　　2. $x = \dfrac{3}{10}$ 4. $r = -11$ 6. $P = -\dfrac{8}{23}$

<hr>

<hr>

3-9   PROPORTIONS WITH NUMERICAL DENOMINATORS

A proportion is a fractional equation that contains only one fraction on each side. In this section, we will discuss a method for solving proportions in which both denominators are numbers.

<hr>

69.   In the proportion below, both denominators are 3. We cleared the fractions by multiplying both sides by 3. Solve the other proportion. Begin by multiplying both sides by 5.

$$\frac{2x}{3} = \frac{5}{3}$$

Check

$$3\left(\frac{2x}{3}\right) = 3\left(\frac{5}{3}\right)$$

$$\frac{3\left(\frac{5}{2}\right)}{3} = \frac{5}{3}$$

$$2x = 5$$

$$\frac{5}{3} = \frac{5}{3}$$

$$x = \frac{5}{2}$$

$5 \cdot \dfrac{7}{5} = \dfrac{4y}{5}$

$\dfrac{35}{5} = 4y$

$7 = 4y$

$\dfrac{7}{4} = y$

$y = \dfrac{7}{4}$

70.   To solve the proportion below, we began by multiplying both sides by 7. Solve the other proportion.

$$\frac{x+2}{7} = \frac{9}{7}$$

$$7\left(\frac{x+2}{7}\right) = 7\left(\frac{9}{7}\right)$$

$$x + 2 = 9$$

$$x + 2 + (-2) = 9 + (-2)$$

$$x = 7$$

$$\frac{2m+3}{4} = \frac{m}{4}$$

[handwritten:] $2m+3 = m$
$3 = -m$
$-3 = m$

---

71.   In the proportion below, the larger denominator is a multiple of the smaller. We cleared the fraction by multiplying both sides by 8. Solve the other proportion. Begin by multiplying both sides by 9.

$$\frac{x}{4} = \frac{7}{8}$$

$$\overset{2}{\cancel{8}}\left(\frac{x}{\cancel{4}}\right) = \cancel{8}\left(\frac{7}{\cancel{8}}\right)$$

$$2x = 7$$

$$x = \frac{7}{2}$$

Check

$$\frac{\frac{7}{2}}{4} = \frac{7}{8}$$

$$\frac{7}{2}\left(\frac{1}{4}\right) = \frac{7}{8}$$

$$\frac{7}{8} = \frac{7}{8}$$

$$\frac{1}{3} = \frac{y}{9}$$

[handwritten:] $\frac{1}{3} \cdot \frac{9}{1} = \frac{y}{9} \cdot \frac{9}{1}$
$3 = y$

[right column:] m = -3

---

72.   When the larger denominator is a multiple of the smaller, we can clear the fractions by multiplying both sides <u>by the larger denominator</u>. Following the example, solve the other proportions.

$$\frac{3m}{5} = \frac{7}{10}$$

$$\overset{2}{\cancel{10}}\left(\frac{3m}{\cancel{5}}\right) = \cancel{10}\left(\frac{7}{\cancel{10}}\right)$$

$$6m = 7$$

$$m = \frac{7}{6}$$

a) $\frac{9R}{4} = \frac{1}{2}$ *[handwritten:]* $\cdot \frac{4}{1}$
$9R = 2$
$R = \frac{2}{9}$

b) $\frac{3}{8} = \frac{5d}{2}$ *[handwritten:]* $\cdot \frac{8}{1}$
$3 = 20d$
$\frac{3}{20} = d$

[right column:] y = 3

---

73.   Following the example, solve the other proportion.

$$\frac{5-x}{12} = \frac{1}{4}$$

$$\cancel{12}\left(\frac{5-x}{\cancel{12}}\right) = \overset{3}{\cancel{12}}\left(\frac{1}{\cancel{4}}\right)$$

$$5 - 1x = 3$$

$$(-5) + 5 - 1x = 3 + (-5)$$

$$-1x = -2$$

$$x = \frac{-2}{-1} = 2$$

[handwritten:] $\frac{5}{1} \overset{10}{\cancel{\frac{1}{2}}} \,\frac{1}{2} = \frac{5-y}{10}$
$5 = 5 - y$
$0 = y$

[right column:]
a) $R = \frac{2}{9}$, from: $9R = 2$

b) $d = \frac{3}{20}$, from: $3 = 20d$

74. Following the example, solve the other proportion.

$$\frac{5}{6} = \frac{m+2}{3}$$

$$\cancel{6}\left(\frac{5}{\cancel{6}}\right) = \overset{2}{\cancel{6}}\left(\frac{m+2}{\cancel{3}}\right)$$

$$5 = 2(m+2)$$

$$5 = 2m + 4$$

$$5 + (-4) = 2m + 4 + (-4)$$

$$1 = 2m$$

$$m = \frac{1}{2}$$

(handwritten)
$$\overset{3}{\cancel{\frac{6}{1}}} \cdot \frac{t+1}{\cancel{2}_1} = \frac{1}{6}$$

$$3t + 3 = 1$$
$$3t = -2$$
$$t = -\frac{2}{3}$$

$y = 0$, from:

$$5 = 5 - 1y$$

---

75. In the proportion below, the larger denominator is not a multiple of the smaller. To clear the fractions, we multiplied both sides by 12, the smallest common multiple of both denominators. To solve the other equation, begin by multiplying both sides by 24.

$$\frac{x}{3} = \frac{3}{4}$$

$$\overset{4}{\cancel{12}}\left(\frac{x}{\cancel{3}}\right) = \overset{3}{\cancel{12}}\left(\frac{3}{\cancel{4}}\right)$$

$$4x = 9$$

$$x = \frac{9}{4}$$

(handwritten)
$$\overset{4}{\cancel{24}} \cdot \frac{1}{\cancel{6}_1} = \frac{y}{\cancel{8}_1} \cdot \overset{3}{\cancel{\frac{24}{1}}}$$

$$4 = 3y$$
$$\frac{4}{3} = y$$

$t = -\frac{2}{3}$, from:

$$3t + 3 = 1$$

---

76. When the larger denominator is not a multiple of the smaller, we can clear the fractions in a proportion by multiplying both sides <u>by the smallest common multiple of both denominators</u>. Identify the multiplier we would use to clear the fractions in each proportion below.

a) $\dfrac{3x}{2} = \dfrac{7}{3}$ _6_   b) $\dfrac{y-1}{7} = \dfrac{4}{5}$ _35_   c) $\dfrac{1}{4} = \dfrac{7-t}{6}$ _12_

$y = \frac{4}{3}$, from:

$$4 = 3y$$

---

77. Solve: a) $\dfrac{2t}{5} = \dfrac{1}{3}$ b) $\dfrac{m+4}{10} = \dfrac{m}{4}$

(handwritten a)
$$\overset{3}{\cancel{\frac{15}{1}}} \cdot \frac{2t}{\cancel{5}_1} = \frac{1}{\cancel{3}_1} \cdot \overset{5}{\cancel{\frac{15}{1}}}$$

$$6t = 5$$
$$t = \frac{5}{6}$$

(handwritten b)
$$\overset{2}{\cancel{\frac{20}{1}}} \cdot \frac{m+4}{\cancel{10}} = \frac{m}{\cancel{4}_1} \cdot \overset{5}{\cancel{\frac{20}{1}}}$$

$$2m + 8 = 5m$$
$$8 = 3m$$
$$\frac{8}{3} = m$$

---

a) 6   b) 35   c) 12

78.   Solve: a) $\dfrac{5}{6} = \dfrac{3y}{10}$ b) $\dfrac{2F-1}{5} = \dfrac{1}{2}$

$\dfrac{\overset{5}{\cancel{30}}}{1} \cdot \dfrac{5}{\cancel{6}} = \dfrac{3y}{\underset{1}{\cancel{10}}} \cdot \dfrac{\overset{3}{\cancel{30}}}{1}$ $\dfrac{\overset{2}{\cancel{10}}}{1} \cdot \dfrac{2F-1}{\underset{1}{\cancel{5}}} = \dfrac{1}{\cancel{2}} \cdot \dfrac{\overset{5}{\cancel{10}}}{1}$

$25 = 9y$ $4F - 2 = 5$
$\dfrac{25}{9} = y$ $4F = 7$
   $F = \dfrac{7}{4}$

a) $t = \dfrac{5}{6}$, from:

   $6t = 5$

b) $m = \dfrac{8}{3}$, from:

   $2m + 8 = 5m$

a) $y = \dfrac{25}{9}$ b) $F = \dfrac{7}{4}$

3-10   MORE-COMPLICATED EQUATIONS WITH NUMERICAL DENOMINATORS

In this section, we will discuss a method for solving fractional equations that contain two terms on a side. The equations are limited to those with numerical denominators.

---

79.   The equation below has two terms on the left side. To clear the fraction, we multiplied both sides by 4. Notice that we multiplied by the distributive principle on the left side. Complete the solution of the other equation.

$\dfrac{x}{4} + 2 = 3$ $1 = \dfrac{5t}{2} - 3$

$4\left(\dfrac{x}{4} + 2\right) = 4(3)$ $2(1) = 2\left(\dfrac{5t}{2} - 3\right)$

$\cancel{4}\left(\dfrac{x}{\cancel{4}}\right) + 4(2) = 12$ $2 = \cancel{2}\left(\dfrac{5t}{\cancel{2}}\right) - 2(3)$

$x + 8 = 12$ $2 = 5t - 6$

$x + 8 + (-8) = 12 + (-8)$ $8 = 5t$

$x = 4$ $\dfrac{8}{5} = t$

---

80.   To solve equations like those in the last frame, we must multiply by the distributive principle on one side. Some examples of multiplications of that type are shown.

$3\left(\dfrac{y}{3} + 5\right) = \cancel{3}\left(\dfrac{y}{\cancel{3}}\right) + 3(5) = y + 15$

$7\left(4 - \dfrac{2m}{7}\right) = 7(4) - \cancel{7}\left(\dfrac{2m}{\cancel{7}}\right) = 28 - 2m$

Following the examples, complete these:

a) $10\left(1 + \dfrac{7d}{10}\right) = \left(10\right)\left(1\right) + \left(10\right)\left(\dfrac{7d}{10}\right) = \underline{10 + 7d}$

b) $8\left(\dfrac{p}{8} - 3\right) = \left(8\right)\left(\dfrac{p}{8}\right) - \left(8\right)\left(-3\right) = \underline{p - 24}$

$t = \dfrac{8}{5}$, from:

   $2 = 5t - 6$

81.   Following the example, solve the other equation. Be sure to multiply <u>both</u> <u>sides</u> by 3.

$$d + \frac{3d}{5} = 2$$

$$5\left(d + \frac{3d}{5}\right) = 5(2)$$

$$5(d) + \cancel{5}\left(\frac{3d}{\cancel{5}}\right) = 10$$

$$5d + 3d = 10$$

$$8d = 10$$

$$d = \frac{5}{4}$$

$$\frac{m}{3} + 1 = 5$$

$$3\left(\frac{m}{3} + 1\right) = 15$$

$$m + 3 = 15$$

$$m = 12$$

a) $10(1) + \cancel{10}\left(\frac{7d}{\cancel{10}}\right)$

$$= 10 + 7d$$

b) $\cancel{8}\left(\frac{p}{\cancel{8}}\right) - 8(3) = p - 24$

---

82.   To clear the fractions below, we multiplied both sides by 2. Solve the other equation. Begin by multiplying both sides by 4.

$$\frac{x}{2} + 5 = \frac{7}{2}$$

$$2\left(\frac{x}{2} + 5\right) = \cancel{2}\left(\frac{7}{\cancel{2}}\right)$$

$$\cancel{2}\left(\frac{x}{\cancel{2}}\right) + 2(5) = 7$$

$$x + 10 = 7$$

$$x + 10 + (-10) = 7 + (-10)$$

$$x = -3$$

$$\frac{3}{4} - t = \frac{5t}{4}$$

$$4\left(\frac{3}{4}\right) - 4(t) = 4\left(\frac{5t}{4}\right)$$

$$3 + 4t = 5t$$

$$3 = 9t$$

$$\frac{3}{9} = t$$

$$\frac{1}{3} = t$$

$m = 12$, from:

$$m + 3 = 15$$

$$\frac{x}{12} = x + \frac{1}{4}$$

$$12\left(\frac{x}{12}\right) = (12)(x)(12)\left(\frac{1}{4}\right)$$

$$x = 12x + 3$$

$$-11x = 3$$

$$x = \frac{3}{11}$$

---

83.   In the equation below, the larger denominator is a multiple of the smaller. We cleared the fractions by multiplying both sides by 12. Complete the other solution.

$$\frac{x}{12} = x + \frac{1}{4}$$

$$\cancel{12}\left(\frac{x}{\cancel{12}}\right) = 12\left(x + \frac{1}{4}\right)$$

$$x = 12(x) + \overset{3}{\cancel{12}}\left(\frac{1}{\cancel{4}}\right)$$

$$x = 12x + 3$$

$$x + (-12x) = (-12x) + 12x + 3$$

$$-11x = 3$$

$$x = -\frac{3}{11}$$

$$1 - \frac{y}{4} = \frac{3y}{8}$$

$$8\left(1 - \frac{y}{4}\right) = \cancel{8}\left(\frac{3y}{\cancel{8}}\right)$$

$$8(1) - \overset{2}{\cancel{8}}\left(\frac{y}{\cancel{4}}\right) = 3y$$

$$8 + 2y = 3y$$

$$8 = 5y$$

$$\frac{8}{5} = y$$

$t = \frac{1}{3}$, from:

$$3 - 4t = 5t$$

$y = \frac{8}{5}$, from:

$$8 - 2y = 3y$$

84.   In the equation below, the larger denominator is not a multiple of the smaller. We cleared the fractions by multiplying both sides by 10, the smallest common multiple of both denominators. Complete the other solution.

$$\frac{3x}{2} - 3 = \frac{x}{5}$$

$$10\left(\frac{3x}{2} - 3\right) = \overset{2}{\cancel{10}}\left(\frac{x}{\cancel{5}}\right)$$

$$\overset{5}{\cancel{10}}\left(\frac{3x}{\cancel{2}}\right) - 10(3) = 2x$$

$$15x - 30 = 2x$$

$$(-15x) + 15x - 30 = 2x + (-15x)$$

$$-30 = -13x$$

$$x = \frac{-30}{-13} = \frac{30}{13}$$

$$\frac{y}{4} = \frac{5y}{6} + 2$$

$$\overset{3}{\cancel{12}}\left(\frac{y}{\cancel{4}}\right) = 12\left(\frac{5y}{6} + 2\right)$$

$$3y = \overset{2}{\cancel{12}}\left(\frac{5y}{\cancel{6}}\right) + 12(2)$$

(handwritten)
$$3y = 10y + 24$$
$$-7y = 24$$
$$y = \frac{-24}{7}$$

---

85.   To clear the fractions when an equation contains two different numerical denominators:

    1) Use the <u>larger</u> <u>denominator</u> if it is a multiple of the smaller.

    2) Use the <u>smallest</u> <u>common</u> <u>multiple</u> of both denominators if the larger denominator is not a multiple of the smaller.

Solve: a) $1 = \frac{3t}{4} - \frac{2t}{3}$

(handwritten)
$$\frac{1}{1} \cdot \frac{12}{1} = \overset{3}{\cancel{12}}\left(\frac{3t}{\cancel{4}_1}\right) - \overset{4}{\cancel{12}}\left(\frac{2t}{\cancel{3}_1}\right)$$
$$12 = 9t - 8t$$
$$12 = t$$

   b) $\frac{m}{20} + 1 = \frac{m}{4}$

(handwritten)
$$m + 20 = 5m$$
$$20 = 4m$$
$$5 = m$$

y = -$\frac{24}{7}$, from:

$$3y = 10y + 24$$

---

86.   Following the example, solve the other equation.

$$\frac{5x}{3} + \frac{1}{2} = 0$$

$$6\left(\frac{5x}{3} + \frac{1}{2}\right) = 6(0)$$

$$\overset{2}{\cancel{6}}\left(\frac{5x}{\cancel{3}}\right) + \overset{3}{\cancel{6}}\left(\frac{1}{\cancel{2}}\right) = 0$$

$$10x + 3 = 0$$

$$10x + 3 + (-3) = 0 + (-3)$$

$$10x = -3$$

$$x = -\frac{3}{10}$$

$$0 = \frac{y}{8} - \frac{1}{2}$$

(handwritten)
$$0 = y - 4$$
$$4 = y$$

a)   t = 12, from:

     12 = 9t - 8t

b)   m = 5, from:

     m + 20 = 5m

87. The equation below contains three fractions whose denominators are identical. To clear the fractions, we multiplied both sides by 6. Notice that we had to multiply by the distributive principle on both sides. Solve the other equation.

$$\frac{x}{6} + 1 = \frac{2x}{6} - \frac{5}{6}$$

$$6\left(\frac{x}{6} + 1\right) = 6\left(\frac{2x}{6} - \frac{5}{6}\right)$$

$$6\left(\frac{x}{6}\right) + 6(1) = 6\left(\frac{2x}{6}\right) - 6\left(\frac{5}{6}\right)$$

$$x + 6 = 2x - 5$$

$$(-x) + x + 6 = (-x) + 2x - 5$$

$$6 = x - 5$$

$$5 + 6 = x - 5 + 5$$

$$x = 11$$

$$1 - \frac{t}{2} = \frac{3t}{2} + \frac{1}{2}$$

$2 - t = 3t + 1$

$1 = 4t$

$\frac{1}{4} = t$

$y = 4$, from:

$$0 = y - 4$$

88. When an equation contains more than two fractions and the denominators are not identical, we can clear the fractions by multiplying both sides by the smallest common multiple of the denominators. That is:

For $\frac{x}{2} + \frac{x}{5} = \frac{3x}{10} + 4$, we multiply both sides by 10.

For $\frac{5y}{7} - 1 = \frac{y}{2} + \frac{1}{2}$, we multiply both sides by 14.

What number would we multiply both sides by to clear the fractions in each equation below?

a) $\frac{m}{2} + \frac{3m}{8} = \frac{m}{8} - 10$ 8   b) $\frac{t}{3} - \frac{4t}{5} = 1 + \frac{t}{3}$ 15

$t = \frac{1}{4}$, from:

$$2 - t = 3t + 1$$

89. Solve each equation.

a) $\frac{d}{2} - \frac{d}{4} = \frac{d-1}{8}$

$4d - 2d = d - 1$

$2d = d - 1$

$d = -1$

b) $\frac{F}{4} + \frac{3}{5} = 2 - \frac{3F}{5}$

$5F + 12 = 40 - 12F$

$17F = 28$

$F = \frac{28}{17}$

a) 8   b) 15

a) $d = -1$, from:

$$4d - 2d = d - 1$$

b) $F = \frac{28}{17}$, from:

$$5F + 12 = 40 - 12F$$

3-11   PROPORTIONS WITH LITERAL DENOMINATORS

In this section, we will discuss a method for solving proportions in which one or both denominators contain a letter.

---

90.   When a proportion contains a number as one denominator and a letter as the other denominator, we can clear the fractions by multiplying both sides by a letter term in which the numerical denominator is the coefficient of the letter. That is:

$$\text{For } \frac{6}{5} = \frac{9}{x} , \text{ we multiply by } 5x.$$

$$\text{For } \frac{y-7}{y} = \frac{1}{2} , \text{ we multiply by } 2y.$$

To clear the fraction below, we multiplied both sides by 8x. Solve the other proportion.

$$\frac{7}{8} = \frac{3}{x} \qquad\qquad\qquad \frac{4}{y} = \frac{3}{2}$$

$$\overset{x}{\cancel{8x}}\left(\frac{7}{\cancel{8}}\right) = \overset{8}{\cancel{8x}}\left(\frac{3}{\cancel{x}}\right)$$

$$7x = 24$$

$$x = \frac{24}{7}$$

[handwritten:] $2y\left(\frac{4}{y}\right) = 2y\left(\frac{3}{2}\right)$

$8 = 3y$

$\dfrac{8}{3} = y$

---

91.   To clear the fractions below, we multiplied both sides by 4t. Solve the other proportion.

$$\frac{t+8}{t} = \frac{3}{4} \qquad\qquad\qquad \frac{4}{3} = \frac{d-1}{d}$$

$$\overset{4}{\cancel{4t}}\left(\frac{t+8}{\cancel{x}}\right) = \overset{t}{\cancel{4t}}\left(\frac{3}{\cancel{4}}\right)$$

$$4(t+8) = 3t$$

$$4t + 32 = 3t$$

$$(-4t) + 4t + 32 = 3t + (-4t)$$

$$32 = -1t$$

$$t = \frac{32}{-1} = -32$$

[handwritten:] $3d\left(\frac{4}{3}\right) = 3d\left(\frac{d-1}{d}\right)$

$4d = 3d - 3$

$d = -3$

$y = \dfrac{8}{3}$, from:

$8 = 3y$

$d = -3$, from:

$4d = 3(d-1)$

92. To clear the fractions in equations like those below, we multiply both sides by a letter term in which the coefficient is the <u>smallest</u> <u>common</u> <u>multiple</u> of the numerical factors. Therefore:

For $\dfrac{5}{3x} = \dfrac{1}{2}$, we multiply by 6x.

For $\dfrac{7}{10} = \dfrac{3}{2d}$, we multiply by 10d.

For $\dfrac{t-2}{4t} = \dfrac{1}{6t}$, we multiply by 12t.

Following the example, solve the other proportion.

$$\dfrac{5}{3x} = \dfrac{1}{2}$$

$$\overset{2}{6x}\left(\dfrac{5}{3x}\right) = \overset{3x}{6x}\left(\dfrac{1}{2}\right)$$

$$10 = 3x$$

$$x = \dfrac{10}{3}$$

$$\dfrac{7}{10} = \dfrac{3}{2d}$$

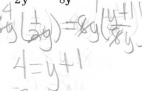

---

93. Following the example, solve the other proportion.

$$\dfrac{t-2}{4t} = \dfrac{1}{3t}$$

$$\overset{3}{12t}\left(\dfrac{t-2}{4t}\right) = \overset{4}{12t}\left(\dfrac{1}{3t}\right)$$

$$3(t-2) = 4$$

$$3t - 6 = 4$$

$$3t - 6 + 6 = 4 + 6$$

$$3t = 10$$

$$t = \dfrac{10}{3}$$

$$\dfrac{1}{2y} = \dfrac{y+1}{8y}$$

$d = \dfrac{15}{7}$, from:

$7d = 15$

---

94. When one denominator is a number and the other is an addition or subtraction, we can clear the fractions by multiplying both sides by the number <u>and</u> the addition or subtraction. For example:

For $\dfrac{2}{5} = \dfrac{8}{x+3}$, we multiply by 5(x + 3).

For $\dfrac{10}{y-7} = \dfrac{5}{3}$, we multiply by 3(y - 7).

$y = 3$, from:

$4 = y + 1$

Continued on following page.

94.   Continued

To clear the fractions in proportions, we must do multiplications like these:

$$5(x + 3)\left(\frac{8}{x + 3}\right) \;=\; \frac{(5)(x + 3)(8)}{x + 3} \;=\; \left(\frac{x + 3}{x + 3}\right)(5)(8) \;=\; (1)(5)(8) \;=\; 40$$

$$3(y - 7)\left(\frac{10}{y - 7}\right) \;=\; \frac{(3)(y - 7)(10)}{y - 7} \;=\; \left(\frac{y - 7}{y - 7}\right)(3)(10) \;=\; (1)(3)(10) \;=\; 30$$

We can use cancelling to perform the same multiplications. That is:

$$5(x + 3)\left(\frac{8}{x + 3}\right) \;=\; 40 \qquad\qquad 3(y - 7)\left(\frac{10}{y - 7}\right) \;=\; \underline{30}$$

---

95.   To clear the fractions in proportions like those in the last frame, we also have to do multiplications like these:

$$\frac{2}{5}(5)(x + 3) \;=\; \frac{(2)(5)(x + 3)}{5} \;=\; \left(\frac{5}{5}\right)(2)(x + 3) \;=\; (1)(2)(x + 3) \;=\; 2(x + 3)$$

$$\frac{5}{3}(3)(y - 7) \;=\; \frac{(5)(3)(y - 7)}{3} \;=\; \left(\frac{3}{3}\right)(5)(y - 7) \;=\; (1)(5)(y - 7) \;=\; 5(y - 7)$$

We can use cancelling to perform the same multiplications. That is:

$$\frac{2}{5}(5)(x + 3) \;=\; 2(x + 3) \qquad\qquad \frac{5}{3}(3)(y - 7) \;=\; \underline{5(y-7)}$$

> 30

---

96.   Following the example, solve the other proportion.

$$\frac{16}{t + 9} = \frac{4}{5} \qquad\qquad\qquad\qquad \frac{3m}{m - 5} = \frac{3}{2}$$

$$5(t + 9)\left(\frac{16}{t + 9}\right) = \frac{4}{5}(5)(t + 9)$$

$$80 = 4(t + 9)$$

$$80 = 4t + 36$$

$$80 + (-36) = 4t + 36 + (-36)$$

$$44 = 4t$$

$$t = 11$$

[Handwritten solution for the second proportion:]

$$2(m-5)\left(\frac{3m}{m-5}\right) = \frac{3}{2}(m-5)$$

$$6m = 3m - 15$$

$$3m = -15$$

$$m = -5$$

> 5(y − 7)

> m = −5, from:
>
> 6m = 3m − 15

97.   Following the example, solve the other proportion.

$$\frac{5}{4} = \frac{1}{3a - 4}$$

$$\frac{5}{4}(4)(3a - 4) = 4(3a - 4)\left(\frac{1}{3a - 4}\right)$$

$$5(3a - 4) = 4$$

$$15a - 20 = 4$$

$$15a - 20 + 20 = 4 + 20$$

$$15a = 24$$

$$a = \frac{24}{15} = \frac{8}{5}$$

$$\frac{4}{5} = \frac{3}{2b - 1}$$

$$5(2b-1)\left(\frac{4}{5}\right) = 5(2b-1)\left(\frac{3}{2b-1}\right)$$

$$8b - 4 = 15$$

$$8b = 19$$

$$b = \frac{19}{8}$$

---

98.   To clear the fractions below, we multiplied by both "x" and (x + 2).
Solve the other proportion.

$$\frac{5}{x + 2} = \frac{3}{x}$$

$$x(x + 2)\left(\frac{5}{x + 2}\right) = \left(\frac{3}{x}\right)(x)(x + 2)$$

$$5x = 3(x + 2)$$

$$5x = 3x + 6$$

$$5x + (-3x) = (-3x) + 3x + 6$$

$$2x = 6$$

$$x = 3$$

$$\frac{4}{m - 3} = \frac{2}{m}$$

$$m(m-3)\left(\frac{4}{m-3}\right) = m(m-3)\left(\frac{2}{m}\right)$$

$$4m = 2m - 6$$

$$2m = -6$$

$$m = -3$$

$b = \dfrac{19}{8}$, from:

$$8b - 4 = 15$$

---

99.   Following the example, solve the other proportion.

$$\frac{3}{t} = \frac{9}{2t - 1}$$

$$\left(\frac{3}{t}\right)(t)(2t - 1) = t(2t - 1)\left(\frac{9}{2t - 1}\right)$$

$$3(2t - 1) = 9t$$

$$6t - 3 = 9t$$

$$(-6t) + 6t - 3 = 9t + (-6t)$$

$$-3 = 3t$$

$$t = \frac{-3}{3} = -1$$

$$\frac{5}{F} = \frac{1}{3F + 4}$$

$m = -3$, from:

$$4m = 2m - 6$$

$F = -\dfrac{10}{7}$, from:

$$15F + 20 = F$$

100. To clear the fraction below, we multiplied both sides by 3t and (t + 1).
      Solve the other proportion.

$$\frac{7}{3t} = \frac{5}{t + 1}$$ $\qquad\qquad$ $$\frac{3}{m - 2} = \frac{1}{4m}$$

$$\left(\frac{7}{\cancel{3t}}\right)(\cancel{3t})(t + 1) = 3t(\cancel{t+1})\left(\frac{5}{\cancel{t+1}}\right)$$

$$7(t + 1) = 15t$$

$$7t + 7 = 15t$$

$$(-7t) + 7t + 7 = 15t + (-7t)$$

$$7 = 8t$$

$$t = \frac{7}{8}$$

---

101. When one denominator is an instance of the distributive principle, we
      can clear the fractions by multiplying both sides by the smallest common
      multiple of the numerical factors, any letter, and the addition or
      subtraction. For example:

      For $\dfrac{5}{2(x + 3)} = \dfrac{2}{3}$, we multiply by 6(x + 3).

      For $\dfrac{1}{4(y - 1)} = \dfrac{7}{8}$, we multiply by 8(y - 1).

      For $\dfrac{2}{3(d + 1)} = \dfrac{1}{d}$, we multiply by 3d(d + 1).

      To clear the fractions, we must multiply each of the following proportions
      by what?

      a) $\dfrac{x}{3(x - 1)} = \dfrac{1}{3}$ $\qquad$ b) $\dfrac{2y - 5}{7(y + 3)} = \dfrac{1}{2}$ $\qquad$ c) $\dfrac{3}{5(V - 2)} = \dfrac{7}{5V}$

      _____   _____   _____

$m = -\dfrac{2}{11}$, from:

$$12m = m - 2$$

---

102. Following the example, solve the other proportion.

$$\frac{2}{3(d + 1)} = \frac{1}{d}$$ $\qquad\qquad$ $$\frac{3}{2(x + 1)} = \frac{1}{x}$$

$$\overset{d}{\cancel{3d(d+1)}}\left[\frac{2}{\cancel{3(d+1)}}\right] = \left(\frac{1}{\cancel{d}}\right)\overset{3}{\cancel{(3d)}}(d + 1)$$

$$2d = 3(d + 1)$$

$$2d = 3d + 3$$

$$2d + (-3d) = (-3d) + 3d + 3$$

$$-1d = 3$$

$$d = \frac{3}{-1} = -3$$

a) 3(x - 1)

b) 14(y + 3)

c) 5V(V - 2)

x = 2 , from:

$$3x = 2x + 2$$

SELF-TEST 11 (pages 113-125)

Solve each equation. Report each solution in lowest terms.

1. $\dfrac{5x}{3} = \dfrac{7}{12}$

2. $\dfrac{2t-3}{6} = \dfrac{t}{9}$

3. $\dfrac{d}{2} = 3d - \dfrac{5}{8}$

4. $\dfrac{2w}{3} - 1 = \dfrac{w}{2} + \dfrac{5}{3}$

5. $\dfrac{5}{6y} = \dfrac{y+2}{2y}$

6. $\dfrac{1}{r+1} = \dfrac{3}{2r}$

ANSWERS: 1. $x = \dfrac{7}{20}$ 2. $t = \dfrac{9}{4}$ 3. $d = \dfrac{1}{4}$ 4. $w = 16$ 5. $y = -\dfrac{1}{3}$ 6. $r = -3$

3-12   MORE-COMPLICATED EQUATIONS WITH LITERAL DENOMINATORS

In this section, we will discuss a method for solving fractional equations that contain two terms on a side when the denominators contain a letter.

103.   The equation below has two terms on the left side. To clear the fraction, we multiplied both sides by "x". Notice that we multiplied by the distributive principle on the left side. Solve the other equation. Begin by multiplying both sides by "y".

$$\frac{5}{x} + 3 = 9 \qquad\qquad 3 = \frac{4}{y} + 1$$

$$x\left(\frac{5}{x} + 3\right) = x(9)$$

$$\cancel{x}\left(\frac{5}{\cancel{x}}\right) + x(3) = 9x$$

$$5 + 3x = 9x$$

$$5 + 3x + (-3x) = 9x + (-3x)$$

$$5 = 6x$$

$$x = \frac{5}{6}$$

104.   To clear the fraction below, we multiplied both sides by 3t. Solve the other equation.

$$2 - \frac{45}{3t} = 5 \qquad\qquad 7 = 10 - \frac{5}{2m}$$

$$3t\left(2 - \frac{45}{3t}\right) = 3t(5)$$

$$3t(2) - \cancel{3t}\left(\frac{45}{\cancel{3t}}\right) = 15t$$

$$6t - 45 = 15t$$

$$(-6t) + 6t - 45 = 15t + (-6t)$$

$$-45 = 9t$$

$$t = -5$$

$y = 2$, from:

$$3y = 4 + 1y$$

---

105.   To solve the equation below, we began by multiplying both sides by "x". Solve the other equation.

$$\frac{2}{x} = \frac{3}{x} + 5 \qquad\qquad 4 - \frac{5}{w} = \frac{1}{w}$$

$$\cancel{x}\left(\frac{2}{\cancel{x}}\right) = x\left(\frac{3}{x} + 5\right)$$

$$2 = \cancel{x}\left(\frac{3}{\cancel{x}}\right) + x(5)$$

$$2 = 3 + 5x$$

$$2 + (-3) = (-3) + 3 + 5x$$

$$-1 = 5x$$

$$x = -\frac{1}{5}$$

$m = \dfrac{5}{6}$, from:

$$14m = 20m - 5$$

---

106.   To clear the fraction below, we multiplied both sides by 5y. Solve the other equation.

$$\frac{17}{5y} = 2 - \frac{11}{5y} \qquad\qquad \frac{1}{3x} + 4 = \frac{7}{3x}$$

$$\cancel{5y}\left(\frac{17}{\cancel{5y}}\right) = 5y\left(2 - \frac{11}{5y}\right)$$

$$17 = 5y(2) - \cancel{5y}\left(\frac{11}{\cancel{5y}}\right)$$

$$17 = 10y - 11$$

$$17 + 11 = 10y - 11 + 11$$

$$28 = 10y$$

$$y = \frac{28}{10} = \frac{14}{5}$$

$w = \dfrac{3}{2}$, from:

$$4m - 5 = 1$$

107. To clear the fractions below, we multiplied both sides by 2x. Solve the other equation.

$$\frac{3}{x} + \frac{7}{2} = 1 \qquad\qquad \frac{1}{3} = \frac{5}{m} - 4$$

$$2x\left(\frac{3}{x} + \frac{7}{2}\right) = 2x(1)$$

$$\overset{2}{\cancel{2x}}\left(\frac{3}{\cancel{x}}\right) + \overset{x}{\cancel{2x}}\left(\frac{7}{\cancel{2}}\right) = 2x$$

$$6 + 7x = 2x$$

$$6 + 7x + (-7x) = 2x + (-7x)$$

$$6 = -5x$$

$$x = -\frac{6}{5}$$

$x = \frac{1}{2}$, from:

$$1 + 12x = 7$$

---

108. To clear the fractions below, we multiplied both sides by 3F. Solve the other equation.

$$\frac{1}{3F} - 2 = \frac{5}{F} \qquad\qquad \frac{15}{4d} = 1 + \frac{5}{d}$$

$$3F\left(\frac{1}{3F} - 2\right) = \overset{3}{\cancel{3F}}\left(\frac{5}{\cancel{F}}\right)$$

$$\cancel{3F}\left(\frac{1}{\cancel{3F}}\right) - 3F(2) = 15$$

$$1 - 6F = 15$$

$$(-1) + 1 - 6F = 15 + (-1)$$

$$-6F = 14$$

$$F = -\frac{14}{6} = -\frac{7}{3}$$

$m = \frac{15}{13}$, from:

$$m = 15 - 12m$$

---

109. To clear the fractions below, we multiplied both sides by 4t. Solve the other equation.

$$\frac{3}{t} + \frac{1}{4} = \frac{5}{4t} \qquad\qquad \frac{3}{5} - \frac{1}{5m} = \frac{4}{m}$$

$$4t\left(\frac{3}{t} + \frac{1}{4}\right) = \cancel{4t}\left(\frac{5}{\cancel{4t}}\right)$$

$$\overset{4}{\cancel{4t}}\left(\frac{3}{\cancel{t}}\right) + \overset{t}{\cancel{4t}}\left(\frac{1}{\cancel{4}}\right) = 5$$

$$12 + t = 5$$

$$(-12) + 12 + t = 5 + (-12)$$

$$t = -7$$

$d = -\frac{5}{4}$, from:

$$15 = 4d + 20$$

$m = 7$, from:

$$3m - 1 = 20$$

110. When the denominators contain various numerical factors and a letter, we can clear the fractions by multiplying by the <u>smallest</u> <u>common</u> <u>multiple</u> of the numbers and the letter. For example:

$$\text{For } \frac{3}{2x} - 1 = \frac{5}{3x}, \text{ we multiply by 6x.}$$

$$\text{For } \frac{1}{4} = \frac{1}{8} + \frac{1}{y}, \text{ we multiply by 8y.}$$

What multiplier would we use to clear the fractions in each equation below?

a) $\frac{7}{2d} = 1 - \frac{3}{10d}$ b) $\frac{5}{x} - \frac{1}{4} = \frac{2}{3x}$ c) $\frac{1}{4} - \frac{1}{y} = \frac{1}{6}$

_____   _____   _____

---

111. Solve each equation.

a) $\frac{3}{2x} - 4 = \frac{1}{10x}$

b) $\frac{1}{3} = \frac{1}{9} + \frac{1}{y}$

a) 10d

b) 12x

c) 12y

---

112. Sometimes we have to multiply by the distributive principle on both sides to clear the fractions. An example is shown. Solve the other equation.

$$\frac{1}{m} - \frac{3}{4} = \frac{1}{4} + \frac{5}{m}$$

$$\frac{1}{2} + \frac{2}{3y} = \frac{1}{y} - \frac{1}{3}$$

$$4m\left(\frac{1}{m} - \frac{3}{4}\right) = 4m\left(\frac{1}{4} + \frac{5}{m}\right)$$

$$\overset{4}{\cancel{4m}}\left(\frac{1}{m}\right) - \overset{m}{\cancel{4m}}\left(\frac{3}{4}\right) = \overset{m}{\cancel{4m}}\left(\frac{1}{4}\right) + \overset{4}{\cancel{4m}}\left(\frac{5}{m}\right)$$

$$4 - 3m = 1m + 20$$

$$(-4) + 4 - 3m = 1m + 20 + (-4)$$

$$-3m = 1m + 16$$

$$-3m + (-1m) = (-1m) + 1m + 16$$

$$-4m = 16$$

$$m = -4$$

a) $x = \frac{7}{20}$, from:

   $15 - 40x = 1$

b) $y = \frac{9}{2}$, from:

   $3y = y + 9$

$y = \frac{2}{5}$, from:

$3y + 4 = 6 - 2y$

3-13   FORMULA EVALUATIONS

In this section, we will discuss formula evaluations with formulas containing fractions.

---

113.   Any fraction containing letters stands for a division. For example:

$\dfrac{E}{R}$ means: E ÷ R $\dfrac{bh}{2}$ means: _____ ÷ _____

---

114.   To perform the evaluation below, we simply divided. Complete the other evaluation.

   In $\boxed{I = \dfrac{E}{R}}$,   find I when   E = 40 and R = 5.

   $I = \dfrac{E}{R} = \dfrac{40}{5} = 8$

   In $\boxed{t = \dfrac{1}{f}}$,   find "t" when f = 100.

   $t = \dfrac{1}{f} =$ _____

bh ÷ 2

---

115.   In the evaluation below, we performed the multiplication in the numerator first. Complete the other evaluation.

   In $\boxed{V = \dfrac{4st}{a}}$,   find V when s = 5, t = 3, and a = 10.

   $V = \dfrac{4st}{a} = \dfrac{4(5)(3)}{10} = \dfrac{60}{10} = 6$

   In $\boxed{P = \dfrac{Fs}{t}}$,   find P when F = 20, s = 100, and t = 10.

   $P = \dfrac{Fs}{t} =$

$t = .01$, from $\dfrac{1}{100}$

---

116.   In the evaluation below, we performed the addition in the numerator first. Complete the other evaluation.

   In $\boxed{R = \dfrac{p+q}{2}}$,   find R when p = 20 and q = 30.

   $R = \dfrac{p+q}{2} = \dfrac{20+30}{2} = \dfrac{50}{2} = 25$

   In $\boxed{t = \dfrac{a-b}{c-d}}$,   find "t" when a = 30, b = 10, c = 10, and d = 6.

   $t = \dfrac{a-b}{c-d} =$ _____

$P = 200$, from $\dfrac{2,000}{10}$

$t = 5$, from $\dfrac{20}{4}$

117.   Following the example, complete the other evaluation.

In $\boxed{\alpha = \dfrac{\beta}{\beta + 1}}$, find α when $\beta = 99$.

$$\alpha = \frac{\beta}{\beta + 1} = \frac{99}{99 + 1} = \frac{99}{100} = .99$$

In $\boxed{S = \dfrac{PT}{P + T}}$, find S when P = 4 and T = 6.

$$S = \frac{PT}{P + T} = \underline{\hspace{4cm}}$$

---

118.   The formula below can be used to convert from degrees-Fahrenheit (F) to degrees-Celsius (C). We used it for one conversion. Complete the other conversion.

Convert 68°F to degrees-Celsius.

$$C = \frac{5F - 160}{9} = \frac{5(68) - 160}{9} = \frac{340 - 160}{9} = \frac{180}{9} = 20°C$$

Convert 32°F to degrees-Celsius.

$$C = \frac{5F - 160}{9} = \underline{\hspace{4cm}}$$

---

119.   The formula below can be used to convert from degrees-Celsius (C) to degrees-Fahrenheit (F). We used it for one conversion. Complete the other conversion.

Convert 50°C to degrees-Fahrenheit.

$$F = \frac{9C}{5} + 32 = \frac{9(50)}{5} + 32 = \frac{450}{5} + 32 = 90 + 32 = 122°F$$

Convert 10°C to degrees-Fahrenheit.

$$F = \frac{9C}{5} + 32 = \underline{\hspace{4cm}}$$

$S = 2.4$, from $\dfrac{24}{10}$

0°C

50°F

3-14   SUBSCRIPTS IN FORMULAS

Some letters in formulas have letters or numbers as subscripts. In this section, we will discuss the meaning of subscripts and do some evaluations with formulas containing subscripts.

120.   When it makes sense to use the same letter more than once in a formula, subscripts are used. Either letters or numbers can be used as the subscripts. Two examples are discussed.

The formula below shows the relationship between average velocity (or speed), original velocity, and final velocity when acceleration is constant.

$$v_{av} = \frac{v_o + v_f}{2}$$

where: v_{av} = average velocity

v_o = original velocity

v_f = final velocity

The formula below shows the relationship between the total resistance in an electric circuit and the three separate resistances in the circuit.

$$R_t = R_1 + R_2 + R_3$$

where: R_t = total resistance

R_1 = first resistance

R_2 = second resistance

R_3 = third resistance

Evaluations with formulas containing subscripts are performed in the usual way. For example:

In $\boxed{v_{av} = \frac{v_o + v_f}{2}}$, find v_{av} when $v_o = 20$ and $v_f = 40$.

$$v_{av} = \frac{v_o + v_f}{2} = \frac{20 + 40}{2} = \frac{60}{2} = 30$$

In $\boxed{R_t = R_1 + R_2 + R_3}$, find R_t when $R_1 = 10$, $R_2 = 15$, and $R_3 = 30$.

$$R_t = R_1 + R_2 + R_3 = \underline{\hspace{4cm}}$$

$R_t = 55$

121.   The abbreviation "sub" is used when naming a letter with a subscript. For example:

$$V_2 \text{ is called "V sub 2".}$$

$$V_t \text{ is called "V sub t".}$$

Sometimes, however, the name is shortened by skipping the abbreviation "sub". For example:

$$V_2 \text{ is called "V2".}$$

$$V_t \text{ is called "Vt".}$$

Write a letter with a subscript for each of these.

a) F sub 1 _____   b) v sub f _____   c) R3 _____

---

122.   a) In $\boxed{E = \dfrac{P_o}{P_i}}$, find E when $P_o = 50$ and $P_i = 10$.

$$E = \frac{P_o}{P_i} = \underline{\hspace{4cm}}$$

b) In $\boxed{V_1 = \dfrac{V_2 T_1}{T_2}}$, find V_1 when $V_2 = 8$, $T_1 = 10$, and $T_2 = 4$.

$$V_1 = \frac{V_2 T_1}{T_2} = \underline{\hspace{4cm}}$$

---

123.   a) In $\boxed{a = \dfrac{V_2 - V_1}{t}}$, find "a" when $V_2 = 90$, $V_1 = 50$, and $t = 5$.

$$a = \frac{V_2 - V_1}{t} = \underline{\hspace{4cm}}$$

b) In $\boxed{R_t = \dfrac{R_1 R_2}{R_1 + R_2}}$, find R_t when $R_1 = 10$ and $R_2 = 15$.

$$R_t = \frac{R_1 R_2}{R_1 + R_2} = \underline{\hspace{4cm}}$$

Answer column:

a) F_1

b) v_f

c) R_3

a) $E = 5$

b) $V_1 = 20$

a) $a = 8$

b) $R_t = 6$

124.

a) In $\boxed{v = \dfrac{s_2 - s_1}{t_2 - t_1}}$, find "v" when $s_2 = 100$, $s_1 = 50$, $t_2 = 5$, and $t_1 = 3$.

$v = \dfrac{s_2 - s_1}{t_2 - t_1} = $ _____

b) In $\boxed{H = \dfrac{AKT(t_2 - t_1)}{L}}$, find H when $A = 2$, $K = 4$, $T = 10$, $t_2 = 10$, $t_1 = 5$, and $L = 40$.

$H = \dfrac{AKT(t_2 - t_1)}{L} = $ _____

---

a) $v = 25$ b) $H = 10$

3-15   FORMULA EVALUATIONS REQUIRING EQUATION-SOLVING

In this section, we will discuss some evaluations with fractional formulas that require solving a fractional equation.

125. To find "s" in the evaluation below, we had to solve an equation. Use the same method to find "P_i" in the other evaluation

In the formula below, find "s" when $v = 50$ and $t = 3$.

$\boxed{v = \dfrac{s}{t}}$

$50 = \dfrac{s}{3}$

$3(50) = \cancel{3}\left(\dfrac{s}{\cancel{3}}\right)$

$s = 150$

In the formula below, find P_i when $E = 20$ and $P_O = 100$.

$\boxed{E = \dfrac{P_O}{P_i}}$

126. In this frame and the following frames, we will give an example that is solved and ask you to complete a similar evaluation.

In the formula below, find F when $P = 20$, $s = 8$, and $t = 2$.

$\boxed{P = \dfrac{Fs}{t}}$

$20 = \dfrac{F(8)}{2}$

$2(20) = \cancel{2}\left(\dfrac{8F}{\cancel{2}}\right)$

$40 = 8F$

$F = 5$

In the formula below, find V_2 when $P_1 = 25$, $P_2 = 5$, and $V_1 = 4$.

$\boxed{P_1 = \dfrac{P_2V_2}{V_1}}$

$P_i = 5$

127. In the formula below, find "v_o" when $v_{av} = 50$ and $v_f = 70$.

$$v_{av} = \frac{v_o + v_f}{2}$$

$$50 = \frac{v_o + 70}{2}$$

$$2(50) = 2\left(\frac{v_o + 70}{2}\right)$$

$$100 = v_o + 70$$

$$v_o = 30$$

In the formula below, find "q" when $R = 10.4$ and $p = 8.2$.

$$R = \frac{p + q}{2}$$

$V_2 = 20$

---

128. In the formula below, find "v_1" when $a = 20$, $v_2 = 100$, and $t = 3$.

$$a = \frac{v_2 - v_1}{t}$$

$$20 = \frac{100 - v_1}{3}$$

$$3(20) = 3\left(\frac{100 - v_1}{3}\right)$$

$$60 = 100 - 1v_1$$

$$-40 = -1v_1$$

$$v_1 = \frac{-40}{-1} = 40$$

In the formula below, find T_R when $E = 0.9$ and $T = 10$.

$$E = \frac{T - T_R}{T}$$

$q = 12.6$

---

129. In the formula below, find T when $S = 4$ and $P = 5$.

$$S = \frac{PT}{P + T}$$

$$4 = \frac{5T}{5 + T}$$

$$4(5 + T) = (5 + T)\left(\frac{5T}{5 + T}\right)$$

$$20 + 4T = 5T$$

$$T = 20$$

In the formula below, find C_1 when $C_2 = 10$ and $C_T = 5$.

$$C_2 = \frac{C_1 C_T}{C_1 - C_T}$$

$T_R = 1$

$C_1 = 10$

130. In the formula below, find "y_2" when $m = 2$, $y_1 = 4$, $x_2 = 12$, and $x_1 = 9$.

$$\boxed{m = \frac{y_2 - y_1}{x_2 - x_1}}$$

$$2 = \frac{y_2 - 4}{12 - 9}$$

$$2 = \frac{y_2 - 4}{3}$$

$$3(2) = \cancel{3}\left(\frac{y_2 - 4}{\cancel{3}}\right)$$

$$6 = y_2 - 4$$

$$y_2 = 10$$

In the formula below, find "t_2" when $v = 50$, $s_2 = 200$, $s_1 = 100$, and $t_1 = 2$.

$$\boxed{v = \frac{s_2 - s_1}{t_2 - t_1}}$$

$t_2 = 4$

---

131. In the formula below, find "d_1" when $F_1 = 12$, $F_2 = 20$, and $d_2 = 40$.

$$\boxed{\frac{F_1}{F_2} = \frac{d_1}{d_2}}$$

$$\frac{12}{20} = \frac{d_1}{40}$$

$$\overset{2}{\cancel{40}}\left(\frac{12}{\cancel{20}}\right) = \cancel{40}\left(\frac{d_1}{\cancel{40}}\right)$$

$$d_1 = 24$$

In the formula below, find P_1 when $P_2 = 12$, $V_2 = 12$, and $V_1 = 18$.

$$\boxed{\frac{P_1}{P_2} = \frac{V_2}{V_1}}$$

$P_1 = 8$

---

132. In the formula below, find T_1 when $P_1 = 2$, $V_1 = 5$, $P_2 = 3$, $V_2 = 4$, and $T_2 = 6$.

$$\boxed{\frac{P_1 V_1}{T_1} = \frac{P_2 V_2}{T_2}}$$

$$\frac{(2)(5)}{T_1} = \frac{(3)(4)}{6}$$

$$\frac{10}{T_1} = \frac{12}{6}$$

$$\overset{6}{\cancel{6T_1}}\left(\frac{10}{\cancel{T_1}}\right) = \cancel{6T_1}\left(\frac{12}{\cancel{6}}\right)$$

$$60 = 12T_1$$

$$T_1 = 5$$

In the formula below, find T_2 when $P_1 = 5$, $V_1 = 6$, $T_1 = 10$, $P_2 = 4$, and $V_2 = 3$.

$$\boxed{\frac{P_1 V_1}{T_1} = \frac{P_2 V_2}{T_2}}$$

$T_2 = 4$

133.   In the formula below, find "d"   In the formula below, find C_T
        when $D = 4$ and $f = 3$.   when $C_1 = 10$ and $C_2 = 2$.

$$\boxed{\frac{1}{D} + \frac{1}{d} = \frac{1}{f}}$$ $$\boxed{\frac{1}{C_1} = \frac{1}{C_T} - \frac{1}{C_2}}$$

$$\frac{1}{4} + \frac{1}{d} = \frac{1}{3}$$

$$12d\left(\frac{1}{4} + \frac{1}{d}\right) = \overset{4d}{\cancel{12d}\left(\frac{1}{\cancel{3}}\right)}$$

$$\overset{3d}{\cancel{12d}}\left(\frac{1}{\cancel{4}}\right) + \overset{12}{\cancel{12d}}\left(\frac{1}{\cancel{d}}\right) = 4d$$

$$3d + 12 = 4d$$

$$d = 12$$

$$C_T = \frac{5}{3} \text{ or } 1.67$$

SELF-TEST 12 (pages 125-136)

Solve each equation. Report each solution in lowest terms.

1. $\dfrac{3}{x} + 1 = \dfrac{5}{2}$

2. $\dfrac{2}{3w} - \dfrac{1}{3} = \dfrac{4}{w}$

3. $\dfrac{3}{5} + \dfrac{1}{2y} = \dfrac{2}{y} - \dfrac{3}{10}$

4. In $\boxed{R_1 = \dfrac{R_2 R_t}{R_2 - R_t}}$, find R_1 when $R_2 = 15$ and $R_t = 10$.

5. In $\boxed{m = \dfrac{y_2 - y_1}{x_2 - x_1}}$, find "$x_1$"
   when $m = 4$, $y_2 = 9$, $y_1 = 1$, and $x_2 = 5$.

6. In $\boxed{F = \dfrac{r - h}{h}}$, find "h"
   when $F = 3$ and $r = 8$.

ANSWERS: 1. $x = 2$ 2. $w = -10$ 3. $y = \dfrac{5}{3}$ 4. $R_1 = 30$ 5. $x_1 = 3$ 6. $h = 2$

SUPPLEMENTARY PROBLEMS - CHAPTER 3

Assignment 9

Multiply.

1. $\left(\dfrac{x}{5}\right)\left(\dfrac{7}{8}\right)$ 2. $\left(\dfrac{3}{4}\right)\left(\dfrac{5}{t}\right)$ 3. $\left(\dfrac{9}{2}\right)\left(\dfrac{3y}{8}\right)$ 4. $\left(\dfrac{1}{2w}\right)\left(\dfrac{7}{3}\right)$

5. $6\left(\dfrac{2p}{5}\right)$ 6. $4r\left(\dfrac{1}{7}\right)$ 7. $\left(\dfrac{4}{3x}\right)(5)$ 8. $\left(\dfrac{3}{R}\right)(1)$

Reduce each fractional expression to lowest terms.

9. $\dfrac{4a}{5a}$ 10. $\dfrac{9y}{9y}$ 11. $\dfrac{4d}{10d}$ 12. $\dfrac{24R}{16R}$

13. $\dfrac{h}{2h}$ 14. $\dfrac{12P}{4}$ 15. $\dfrac{7w}{w}$ 16. $\dfrac{3x}{9}$

Multiply. Reduce each product to lowest terms.

17. $\left(\dfrac{2m}{3}\right)\left(\dfrac{9}{10m}\right)$ 18. $\left(\dfrac{5}{h}\right)\left(\dfrac{4h}{15}\right)$ 19. $\left(\dfrac{9t}{2}\right)\left(\dfrac{8}{3}\right)$ 20. $\left(\dfrac{1}{4p}\right)\left(\dfrac{2p}{3}\right)$

21. $\left(\dfrac{r}{5}\right)\left(\dfrac{5}{r}\right)$ 22. $\left(\dfrac{10}{7b}\right)\left(\dfrac{7b}{10}\right)$ 23. $\left(\dfrac{5w}{2}\right)\left(\dfrac{8}{w}\right)$ 24. $\left(\dfrac{3}{4}\right)\left(\dfrac{4x}{3}\right)$

25. $y\left(\dfrac{7}{3y}\right)$ 26. $2P\left(\dfrac{6}{P}\right)$ 27. $12a\left(\dfrac{1}{8a}\right)$ 28. $6t\left(\dfrac{2}{3t}\right)$

29. $5\left(\dfrac{3R}{5}\right)$ 30. $12\left(\dfrac{7x}{4}\right)$ 31. $6m\left(\dfrac{4}{3}\right)$ 32. $5h\left(\dfrac{1}{5h}\right)$

Assignment 10

Solve each equation. Report each solution in lowest terms.

1. $\dfrac{w}{6} = 9$ 2. $\dfrac{24}{x} = 8$ 3. $15 = \dfrac{3y}{5}$

4. $\dfrac{36}{4R} = 3$ 5. $4 = \dfrac{5G}{3}$ 6. $\dfrac{1}{2v} = 5$

7. $1 = \dfrac{7}{2d}$ 8. $\dfrac{3p}{8} = 1$ 9. $\dfrac{5t}{8} = 0$

10. $\dfrac{h + 9}{2} = 3$ 11. $x = \dfrac{3x - 1}{4}$ 12. $\dfrac{4 - 2y}{3} = 2y$

13. $\dfrac{4w + 3}{6} = 0$ 14. $\dfrac{8}{F + 3} = 2$ 15. $\dfrac{a}{a - 5} = 6$

16. $\dfrac{2k + 1}{2k - 1} = 5$ 17. $2 = \dfrac{1 - x}{3}$ 18. $\dfrac{3 - 2t}{t + 4} = 0$

Continued on following page.

19. $\dfrac{4(x + 5)}{3} = 7$

20. $6 = \dfrac{3(H - 1)}{2}$

21. $\dfrac{3(v - 5)}{4v} = 2$

22. $5 = \dfrac{6(P + 3)}{P}$

23. $\dfrac{2(y + 1)}{5y} = 0$

24. $1 = \dfrac{2}{5(B - 1)}$

25. $\dfrac{t}{3(t + 2)} = 1$

26. $6 = \dfrac{2k}{3(k - 1)}$

27. $\dfrac{3s}{5(s + 4)} = 0$

Assignment 11

Solve each equation. Report each solution in lowest terms.

1. $\dfrac{2t}{5} = \dfrac{7}{5}$

2. $\dfrac{5y}{9} = \dfrac{1}{3}$

3. $\dfrac{5x - 2}{6} = \dfrac{x}{2}$

4. $\dfrac{1}{5} = \dfrac{w + 8}{10}$

5. $\dfrac{2r}{3} = \dfrac{5}{4}$

6. $\dfrac{d - 1}{2} = \dfrac{3}{5}$

7. $\dfrac{9 - a}{6} = \dfrac{7}{8}$

8. $\dfrac{h - 1}{6} = \dfrac{3h}{4}$

9. $\dfrac{3P + 5}{10} = \dfrac{1}{4}$

10. $\dfrac{3x}{4} - 1 = 5$

11. $y + \dfrac{2y}{3} = 10$

12. $\dfrac{t}{6} = \dfrac{3t}{2} + 4$

13. $\dfrac{h}{3} + 1 = \dfrac{h}{9}$

14. $\dfrac{4p}{3} - \dfrac{3p}{2} = 1$

15. $\dfrac{7w}{5} - \dfrac{3}{4} = 0$

16. $\dfrac{2m}{3} + \dfrac{4}{3} = 2 - \dfrac{m}{3}$

17. $\dfrac{5k}{3} - \dfrac{1}{6} = \dfrac{3k}{2} + 1$

18. $1 - \dfrac{r}{4} = \dfrac{2r - 3}{3}$

19. $\dfrac{3}{2x} = \dfrac{5}{8}$

20. $\dfrac{5}{2} = \dfrac{b - 1}{2b}$

21. $\dfrac{2E + 3}{5E} = \dfrac{1}{4E}$

22. $\dfrac{7}{3} = \dfrac{2}{3t - 1}$

23. $\dfrac{3}{2R + 5} = \dfrac{1}{R}$

24. $\dfrac{2}{4y} = \dfrac{3}{y - 5}$

25. $\dfrac{F}{F + 1} = \dfrac{3}{8}$

26. $\dfrac{2w + 3}{2(w - 4)} = \dfrac{3}{4}$

27. $\dfrac{3}{x} = \dfrac{1}{5(x + 2)}$

Assignment 12

Solve each equation. Report each solution in lowest terms.

1. $\dfrac{3}{y} - 6 = 15$

2. $8 = \dfrac{3}{5R} + 2$

3. $\dfrac{1}{t} = 2 + \dfrac{3}{t}$

4. $\dfrac{7}{4x} = 5 - \dfrac{3}{4x}$

5. $\dfrac{2}{w} + \dfrac{4}{3} = 8$

6. $\dfrac{5}{2d} - 3 = \dfrac{4}{d}$

7. $\dfrac{2}{F} - \dfrac{1}{3} = \dfrac{5}{3F}$

8. $\dfrac{5}{6p} + 1 = \dfrac{2}{3p}$

9. $\dfrac{1}{2} = \dfrac{1}{R} - \dfrac{1}{5}$

10. $\dfrac{3}{4} + \dfrac{2}{m} = \dfrac{1}{3m}$

11. $5 - \dfrac{2}{h} = \dfrac{3}{2h} + \dfrac{7}{4}$

12. $\dfrac{1}{4y} + \dfrac{5}{6} = \dfrac{2}{y} - \dfrac{5}{2}$

Continued on following page.

Do these evaluations.

13. In $\boxed{a = \dfrac{F}{m}}$, find "a"

when $F = 40$ and $m = 8$.

14. In $\boxed{W = \dfrac{P - 2L}{2}}$, find W

when $P = 84$ and $L = 30$.

15. In $\boxed{I = \dfrac{E}{R - r}}$, find I

when $E = 120$, $R = 50$, and $r = 20$.

16. In $\boxed{f = \dfrac{Dd}{D + d}}$, find "f"

when $D = 12$ and $d = 6$.

Do these evaluations.

17. In $\boxed{L = \dfrac{A}{W}}$, find W

when $L = 8$ and $A = 48$.

18. In $\boxed{P = \dfrac{pv}{V}}$, find "v"

when $p = 9$, $V = 3$, and $P = 12$.

19. In $\boxed{E = \dfrac{I - i}{R}}$, find "i"

when $I = 5$, $R = 2$, and $E = 1$.

20. In $\boxed{v = \dfrac{s}{t_2 - t_1}}$, find "$t_2$"

when $v = 20$, $s = 60$, and $t_1 = 5$.

21. In $\boxed{F = \dfrac{m(v_2 - v_1)}{t}}$, find "$v_1$"

when $F = 12$, $m = 6$, $v_2 = 10$, and $t = 3$.

22. In $\boxed{d = h + \dfrac{r}{w}}$, find "r"

when $d = 10$, $h = 8$, and $w = 3$.

23. In $\boxed{G = \dfrac{r}{w - r}}$, find "r"

when $G = 5$ and $w = 12$.

24. In $\boxed{P = \dfrac{k}{s(a + b)}}$, find "b" when

$P = 5$, $k = 40$, $s = 2$, and $a = 3$.

25. In $\boxed{N = \dfrac{AB}{A + B}}$, find A

when $B = 10$ and $N = 6$.

26. In $\boxed{\dfrac{1}{R_t} = \dfrac{1}{R_1} + \dfrac{1}{R_2}}$, find R_1

when $R_t = 2$ and $R_2 = 6$.

Chapter 4 SPECIAL PRODUCTS AND FACTORING

In this chapter, we will discuss some special products and factorings that occur frequently. The "special products" include multiplication of two monomials, of a monomial and a binomial, and of two binomials. The "special factorings" include factoring monomials, binomials, and trinomials. The procedure for squaring binomials is also included.

4-1   MULTIPLYING AND SQUARING MONOMIALS

Any number, letter, or letter term is a monomial. In this section, we will discuss the procedure for multiplying and squaring monomials.

1. A monomial is an expression like 5, x, or 3y. We have already discussed some multiplications of monomials. For example: $x(5) = 5x$ a) $3(4y) =$ _____ b) $(9m)(1) =$ _____	

2.   To "square" a number, we "multiply the number by itself". A small raised "2", called an exponent, is used as the symbol for squaring. That is:
 	
    $4^2 = (4)(4) = 16$ $7^2 = (7)(7) = 49$
 	
Complete these:
 	
    a) $3^2 =$ _____   b) $1^2 =$ _____   c) $12^2 =$ _____
	a) 12y b) 9m

3.   When a number is squared, the new number is called the "square" of the original number. For example:
 	
    Since $4^2 = 16$,   16 is called the "square" of 4.
 	
Find the square of each number.
 	
    a) $10^2 =$ ____   b) $0^2 =$ _____   c) $11^2 =$ ____
	a) 9
	
	b) 1
	
	c) 144

4. The word names for expressions like 6^2 and 9^2 are "6 squared" and "9 squared". Therefore: a) 6 squared $= 6^2 =$ _____ b) 9 squared $= 9^2 =$ _____	a) 100 b) 0 c) 121

5. Any number that is the square of a whole number is called a "<u>perfect square</u>". For example:

$$\text{Since } 7^2 = 49, \quad 49 \text{ is a perfect square.}$$

The first nine perfect-square whole numbers are listed below. Write the next three in the blanks.

$$1, \ 4, \ 9, \ 16, \ 25, \ 36, \ 49, \ 64, \ 81, \ \underline{\hspace{1cm}}, \ \underline{\hspace{1cm}}, \ \underline{\hspace{1cm}}$$

a) 36   b) 81

---

6. To square a letter, we multiply the letter by itself. That is:

$$x^2 = (x)(x) \qquad\qquad y^2 = (y)(y)$$

Expressions like x^2 and y^2 are called indicated squares. The word names "x squared" and "y squared" are used for them. Write the indicated square for each word name.

a) t squared = $\underline{\hspace{1cm}}$ b) F squared = $\underline{\hspace{1cm}}$

100, 121, 144

---

7. When a letter is multiplied by itself, we can write it as an indicated square. For example:

$$(m)(m) = m^2 \qquad\qquad (V)(V) = V^2$$

Therefore, to multiply a letter term by the same letter, we simply write the letter as an indicated square. For example:

$$x(5x) = (5)(x)(x) = 5x^2$$
$$(2y)(y) = (2)(y)(y) = 2y^2$$

Following the examples, complete these:

a) t(7t) = $\underline{\hspace{1.5cm}}$ b) (9F)(F) = $\underline{\hspace{1.5cm}}$

a) t^2 b) F^2

---

8. To perform the multiplications below, we multiplied the numerical factors and wrote the indicated square of the letter.

$$(2x)(5x) = (2)(5)(x)(x) = 10x^2$$
$$(7y)(6y) = (7)(6)(y)(y) = 42y^2$$

Following the examples, complete these:

a) (3t)(4t) = $\underline{\hspace{1.5cm}}$ b) (9d)(10d) = $\underline{\hspace{1.5cm}}$

a) $7t^2$ b) $9F^2$

---

a) $12t^2$ b) $90d^2$

9.   To square a letter term, we also multiply the letter term by itself.
      For example:

$$(3x)^2 = (3x)(3x) = 9x^2$$

$$(8y)^2 = (8y)(8y) = 64y^2$$

As you can see, we can obtain the squares above by squaring the coefficient
and writing the indicated square of the letter.   Complete these:

   a) $(2P)^2 = $ _____   b) $(10m)^2 = $ _____

---

10.   Any letter term that is the square of a letter term is called a "perfect
      square".   That is:

      Since $(4x)^2 = 16x^2$, $16x^2$ is a perfect square.

      Since $(9y)^2 = 81y^2$, $81y^2$ is a perfect square.

As you can see, a squared-letter term is a perfect square only if the
numerical coefficient is a perfect square.   Which of the following are
perfect squares? _____

   a) $25d^2$ b) $47F^2$ c) $61V^2$ d) $4t^2$

a) $4P^2$ b) $100m^2$

Only (a) and (d)

4-2   MULTIPLYING A BINOMIAL BY A MONOMIAL

In this section, we will define what is meant by a monomial, a binomial, and a trinomial.   Then we will
show how the distributive principle is used to multiply a binomial by a monomial.

---

11.   There are special names for algebraic expressions containing one, two,
      or three terms.

      Any expression with one term is a "monomial".   For example:

          5   3x y^2

      Any expression with two terms is a "binomial".   For example:

      2x + 7   y - 1 $4x^2 + 5$

      Any expression with three terms is a "trinomial".   For example:

      $x^2 + 5x + 6$ $2y^2 - 4y + 7$

State whether each expression is a monomial, binomial, or trinomial.

   a) $t^2 - 25$ _____   c) $m^2 - 7m - 8$ _____

   b) 10d _____   d) 8x + 12 _____

12. We have already used the distributive principle to multiply a binomial by a monomial when the monomial is a number.   For example:

$$3(x + 5) = 3(x) + 3(5) = 3x + 15$$
$$7(4y - 1) = 7(4y) - 7(1) = 28y - 7$$

The distributive principle is also used when the monomial is a letter or letter term.   For example:

$$x(x + 2) = x(x) + x(2) = x^2 + 2x$$
$$y(y - 7) = y(y) - y(7) = y^2 - 7y$$

Following the examples, complete these multiplications.

a) m(m + 5) = _____   b) t(t - 6) = _____

a) binomial
b) monomial
c) trinomial
d) binomial

---

13. Notice that we did not write the coefficient "1" in each product below.

$$x(x + 1) = x(x) + x(1) = x^2 + x$$
$$y(y - 1) = y(y) - y(1) = y^2 - y$$

Following the examples, complete these multiplications.

a) p(p + 1) = _____   b) t(t - 1) = _____

a) $m^2 + 5m$
b) $t^2 - 6t$

---

14. Following the examples, complete the other multiplications.

$$b(3b + 4) = b(3b) + b(4) = 3b^2 + 4b$$
$$t(5t - 1) = t(5t) - t(1) = 5t^2 - t$$

a) x(7x + 1) = _____   b) y(2y - 8) = _____

a) $p^2 + p$
b) $t^2 - t$

---

15. Following the examples, complete the other multiplications.

$$2x(x + 5) = 2x(x) + 2x(5) = 2x^2 + 10x$$
$$6y(y - 1) = 6y(y) - 6y(1) = 6y^2 - 6y$$

a) 3m(m + 1) = _____   b) 7d(d - 4) = _____

a) $7x^2 + x$
b) $2y^2 - 8y$

---

16. Following the examples, complete the other multiplications.

$$4x(2x + 1) = 4x(2x) + 4x(1) = 8x^2 + 4x$$
$$3y(5y - 6) = 3y(5y) - 3y(6) = 15y^2 - 18y$$

a) 7x(3x + 8) = _____   b) 9y(4y - 1) = _____

a) $3m^2 + 3m$
b) $7d^2 - 28d$

---

a) $21x^2 + 56x$
b) $36y^2 - 9y$

17.   Complete each multiplication.

a) $t(t + 1)$ = _____   c) $4m(m + 3)$ = _____

b) $y(2y - 5)$ = _____   d) $9a(2a - 7)$ = _____

a) $t^2 + t$ b) $2y^2 - 5y$ c) $4m^2 + 12m$ d) $18a^2 - 63a$

4-3   FACTORING MONOMIALS

In this section, we will discuss the factoring process for monomials.

18.   Factoring a number is the same as writing the number as a multiplication. For example:

15 can be factored into (3)(5)

42 can be factored into (7)(6)

The two numbers in the multiplication are called "<u>factors</u>" of the original number. That is:

"3" and "5" are called "factors" of 15.

"7" and "6" are called "_____" of 42.

19.   A first number is a factor of a second number if the second is <u>divisible</u> by the first.   By "divisible", we mean that the division comes out exact.   For example:

8 is a factor of 32, since $32 \div 8 = 4$

5 is not a factor of 17, since $17 \div 5 = 3\ r2$

a) Is 3 a factor of 36? _____   c) Is 6 a factor of 18? _____

b) Is 7 a factor of 52? _____   d) Is 9 a factor of 40? _____

factors

20.   To show that 8 has four factors (1, 2, 4, and 8), we can use these steps:

1) Since any number is divisible by "1" and itself, "1" and the number itself are factors of any number.   Therefore:

Two factors of 8 are "1" and "8".

2) To find the other factors of 8, we test whole numbers up to 4 (one-half of 8).   That is:

Is 8 divisible by 2?   Yes
Is 8 divisible by 3?   No
Is 8 divisible by 4?   Yes

Using the same steps, list all of the factors for each number.

a) 7 _____   c) 32 _____

b) 25 _____   d) 36 _____

a) Yes   c) Yes

b) No   d) No

21. Some numbers can be factored into only one pair of factors. For example:

$$3 = (1)(3) \text{ or } (3)(1) \qquad 7 = (1)(7) \text{ or } (7)(1)$$

Most numbers, however, can be factored into more than one pair of factors. For example, 18 can be factored into three pairs of factors. They are:

$$18 = (1)(18) \text{ or } (18)(1)$$
$$18 = (2)(9) \text{ or } (9)(2)$$
$$18 = (3)(6) \text{ or } (6)(3)$$

24 can be factored into four pairs of factors. Do so below.

24 = (   )(   )   24 = (   )(   )

24 = (   )(   )   24 = (   )(   )

a) 1, 7

b) 1, 5, 25

c) 1, 2, 4, 8, 16, 32

d) 1, 2, 3, 4, 6, 9, 12, 18, 36

22. To factor a letter with a coefficient, we can write the factors explicitly in either order. That is:

$$3x = (3)(x) \text{ or } (x)(3) \qquad 7y = (   )(   ) \text{ or } (   )(   )$$

(1)(24) or (24)(1)
(2)(12) or (12)(2)
(3)(8)   or (8)(3)
(4)(6)   or (6)(4)

23. When the coefficient is not explicitly shown, the coefficient is "1". Therefore:

$$x = (1)(x) \text{ or } (x)(1) \qquad t = (   )(   ) \text{ or } (   )(   )$$

(7)(y) or (y)(7)

24. We factored 12x and 18y in various ways below.

$$12x = (2)(6x) \qquad 18y = (2)(9y)$$
$$12x = (3)(4x) \qquad 18y = (3)(6y)$$
$$12x = (4)(3x) \qquad 18y = (6)(3y)$$
$$12x = (6)(2x) \qquad 18y = (9)(2y)$$

Write the missing factor in each blank.

a) 6t = (2)(   )   b) 10t = (   )(2t)   c) 16x = (4)(   )

(1)(t) or (t)(1)

25. We factored 16x and 20y in various ways below.

$$16x = (2x)(8) \qquad 20y = (2y)(10)$$
$$16x = (4x)(4) \qquad 20y = (4y)(5)$$
$$16x = (8x)(2) \qquad 20y = (5y)(4)$$
$$\qquad\qquad\qquad 20y = (10y)(2)$$

Write the missing factor in each blank.

a) 30d = (5d)(   )   b) 14m = (   )(2)   c) 32p = (4p)(   )

a) (2)(3t)

b) (5)(2t)

c) (4)(4x)

26. Any letter with a coefficient can be factored into itself and "1". That is:

$$3x = (3x)(1) \text{ or } (1)(3x) \qquad 8t = (   )(   ) \text{ or } (   )(   )$$

a) (5d)(6)

b) (7m)(2)

c) (4p)(8)

27.   Any squared letter can be factored by multiplying the letter by itself.
        That is:

$$x^2 = (x)(x) \qquad y^2 = (y)(y) \qquad p^2 = (\quad)(\quad)$$

| | (8t)(1) or (1)(8t) |

---

28.   We factored $3x^2$ and $7y^2$ in two different ways below.

$$3x^2 = (3x)(x) \qquad\qquad 7y^2 = (7y)(y)$$
$$3x^2 = (x)(3x) \qquad\qquad 7y^2 = (y)(7y)$$

        Write the missing factor in each blank.

        a) $2t^2 = (2t)(\quad)$ b) $5d^2 = (d)(\quad)$ c) $8m^2 = (\quad)(m)$

(p)(p)

---

29.   We factored $12x^2$ and $18y^2$ in various ways below.

$$12x^2 = (2x)(6x) \qquad\qquad 18y^2 = (2y)(9y)$$
$$12x^2 = (3x)(4x) \qquad\qquad 18y^2 = (3y)(6y)$$
$$12x^2 = (4x)(3x) \qquad\qquad 18y^2 = (6y)(3y)$$
$$12x^2 = (6x)(2x) \qquad\qquad 18y^2 = (9y)(2y)$$

        Write the missing factor in each blank.

        a) $20t^2 = (2t)(\quad)$ b) $36d^2 = (9d)(\quad)$ c) $48p^2 = (\quad)(6p)$

a) (2t)(t̲)

b) (d)(5̲d̲)

c) (8̲m̲)(m)

---

a) (2t)(1̲0̲t̲)   b) (9d)(4̲d̲)   c) (8̲p̲)(6p)

4-4   FACTORING BINOMIALS OF THE FORM: ax + b

In this section, we will show how the distributive principle can be used to factor some binomials of
the form: ax + b.

---

30.   We used the distributive principle for each multiplication below.

$$2(x + 3) = 2(x) + 2(3) = 2x + 6$$
$$5(y - 4) = 5(y) - 5(4) = 5y - 20$$

        By simply reversing the process, we can use the distributive principle to
        factor each binomial product. That is:

$$2x + 6 = 2(x) + 2(3) = 2(x + 3)$$
$$5y - 20 = 5(y) - 5(4) = 5(y - 4)$$

        Complete each of these factorings.

        a) $3t + 18 = 3(t) + 3(6) =$ _____

        b) $4d - 36 = 4(d) - 4(9) =$ _____

---

a) $3(t + 6)$

b) $4(d - 9)$

31.   Two more factorings of binomials are shown below.

$$8x + 10 = 2(4x) + 2(5) = 2(4x + 5)$$
$$15y - 10 = 5(3y) - 5(2) = 5(3y - 2)$$

Complete each of these factorings.

   a) $9t + 12 = 3(3t) + 3(4) =$ _____

   b) $35d - 30 = 5(7d) - 5(6) =$ _____

---

32.   When factoring a binomial of the form "ax + b", we "factor out" a number that is a common factor of the coefficient and the number.   For example:

   In $3x + 6$, the common factor is "3".   Therefore:

   $$3x + 6 = 3(x + 2)$$

   In $6y + 4$, the common factor is "2".   Therefore:

   $$6y + 4 = 2(3y + 2)$$

Factor out the common factor from each binomial below.

   a) $7x + 14 =$ _____   c) $6a + 21 =$ _____

   b) $5y - 45 =$ _____   d) $10t - 8 =$ _____

a) $3(3t + 4)$

b) $5(7d - 6)$

---

33.   When factoring a binomial, we can sometimes "factor out" more than one number.   For example, we can factor out 2, 3, or 6 from the binomial below.

   $$6x + 12 = 2(3x) + 2(6) = 2(3x + 6)$$
   $$6x + 12 = 3(2x) + 3(4) = 3(2x + 4)$$
   $$6x + 12 = 6(x) \;\; + 6(2) = 6(x + 2)$$

Ordinarily, we factor out <u>the largest possible number</u>.   Therefore, the ordinary factoring for $6x + 12$ is
_____ .

a) $7(x + 2)$

b) $5(y - 9)$

c) $3(2a + 7)$

d) $2(5t - 4)$

---

34.   Factoring out the largest possible number from a binomial is called "<u>factoring completely</u>".   Factor each binomial below completely.

   a) $4x + 12 =$ _____   c) $12t + 18 =$ _____

   b) $9b - 45 =$ _____   d) $16y - 12 =$ _____

$6(x + 2)$

---

35.   When the largest possible factor is identical to the number term, we get "1" as the second term within the parentheses when factoring.   For example:

   $$6x + 2 = 2(3x) + 2(1) = 2(3x + 1)$$
   $$5y - 5 = 5(y) - 5(1) = 5(y - 1)$$

a) $4(x + 3)$

b) $9(b - 5)$

c) $6(2t + 3)$

d) $4(4y - 3)$

Continued on following page.

35.   Continued

Factor each binomial below completely.

a) 3m + 3 = _____   b) 20p - 10 = _____

---

36.   To check a factoring of a binomial, we simply multiply by the distributive principle.   That is:

$$8x + 12 = 4(2x + 3), \text{ since } 4(2x + 3) = 8x + 12$$
$$14y - 7 = 7(2y - 1), \text{ since } 7(2y - 1) = 14y - 7$$

Factor each binomial completely and check your results by multiplying mentally.

a) 16x + 24 = _____   c) 9m + 9 = _____

b) 36y - 12 = _____   d) 15d - 3 = _____

a) 3(m + 1)

b) 10(2p - 1)

---

37.   A binomial containing a letter term and a number term can be factored <u>only if the coefficient and the number term have a common factor</u>. Therefore, the binomials below <u>cannot</u> be factored.

$$3x + 7 \qquad 9y - 5 \qquad 12t + 17$$

Factor each binomial below completely if possible.

a) 7t + 9 = _____   c) 4y + 2 = _____

b) 6d - 6 = _____   d) 5m - 2 = _____

a) 8(2x + 3)

b) 12(3y - 1)

c) 9(m + 1)

d) 3(5d - 1)

---

a) Not possible   b) 6(d - 1)   c) 2(2y + 1)   d) Not possible

---

4-5   FACTORING BINOMIALS OF THE FORM: $ax^2 + bx$

In this section, we will show how the distributive principle can be used to factor all binomials of the form: $ax^2 + bx$.

38.   We used the distributive principle for each multiplication below.

$$x(x + 5) = x(x) + x(5) = x^2 + 5x$$
$$y(y - 3) = y(y) - y(3) = y^2 - 3y$$

By simply reversing the process, we can use the distributive principle to factor each binomial product.   That is:

$$x^2 + 5x = x(x) + x(5) = x(x + 5)$$
$$y^2 - 3y = y(y) - y(3) = y(y - 3)$$

Complete each of these factorings.

a) $m^2 + 2m = m(m) + m(2) = $ _____

b) $t^2 - 10t = t(t) - t(10) = $ _____

39. Notice how we got a "1" as the second term within the parentheses in each factoring below.

$$x^2 + x = x(x) + x(1) = x(x + 1)$$
$$y^2 - y = y(y) - y(1) = y(y - 1)$$

Following the examples, complete these factorings.

a) $t^2 + t = t(t) + t(1) = $ _____

b) $m^2 - m = m(m) - m(1) = $ _____

a) m(m + 2)

b) t(t - 10)

---

40. Two more examples of "factoring out" a letter from a binomial are shown below.

$$3b^2 + 4b = b(3b) + b(4) = b(3b + 4)$$
$$5t^2 - t = t(5t) - t(1) = t(5t - 1)$$

Following the examples, complete these factorings.

a) $7x^2 + x = x(7x) + x(1) = $ _____

b) $2y^2 - 5y = y(2y) - y(5) = $ _____

a) t(t + 1)

b) m(m - 1)

---

41. To check a factoring, we simply multiply by the distributive principle. That is:

$t^2 + t = t(t + 1)$ is correct, since $t(t + 1) = t^2 + t$

$2p^2 - 5p = p(2p - 5)$ is correct, since $p(2p - 5) = 2p^2 - 5p$

Factor out a letter from each binomial below. Check your results by multiplying mentally.

a) $b^2 + 9b = $ _____   c) $8p^2 + p = $ _____

b) $d^2 - d = $ _____   d) $7x^2 + 8x = $ _____

a) x(7x + 1)

b) y(2y - 5)

---

42. When the coefficients of the terms have a common factor, we can factor out a number and a letter. For example:

$$3x^2 + 6x = 3x(x) + 3x(2) = 3x(x + 2)$$
$$4y^2 - 10y = 2y(2y) - 2y(5) = 2y(2y - 5)$$

Following the examples, complete these factorings.

a) $7t^2 + 21t = 7t(t) + 7t(3) = $ _____

b) $8m^2 - 6m = 2m(4m) - 2m(3) = $ _____

a) b(b + 9)

b) d(d - 1)

c) p(8p + 1)

d) x(7x + 8)

---

43. Sometimes we get a "1" as the second term in the parentheses when factoring out a number and a letter. That is:

$$4x^2 + 2x = 2x(2x) + 2x(1) = 2x(2x + 1)$$
$$15y^2 - 5y = 5y(3y) - 5y(1) = $$ _____

a) 7t(t + 3)

b) 2m(4m - 3)

44.   In order to factor completely, we must factor out the largest possible
number. Factor these completely and check your results.

$5y(3y - 1)$

   a) $5x^2 + 10x$ = _____   c) $12t^2 - 18t$ = _____

   b) $15y^2 - 3y$ = _____   d) $16m^2 - 8m$ = _____

45.   Factor each binomial completely and check your results.

a) $5x(x + 2)$

   a) $2p^2 - p$ = _____   c) $14m^2 + 7m$ = _____

b) $3y(5y - 1)$

   b) $4d^2 + 3d$ = _____   d) $18h^2 - 24h$ = _____

c) $6t(2t - 3)$

d) $8m(2m - 1)$

a) $p(2p - 1)$ b) $d(4d + 3)$ c) $7m(2m + 1)$ d) $6h(3h - 4)$

SELF-TEST 13 (pp. 140-150)

Do these multiplications.

Do these squarings.

1. $(w)(8w)$ = _____   2. $(3P)(7P)$ = _____   3. $(5d)^2$ = _____   4. $(9x)^2$ = _____

Do these multiplications.

5. $d(2d - 1)$ = _____   6. $5m(m + 2)$ = _____   7. $7t(3t - 9)$ = _____

Write the missing factor in each blank.

8. $14x = (\ \ \ )(2x)$ 9. $R = (\ \ \ )(R)$ 10. $56a = (8)(\ \ \ )$ 11. $28y^2 = (7y)(\ \ \ )$

Factor each binomial completely.

12. $6h + 15$ = _____   14. $10x + 10$ = _____   16. $5b^2 - 3b$ = _____

13. $8t - 2$ = _____   15. $2y^2 + y$ = _____   17. $36p^2 - 9p$ = _____

ANSWERS: 1. $8w^2$ 5. $2d^2 - d$ 9. $(\underline{1})(R)$ 13. $2(4t - 1)$

2. $21P^2$ 6. $5m^2 + 10m$ 10. $(8)(\underline{7a})$ 14. $10(x + 1)$

3. $25d^2$ 7. $21t^2 - 63t$ 11. $(7y)(\underline{4y})$ 15. $y(2y + 1)$

4. $81x^2$ 8. $(\underline{7})(2x)$ 12. $3(2h + 5)$ 16. $b(5b - 3)$

17. $9p(4p - 1)$

4-6   MULTIPLYING TWO BINOMIALS: THE DISTRIBUTIVE-PRINCIPLE METHOD

In this section, we will show how the distributive principle can be used to multiply two binomials.

46. Before showing how the distributive principle can be used to multiply two binomials, we will show how it can be used to multiply a monomial and binomial when the binomial is on the left side.   Two examples are discussed below.

$$(x + 4)(2) = x(2) + 4(2) = 2x + 8$$

      Note: The arrows show that "2" is multiplied by each term ("x" and "4") in the binomial.

$$(x + 8)(x) = x(x) + 8(x) = x^2 + 8x$$

      Note: The arrows show that "x" is multiplied by each term ("x" and "8") in the binomial.

Following the examples, complete each multiplication.

   a) $(y + 3)(5)$ = _____

   b) $(2t + 1)(4t)$ = _____

47. The same form of the distributive principle can be used to multiply a binomial by a binomial.   An example is shown.   The arrows show that $(x + 3)$ is multiplied by each term ("x" and "2") in the first binomial.

$$(x + 2)(x + 3) = x(x + 3) + 2(x + 3)$$

      Notice these two points about using the distributive principle above:

    1) The first binomial $(x + 2)$ is broken up.

    2) The right side contains two terms, $x(x + 3)$ and $2(x + 3)$, each of which is a multiplication of a binomial by a monomial.

Using the same pattern, complete these:

   a) $(y + 4)(y + 5) = (\quad)(y + 5) + (\quad)(y + 5)$

   b) $(2p + 3)(p + 7) = (\quad)(p + 7) + (\quad)(p + 7)$

   c) $(3t + 1)(4t + 9) = (\quad)(4t + 9) + (\quad)(4t + 9)$

Answers (right margin):

a) $5y + 15$

b) $8t^2 + 4t$

a) $\underline{(y)}(y + 5) + \underline{(4)}(y + 5)$

b) $\underline{(2p)}(p + 7) + \underline{(3)}(p + 7)$

c) $\underline{(3t)}(4t + 9) + \underline{(1)}(4t + 9)$

48.   The three steps needed to multiply two binomials are shown and discussed below.

Step 1: $(x + 2)(x + 3) = \underline{x(x + 3)} + \underline{2(x + 3)}$

Step 2: $= x^2 + \underline{3x} + \underline{2x} + 6$

Step 3: $= x^2 + \quad 5x \quad + 6$

In Step 1, we applied the distributive principle to break up the first binomial.

In Step 2, we performed the two multiplications of a binomial by a monomial.

In Step 3, we combined 3x and 2x to get 5x.

Following the steps above, complete this multiplication.

$(y + 5)(y + 4) = \quad y(y + 4) + 5(y + 4)$

$= \underline{\hspace{5cm}}$

$= \underline{\hspace{5cm}}$

---

49.   Use the distributive-principle method to complete this multiplication.

$(a + 6)(a + 4) = \underline{\hspace{5cm}}$

$= \underline{\hspace{5cm}}$

$= \underline{\hspace{5cm}}$

$y^2 + 4y + 5y + 20$

$y^2 + 9y + 20$

---

50.   Use the distributive-principle method to complete this multiplication.

$(2x + 1)(x + 5) = \underline{\hspace{5cm}}$

$= \underline{\hspace{5cm}}$

$= \underline{\hspace{5cm}}$

$a(a + 4) + 6(a + 4)$

$a^2 + 4a + 6a + 24$

$a^2 + 10a + 24$

$2x(x + 5) + 1(x + 5)$

$2x^2 + 10x + x + 5$

$2x^2 + 11x + 5$

4-7   MULTIPLYING TWO BINOMIALS: THE FOIL METHOD

In this section, we will show a shorter method for multiplying two binomials. The shorter method is called the "FOIL" method.

51.   The first two steps used to multiply $(x + 2)$ and $(x + 6)$ by the distributive-principle method are shown below.

$$(x + 2)(x + 6) = x(x + 6) + 2(x + 6)$$
$$= x^2 + 6x + 2x + 12$$

In the second step above, there are four terms in the product. We can skip the first step and write those four terms directly as we have done below.

$$\begin{array}{c} \text{F} \quad \text{O} \qquad \text{F} \quad \text{O} \quad \text{I} \quad \text{L} \\ (x + 2)(x + 6) = x^2 + 6x + 2x + 12 \\ \text{I} \quad \text{L} \end{array}$$

Note: 1) To get $\underline{F}$ (or "x^2"),   we multiplied the $\underline{first}$ terms of the binomials.
 2) To get $\underline{O}$ (or "6x"), we multiplied the $\underline{outside}$ terms of the binomials.
 3) To get $\underline{I}$ (or "2x"), we multiplied the $\underline{inside}$ terms of the bionmials.
 4) To get $\underline{L}$ (or "12"), we multiplied the $\underline{last}$ terms of the binomials.

Using the FOIL method, write the four terms of each product below.

a) $\begin{array}{c} \text{F} \quad \text{O} \qquad \underline{\text{F}} \qquad \underline{\text{O}} \qquad \underline{\text{I}} \qquad \underline{\text{L}} \\ (y + 3)(y + 5) = \underline{\quad} + \underline{\quad} + \underline{\quad} + \underline{\quad} \\ \text{I} \quad \text{L} \end{array}$

b) $\begin{array}{c} \text{F} \quad \text{O} \qquad \underline{\text{F}} \qquad \underline{\text{O}} \qquad \underline{\text{I}} \qquad \underline{\text{L}} \\ (2t + 3)(4t + 1) = \underline{\quad} + \underline{\quad} + \underline{\quad} + \underline{\quad} \\ \text{I} \quad \text{L} \end{array}$

52.   In the FOIL method, we multiply $\underline{both}$ $\underline{terms}$ in the $\underline{second}$ bionmial:

 1) by the $\underline{first}$ $\underline{term}$ in the first binomial
 2) by the $\underline{second}$ $\underline{term}$ in the first binomial

Using the FOIL method, write each four-term product.

 a) $(x + 7)(x + 5) = \underline{\quad} + \underline{\quad} + \underline{\quad} + \underline{\quad}$

 b) $(y + 6)(y + 1) = \underline{\quad} + \underline{\quad} + \underline{\quad} + \underline{\quad}$

---

a) $y^2 + 5y + 3y + 15$

b) $8t^2 + 2t + 12t + 3$

a) $x^2 + 5x + 7x + 35$

b) $y^2 + y + 6y + 6$

53.   Using the FOIL method, write each four-term product.

    a) $(3t + 2)(t + 4)$ = _____ + _____ + _____ + _____

    b) $(5x + 1)(2x + 8)$ = _____ + _____ + _____ + _____

---

54.   After writing the four-term product below, we combined the "like" terms $(4x$ and $8x)$ to get a trinomial product.

$$(x + 8)(x + 4) = x^2 + 4x + 8x + 32$$
$$= x^2 + 12x + 32$$

Following the example, write each four-term product and then combine the "like" terms to get a trinomial product.

    a) $(y + 5)(y + 9)$ = _____ + _____ + _____ + _____

                   = _____

    b) $(2t + 1)(4t + 5)$ = _____ + _____ + _____ + _____

                   = _____

Answers (54):
a) $3t^2 + 12t + 2t + 8$
b) $10x^2 + 40x + 2x + 8$

---

55.   Find each trinomial product.

    a) $(y + 1)(y + 10)$ = _____

    b) $(4d + 3)(5d + 1)$ = _____

Answers (55):
a) $y^2 + 9y + 5y + 45$
   $y^2 + 14y + 45$
b) $8t^2 + 10t + 4t + 5$
   $8t^2 + 14t + 5$

Answers:
a) $y^2 + 11y + 10$ b) $20d^2 + 19d + 3$

4-8   MULTIPLICATIONS INVOLVING "DIFFERENCE" BINOMIALS

In this section, we will discuss multiplications in which one or both binomials is a difference.

56.   The FOIL method can also be used when one or both binomials is a difference. An example is shown below.

$$(x + 5)(x - 2) = x^2 - 2x + 5x - 10$$

    Note: 1) To get $\underline{O}$ (or $-2x$), we multiplied "x" and "-2".

            2) To get $\underline{L}$ (or -10), we multiplied "5" and "-2".

Continued on following page.

56.   Continued

Using the FOIL method, write each four-term product below.

a) $(y + 3)(y - 4)$ = _____

b) $(2t + 1)(t - 6)$ = _____

---

57.   The FOIL method was used to get the four-term product below.

$$\begin{array}{ccccccc} & F & O & F & O & I & L \\ (x & - & 4)(x & + & 9) & = & x^2 + 9x - 4x - 36 \\ & & I & & L & & \end{array}$$

Note: 1) To get I (or –4x), we multiplied "–4" and "x".

2) To get L (or –36), we multiplied "–4" and "9".

Using the FOIL method, write each four-term product below.

a) $(y - 7)(y + 2)$ = _____

b) $(t - 1)(4t + 5)$ = _____

a) $y^2 - 4y + 3y - 12$

b) $2t^2 - 12t + t - 6$

---

58.   The FOIL method was used to get the four-term product below.

$$\begin{array}{ccccccc} & F & O & F & O & I & L \\ (m & - & 2)(m & - & 3) & = & m^2 - 3m - 2m + 6 \\ & & I & & L & & \end{array}$$

Note: 1) To get O (or –3m), we multiplied "m" and "–3".

2) To get I (or –2m), we multiplied "–2" and "m".

3) To get L (or +6),   we multiplied "–2" and "–3".

Using the FOIL method, write each four-term product below.

a) $(p - 8)(p - 1)$ = _____

b) $(2t - 3)(2t - 4)$ = _____

a) $y^2 + 2y - 7y - 14$

b) $4t^2 + 5t - 4t - 5$

---

59.   After writing the four-term product below, we combined like terms
(–5x and 3x) to get a trinomial product.

$$(x + 3)(x - 5) = x^2 - 5x + 3x - 15$$
$$= x^2 - 2x - 15$$

a) $p^2 - p - 8p + 8$

b) $4t^2 - 8t - 6t + 12$

Continued on following page.

59.   Continued

Following the example, write each four-term product and then combine the like terms to get a trinomial product.

a) $(y + 2)(y - 8)$ = _____

        = _____

b) $(5m + 1)(m - 4)$ = _____

        = _____

---

60.   After writing the four-term product, we combined like terms (6y and -4y) to get a trinomial product.

$$(y - 4)(y + 6) = y^2 + 6y - 4y - 24$$
$$= y^2 + 2y - 24$$

Following the example, write each four-term product and then combine like terms to get a trinomial product.

a) $(x - 1)(x + 2)$ = _____

        = _____

b) $(3d - 4)(4d + 3)$ = _____

        = _____

a) $y^2 - 8y + 2y - 16$
    $y^2 - 6y - 16$

b) $5m^2 - 20m + m - 4$
    $5m^2 - 19m - 4$

---

61.   After writing the four-term product below, we combined like terms (-4t and -3t) to get a trinomial product.

$$(t - 3)(t - 4) = t^2 - 4t - 3t + 12$$
$$= t^2 - 7t + 12$$

Following the example, write each four-term product and then combine like terms to get a trinomial product.

a) $(x - 9)(x - 1)$ = _____

        = _____

b) $(3y - 1)(2y - 4)$ = _____

        = _____

a) $x^2 + 2x - x - 2$
    $x^2 + x - 2$

b) $12d^2 + 9d - 16d - 12$
    $12d^2 - 7d - 12$

---

62.   Find each trinomial product.

a) $(h + 7)(h - 5)$ = _____

b) $(t - 6)(2t + 1)$ = _____

a) $x^2 - x - 9x + 9$
    $x^2 - 10x + 9$

b) $6y^2 - 12y - 2y + 4$
    $6y^2 - 14y + 4$

63.   Find each trinomial product.

a) $(x - 2)(x - 2)$ = _____

b) $(2y - 5)(2y - 5)$ = _____

a) $h^2 + 2h - 35$

b) $2t^2 - 11t - 6$

---

a) $x^2 - 4x + 4$ b) $4y^2 - 20y + 25$

4-9 MULTIPLYING THE SUM AND DIFFERENCE OF TWO TERMS

In this section, we will discuss multiplications of the sum and difference of two terms.

64.   In the multiplication below, both binomials contain an "x" and a "3". One binomial is a sum; the other binomial is a difference. Since $-3x + 3x = 0$, the product simplifies to a binomial.

$$(x + 3)(x - 3) = x^2 - 3x + 3x - 9$$
$$= x^2 - 9$$

Following the example, complete this multiplication.

$(y + 6)(y - 6)$ = _____

= _____

65.   In the multiplication below, both binomials contain a "2t" and a "5". One binomial is a sum; the other binomial is a difference. Since $-10t + 10t = 0$, the product simplifies to a binomial.

$$(2t + 5)(2t - 5) = 4t^2 - 10t + 10t - 25$$
$$= 4t^2 - 25$$

Following the example, complete this multiplication.

$(3m + 2)(3m - 2)$ = _____

= _____

$y^2 - 6y + 6y - 36$
$y^2 - 36$

66.   Here is a multiplication performed earlier.

$$(y + 6)(y - 6) = y^2 - 36$$

Notice these points about the product $y^2 - 36$.

1) y^2 is the square of "y".

2) 36 is the square of "6".

3) The two squares are <u>subtracted</u>.

Following the pattern above, write each binomial product directly.

a) $(x + 2)(x - 2)$ = ____ - ____   b) $(p + 9)(p - 9)$ = ____ - ____

$9m^2 - 6m + 6m - 4$
$9m^2 - 4$

67.   Here is another multiplication performed earlier.

$$(3m + 2)(3m - 2) = 9m^2 - 4$$

Notice these points about the product $9m^2 - 4$.

     1) $9m^2$ is the square of $3m$.

     2) 4 is the square of 2.

     3) The two squares are <u>subtracted</u>.

Following the pattern above, write each binomial product directly.

a) $(4x + 1)(4x - 1)$ = _____   b) $(5t + 7)(5t - 7)$ = _____

a) $x^2 - 4$

b) $p^2 - 81$

---

68.   Write each binomial product directly.

a) $(x + 1)(x - 1)$ = _____   c) $(2t + 9)(2t - 9)$ = _____

b) $(y + 10)(y - 10)$ = _____   d) $(6p + 1)(6p - 1)$ = _____

a) $16x^2 - 1$

b) $25t^2 - 49$

a) $x^2 - 1$ b) $y^2 - 100$ c) $4t^2 - 81$ d) $36p^2 - 1$

4-10   SQUARING BINOMIALS

In this section, we will show how the FOIL method can be used to square binomials. Then we will show a shortcut for the same operation.

---

69.   To square a binomial, we multiply the binomial by itself. That is:

$$(x + 4)^2 = (x + 4)(x + 4)$$
$$(y - 9)^2 = (y - 9)(y - 9)$$

Notice that squaring a binomial is the same as multiplying two identical binomials. We can use the FOIL method to do so. For example:

$$(x + 4)^2 = (x + 4)(x + 4)$$
$$= x^2 + 4x + 4x + 16$$
$$= x^2 + 8x + 16$$

Following the example, complete the squaring of this binomial.

$$(2m + 5)^2 = (2m + 5)(2m + 5)$$
$$= \text{_____}$$
$$= \text{_____}$$

$4m^2 + 10m + 10m + 25$
$4m^2 + 20m + 25$

70. We used the FOIL method to square the binomial below.

$$(y - 9)^2 = (y - 9)(y - 9)$$
$$= y^2 - 9y - 9y + 81$$
$$= y^2 - 18y + 81$$

Following the example, complete the squaring of this binomial.

$$(6d - 1)^2 = (6d - 1)(6d - 1)$$
$$= \underline{\hspace{5cm}}$$
$$= \underline{\hspace{4cm}}$$

---

71. If we use the FOIL method to square the binomial $(x + 5)$, we get:

$$(x + 5)^2 = x^2 + 10x + 25$$

There is a shortcut that can be used to square the binomial $(x + 5)$.
To see the shortcut, let's examine $x^2 + 10x + 25$.

1) The <u>first</u> <u>term</u> (x^2) is the square of "x".

2) The <u>second</u> <u>term</u> (10x) is double the product of the two
    terms of the binomial. That is: $10x = 2(x)(5)$

3) The <u>third</u> <u>term</u> (25) is the square of "5".

Let's use the shortcut to square $(y + 8)$.

a) Squaring "y", we get _____.

b) Doubling the product of "y" and "8", we get _____.

c) Squaring "8", we get _____.

d) Therefore, $(y + 8)^2 = \underline{\hspace{4cm}}$

[right column:]
$36d^2 - 6d - 6d + 1$
$36d^2 - 12d + 1$

---

72. Let's use the shortcut to square $(2x + 7)$.

a) Squaring "2x", we get _____.

b) Doubling the product of "2x" and "7", we get _____.

c) Squaring "7", we get _____.

d) Therefore, $(2x + 7)^2 = \underline{\hspace{4cm}}$

[right column:]
a) y^2

b) 16y

c) 64

d) $y^2 + 16y + 64$

---

73. Use the shortcut for these. Be sure to <u>double</u> the product of the two terms
to get the middle term of the trinomial.

a) $(m + 3)^2 = \underline{\hspace{5cm}}$

b) $(5d + 2)^2 = \underline{\hspace{5cm}}$

[right column:]
a) $4x^2$

b) 28x

c) 49

d) $4x^2 + 28x + 49$

74.   If we use the FOIL method to square $(y - 4)$, we get:

$$(y - 4)^2 = y^2 - 8y + 16$$

Notice that we can also use the shortcut to get $y^2 - 8y + 16$. However, when using it, <u>we</u> <u>must</u> <u>remember</u> <u>to</u> <u>subtract</u> <u>the</u> <u>second</u> <u>term</u>.

Let's use the shortcut to square $(t - 6)$.

    a) Squaring "t", we get _____.

    b) Doubling the product of "t" and "6", we get _____.

    c) Squaring "6", we get _____.

    d) Therefore, $(t - 6)^2 =$ _____

a) $m^2 + 6m + 9$

b) $25d^2 + 20d + 4$

---

75.   Let's use the shortcut to square $(4x - 5)$.

    a) Squaring "4x", we get _____.

    b) Doubling the product of "4x" and "5", we get _____.

    c) Squaring "5", we get _____.

    d) Therefore, $(4x - 5)^2 =$ _____

a) t^2

b) $12t$

c) 36

d) $t^2 - 12t + 36$

---

76.   Use the shortcut for these. Be sure to <u>subtract</u> the middle term of the trinomial.

    a) $(m - 2)^2 =$ _____

    b) $(3y - 4)^2 =$ _____

a) $16x^2$

b) $40x$

c) 25

d) $16x^2 - 40x + 25$

---

77.   Use the shortcut for these:

    a) $(x + 1)^2 =$ _____   c) $(5m + 1)^2 =$ _____

    b) $(y - 1)^2 =$ _____   d) $(7d - 1)^2 =$ _____

a) $m^2 - 4m + 4$

b) $9y^2 - 24y + 16$

---

78.   In the example below, the number is the first term in the binomial. The same method is used.

$$(3 + 5d)^2 = 9 + 30d + 25d^2$$

Following the example, complete these:

    a) $(12 + t)^2 =$ _____

    b) $(1 - 4m)^2 =$ _____

a) $x^2 + 2x + 1$

b) $y^2 - 2y + 1$

c) $25m^2 + 10m + 1$

d) $49d^2 - 14d + 1$

---

a) $144 + 24t + t^2$

b) $1 - 8m + 16m^2$

SELF-TEST 14 (pp. 151-161)

Do these multiplications.

1. $(7x + 5)(3x + 1)$

2. $(d - 8)(d + 2)$

3. $(2m + 2)(5m - 3)$

4. $(9b - 7)(b - 1)$

5. $(y + 6)(y - 6)$

6. $(4t + 9)(4t - 9)$

Do these squarings.

7. $(5h - 1)^2$

8. $(3 + 2w)^2$

ANSWERS:
1. $21x^2 + 22x + 5$

2. $d^2 - 6d - 16$

3. $10m^2 + 4m - 6$

4. $9b^2 - 16b + 7$

5. $y^2 - 36$

6. $16t^2 - 81$

7. $25h^2 - 10h + 1$

8. $9 + 12w + 4w^2$

4-11   FACTORING THE DIFFERENCE OF TWO PERFECT SQUARES

In this section, we will discuss the procedure for factoring the difference of two perfect squares. Before doing so, the meaning of a "square root" is briefly introduced.

79.   A "square root" of a number N is "a number whose square is N". That is:

Since $5^2 = (5)(5) = 25$,   5 is a square root of 25.

Since $8^2 = (8)(8) = 64$,   8 is a square root of 64.

Instead of saying "the square root of", we use the symbol $\sqrt{}$.
That is:

Since $4^2 = 16$, $\sqrt{16} = 4$ Since $10^2 = 100$, $\sqrt{100} =$ _____

10

80. A whole number whose square root is a whole number is a $\underline{\text{perfect square}}$. For example:

$$49 \text{ is a perfect square, since } \sqrt{49} = 7$$

   a) 81 is a perfect square, since $\sqrt{81}$ = _____

   b) 1 is a perfect square, since $\sqrt{1}$ = _____

---

81. The square root of a squared letter is "$\underline{\text{a quantity whose square equals}}$ $\underline{\text{the squared letter}}$". For example:

$$\text{Since } (x)^2 = (x)(x) = x^2, \ \sqrt{x^2} = x$$
$$\text{Since } (y)^2 = (y)(y) = y^2, \ \sqrt{y^2} = y$$

As you can see, any squared letter is a $\underline{\text{perfect square}}$. That is:

$$t^2 \text{ is a perfect square, since } \sqrt{t^2} = t$$
$$F^2 \text{ is a perfect square, since } \sqrt{F^2} = \text{\_\_\_\_\_}$$

a) 9   b) 1

---

82. The square root of a squared-letter term is also "$\underline{\text{a quantity whose square}}$ $\underline{\text{equals}}$ $\underline{\text{the letter term}}$". For example:

$$\text{Since } (3x)^2 = (3x)(3x) = 9x^2, \ \sqrt{9x^2} = 3x$$
$$\text{Since } (8b)^2 = (8b)(8b) = 64b^2, \ \sqrt{64b^2} = 8b$$

Any squared-letter term with a perfect square coefficient is also a $\underline{\text{perfect}}$ $\underline{\text{square}}$. That is:

$$4d^2 \text{ is a perfect square, since } \sqrt{4d^2} = 2d$$
$$36P^2 \text{ is a perfect square, since } \sqrt{36P^2} = \text{\_\_\_\_\_}$$

F

---

83. In the multiplication below, the product is the difference of two perfect squares.

$$(x + 4)(x - 4) = x^2 - 16$$

By simply reversing the two sides above, we can factor $x^2 - 16$. That is:

$$x^2 - 16 = (x + 4)(x - 4)$$

   Notice these points about the factoring:

   1) "x" is the square root of "x^2".

   2) "4" is the square root of "16".

   3) One factor is a sum; the other factor is a difference.

Following the pattern above, complete these factorings.

   a) $y^2 - 36 = (y + 6)($ _____ $)$ b) $m^2 - 1 = ($ _____ $)(m - 1)$

6P

84. In the multiplication below, the product is the difference of two perfect squares.

$$(3y + 2)(3y - 2) = 9y^2 - 4$$

By simply reversing the two sides, we can factor $9y^2 - 4$. That is:

$$9y^2 - 4 = (3y + 2)(3y - 2)$$

Notice these points about the factoring:

1) "3y" is the square root of "$9y^2$".

2) "2" is the square root of "4".

3) One factor is a sum; the other factor is a difference.

Following the pattern above, complete these factorings.

a) $25t^2 - 64 = (5t + 8)($ $)$ b) $49y^2 - 1 = ($ $)(7y - 1)$

---

a) (y + 6)(y – 6)

b) (m + 1)(m – 1)

---

85. Following the pattern in the last two frames, factor these:

a) $p^2 - 16 = $ _____

b) $t^2 - 100 = $ _____

c) $36x^2 - 81 = $ _____

d) $64h^2 - 9 = $ _____

---

a) (5t + 8)(5t – 8)

b) (7y + 1)(7y – 1)

---

86. The factoring pattern we have been using applies only to the difference of two perfect squares. It does not apply to the sum of two perfect squares. Therefore, it does not apply to either binomial below.

$$x^2 + 100 \qquad\qquad 25y^2 + 81$$

Use the pattern to factor these if it applies.

a) $m^2 - 1 = $ _____

b) $d^2 + 9 = $ _____

c) $64a^2 + 9 = $ _____

d) $4t^2 - 1 = $ _____

---

a) (p + 4)(p – 4)

b) (t + 10)(t – 10)

c) (6x + 9)(6x – 9)

d) (8h + 3)(8h – 3)

---

87. The factoring pattern we have been using applies only when both terms are perfect squares. Therefore, it does not apply to either binomial below.

$$t^2 - 39 \qquad\qquad 17F^2 - 49$$

Use the pattern to factor these if it applies.

a) $y^2 - 84 = $ _____

b) $h^2 - 144 = $ _____

c) $4x^2 - 9 = $ _____

d) $23V^2 - 1 = $ _____

---

a) (m + 1)(m – 1)

b) Does not apply

c) Does not apply

d) (2t + 1)(2t – 1)

---

a) Does not apply

b) (h + 12)(h – 12)

c) (2x + 3)(2x – 3)

d) Does not apply

4-12   COMPUTING "b" AND "c" IN PRODUCTS OF THE FORM: $x^2 + bx + c$

In this section, we will discuss a shortcut for computing "b" and "c" in products of the form: $x^2 + bx + c$.

88. In any trinomial product of the form: $x^2 + bx + c$, "b" is the coefficient of the letter term and "c" is the number term. Both "b" and "c" can be either positive or negative. For example: $\qquad$ In $x^2 + 7x + 12$, b = 7 and c = 12 $\qquad$ In $y^2 - 8y + 16$, b = -8 and c = 16 $\quad$ a) In $m^2 + 3m - 10$, b = _____ and c = _____ $\quad$ b) In $d^2 - 5d - 14$, b = _____ and c = _____	
89. The trinomial product for the multiplication below is given. $\qquad (x + 2)(x + 3) = x^2 + 5x + 6$ Notice these points about "b" and "c" in the product. $\quad$ 1) "b" is 5. It is the <u>sum</u> of 2 and 3. $\quad$ 2) "c" is 6. It is the <u>product</u> of 2 and 3. Using the facts above, we can directly write the product below. $\qquad (y + 5)(y + 4)$ $\quad$ a) Since $5 + 4 = 9$, b = _____ $\quad$ b) Since $(5)(4) = 20$, c = _____ $\quad$ c) Therefore, $(y + 5)(y + 4) =$ _____	a) b = 3, c = -10 b) b = -5, c = -14
90. Using the method in the last frame, write each trinomial product. $\quad$ a) $(m + 1)(m + 7) =$ _____ $\quad$ b) $(p + 6)(p + 8) =$ _____	a) 9 b) 20 c) $y^2 + 9y + 20$
91. The trinomial product for the multiplication below is given. $\qquad (y - 3)(y - 5) = y^2 - 8y + 15$ Notice these points about "b" and "c" in the product. $\quad$ 1) "b" is -8. It is the <u>sum</u> of -3 and -5. $\quad$ 2) "c" is 15. It is the <u>product</u> of -3 and -5.	a) $m^2 + 8m + 7$ b) $p^2 + 14p + 48$

Continued on following page.

91.   Continued

Using these facts, we can directly write the product below.

$$(t - 8)(t - 4)$$

a) Since $(-8) + (-4) = -12$, $b = $ _____

b) Since $(-8)(-4) = 32$, $c = $ _____

c) Therefore, $(t - 8)(t - 4) = $ _____

---

92.   Using the method in the last frame, write each trinomial product.

a) $(t - 1)(t - 2) = $ _____

b) $(m - 7)(m - 5) = $ _____

a) -12

b) 32

c) $t^2 - 12t + 32$

---

93.   The trinomial product for the multiplication below is given.

$$(x + 2)(x - 5) = x^2 - 3x - 10$$

Notice the same points about "b" and "c" in the product.

    1) "b" is -3. It is the <u>sum</u> of 2 and -5.

    2) "c" is -10. It is the <u>product</u> of 2 and -5.

Using the facts above, we can directly write the product below.

$$(y - 3)(y + 4)$$

a) Since $(-3) + 4 = 1$, $b = $ _____

b) Since $(-3)(4) = -12$, $c = $ _____

c) Therefore, $(y - 3)(y + 4) = $ _____

a) $t^2 - 3t + 2$

b) $m^2 - 12m + 35$

---

94.   Using the same method, write each trinomial product.

a) $(x + 7)(x - 1) = $ _____

b) $(y + 3)(y - 5) = $ _____

c) $(m - 3)(m + 6) = $ _____

d) $(p - 2)(p + 1) = $ _____

a) 1

b) -12

c) $y^2 + y - 12$

---

a) $x^2 + 6x - 7$

b) $y^2 - 2y - 15$

c) $m^2 + 3m - 18$

d) $p^2 - p - 2$

4-13   FACTORING TRINOMIALS OF THE FORM: $x^2 + bx + c$

In this section, we will discuss the procedure for factoring trinomials in which the coefficient of the squared letter is "1".

95. In the trinomial product below, both "b" and "c" are positive.
 	
$$(x + 4)(x + 2) = x^2 + 6x + 8$$
 	
To factor $x^2 + 6x + 8$, we can reverse the two sides of the equation. We get:
 	
$$x^2 + 6x + 8 = (x + 4)(x + 2)$$
 	
Notice these points about 4 and 2, the numbers in the factors.
 	
   1) Their <u>sum</u> is 6, which is "b" in the trinomial.
 	
   2) Their <u>product</u> is 8, which is "c" in the trinomial.
 	
Using the facts above, we can factor the trinomial below.
 	
$$y^2 + 5y + 6$$
 	
   1) The <u>product</u> of the numbers in the binomials must be 6.
      The possible pairs of factors are (1 and 6) and (2 and 3).
 	
   2) The <u>sum</u> of the numbers in the binomials must be 5.
      Therefore, the correct pair of factors is (2 and 3).
 	
    Therefore: $y^2 + 5y + 6 = ($ $)($ $)$	

96. Let's factor $y^2 + 5y + 4$. The possible pairs of factors of "4" are (1 and 4) and (2 and 2).
 	
   a) The sum of which pair of factors is 5? _____
 	
   b) Therefore: $y^2 + 5y + 4 =$ _____
	$(y + 2)(y + 3)$

97. Let's factor $t^2 + 8t + 12$. The possible pairs of factors of "12" are (1 and 12), (2 and 6), and (3 and 4).
 	
   a) The sum of which pair of factors is 8? _____
 	
   b) Therefore: $t^2 + 8t + 12 =$ _____
	a) 1 and 4
	
	b) $(y + 1)(y + 4)$

98. Using the same method, factor each trinomial below.
 	
   a) $x^2 + 6x + 5 =$ _____
 	
   b) $p^2 + 10p + 16 =$ _____
	a) 2 and 6
	
	b) $(t + 2)(t + 6)$

	a) $(x + 1)(x + 5)$ b) $(p + 2)(p + 8)$

99. In the trinomial below, "b" is negative and "c" is positive.

$$x^2 - 7x + 10$$

Since the product of the numbers in the binomials must be +10 and their sum must be -7, both numbers must be negative. The possible pairs of negative factors for +10 are (-1 and -10) and (-2 and -5).

a) The sum of which pair of factors is -7? _____

b) Therefore: $x^2 - 7x + 10$ = (   )(   )

---

100. When factoring the trinomial below, the numbers in both binomials must also be negative.

$$y^2 - 7y + 6$$

The possible pairs of negative factors for +6 are (-1 and -6) and (-2 and -3).

a) The sum of which pair of factors is -7? _____

b) Therefore: $y^2 - 7y + 6$ = (   )(   )

a) -2 and -5

b) (x - 2)(x - 5)

---

101. Using the same method, factor each trinomial below.

a) $t^2 - 3t + 2$ = _____

b) $m^2 - 8m + 16$ = _____

a) -1 and -6

b) (y - 1)(y - 6)

---

102. When the number term "c" of a trinomial is positive, the numbers in the binomial factors are either both positive or both negative. To decide whether they are both positive or both negative, we look at "b".

If "b" is positive, both numbers must be positive. That is:

$$x^2 + 5x + 6 = (x + 2)(x + 3)$$

If "b" is negative, both numbers must be negative. That is:

$$x^2 - 5x + 6 = (x - 2)(x - 3)$$

Using the facts above, factor each trinomial below.

a) $y^2 + 6y + 5$ = _____

b) $y^2 - 6y + 5$ = _____

c) $m^2 + 9m + 14$ = _____

d) $m^2 - 9m + 14$ = _____

a) (t - 1)(t - 2)

b) (m - 4)(m - 4)

---

a) (y + 1)(y + 5)

b) (y - 1)(y - 5)

c) (m + 2)(m + 7)

d) (m - 2)(m - 7)

103.   In the trinomial below, "c" is negative. Let's factor the trinomial.

$$x^2 + 5x - 6$$

Since the product of the numbers in the binomials must be –6, one number must be positive and the other negative. The possible pairs of factors for –6 are (–1 and 6), (–6 and 1), (–2 and 3), and (–3 and 2).

a) The sum of which pair of factors is +5? _____

b) Therefore: $x^2 + 5x - 6$ = (   )(   )

---

104.   In the trinomial below, "c" is negative. Let's factor the trinomial.

$$y^2 - 2y - 8$$

Since the product of the numbers in the binomials must be –8, one must be positive and the other negative. The possible pairs of factors for –8 are (–1 and 8), (–8 and 1), (–2 and 4), and (2 and –4).

a) The sum of which pair of factors is –2? _____

b) Therefore: $y^2 - 2y - 8$ = (   )(   )

a) –1 and 6

b) (x – 1)(x + 6)

---

105.   If "c" in a trinomial is negative, the numbers in the two binomials must have different signs. Using that fact, factor each trinomial below.

a) $m^2 + 4m - 5$ = _____

b) $t^2 - 2t - 3$ = _____

c) $x^2 + 3x - 10$ = _____

d) $y^2 - 5y - 14$ = _____

a) 2 and –4

b) (y + 2)(y – 4)

---

106.   To check the factoring of a trinomial, multiply the two binomial factors to see whether you obtain the original trinomial.   For example:

$x^2 + 8x + 15$ = (x + 3)(x + 5) is correct, since:

(x + 3)(x + 5) = $x^2 + 8x + 15$

$y^2 + 7y - 18$ = (y – 2)(y + 9) is correct, since:

(y – 2)(y + 9) = $y^2 + 7y - 18$

Check each factoring.   State whether it is "correct" or "incorrect".

a) $x^2 - 7x + 10$ = (x – 2)(x – 5) _____

b) $p^2 + 2p - 15$ = (p – 3)(p + 5) _____

c) $t^2 - 5t - 14$ = (t – 2)(t + 7) _____

a) (m – 1)(m + 5)

b) (t + 1)(t – 3)

c) (x – 2)(x + 5)

d) (y + 2)(y – 7)

107.   Remember that "c" is the key to factoring a trinomial. That is:

1) If the number term is <u>positive</u>, the numbers in the binomials have <u>the same sign.</u>

2) If the number term is <u>negative</u>, the numbers in the binomials have <u>different signs.</u>

Factor each trinomial below and check your results.

a) $x^2 + 9x + 20$ = _____

b) $y^2 - 3y - 18$ = _____

c) $t^2 - 9t + 8$ = _____

d) $m^2 + 4m - 12$ = _____

a) Correct

b) Correct

c) Incorrect, since:
$(t - 2)(t + 7)$ =
$t^2 + 5t - 14$

---

108.   When the trinomial below is factored, the two binomials are identical.

$$x^2 + 4x + 4 = (x + 2)(x + 2)$$

Factor these trinomials.

a) $y^2 + 10y + 25$ = _____

b) $t^2 - 2t + 1$ = _____

a) $(x + 4)(x + 5)$

b) $(y + 3)(y - 6)$

c) $(t - 1)(t - 8)$

d) $(m - 2)(m + 6)$

---

109.   Some trinomials of the form $x^2 + bx + c$ cannot be factored into binomials in which the numbers are whole numbers. An example is discussed below.

$$x^2 + 4x + 6$$

The possible pairs of factors are (1 and 6) and (2 and 3).
Neither pair has 4 as its sum.

Factor if possible.

a) $t^2 + 7t + 12$ = _____

b) $x^2 - 11x + 16$ = _____

c) $R^2 - R - 11$ = _____

d) $m^2 + m - 2$ = _____

a) $(y + 5)(y + 5)$

b) $(t - 1)(t - 1)$

---

a) $(t + 3)(t + 4)$

b) Not possible

c) Not possible

d) $(m - 1)(m + 2)$

4-14   FACTORING TRINOMIALS OF THE FORM: $ax^2 + bx + c$

In this section, we will discuss the procedure for factoring trinomials in which the coefficient of the squared letter is a number other than "1".

---

110.   The steps needed to multiply $(4x + 3)$ and $(x + 2)$ are shown below.

$$(4x + 3)(x + 2) = 4x^2 + 8x + 3x + 6 = 4x^2 + 11x + 6$$

Notice these points about the trinomial product:

"$4x^2$" is the product of "$4x$" and "x", the letter terms.

"6" is the product of "3" and "2", the number terms.

Using the facts above, if we multiplied $(2y + 4)$ and $(3y + 1)$:

a) The first term of the trinomial product would be _____ .

b) The last term of the trinomial product would be _____ .

---

111.   Here are the steps needed to factor the trinomial below.

$$3x^2 + 14x + 8$$

1) Since the product of the letter terms of the binomials must be "$3x^2$", the only possible pair of letter terms is $(3x$ and $x)$.

2) Since the product of the number terms of the binomials must be 8, the possible pairs of number terms are (1 and 8) and (2 and 4).

3) Using the factors in all combinations, we get four possible pairs of binomials:

> A: $(3x + 1)(x + 8)$
>
> B: $(3x + 8)(x + 1)$
>
> C: $(3x + 2)(x + 4)$
>
> D: $(3x + 4)(x + 2)$

4) Only one pair of binomials is correct. It is the pair that produces "$14x$" as the middle term of the trinomial product.

a) Which pair of binomial factors has "$14x$" as the middle term of its product? Pair _____

b) Therefore: $3x^2 + 14x + 8 = ($ $)($ $)$

---

a) $6y^2$, from $(2y)(3y)$

b) 4, from $(4)(1)$

---

a) Pair C

b) $(3x + 2)(x + 4)$

112.   Let's factor the trinomial: $2y^2 - 13y + 15$

    1) The only possible pair of letter terms is (2y and y).

    2) The possible pairs of number terms are (-1 and -15) and (-3 and -5).

    3) The possible pairs of binomial factors are:

            A: (2y - 1)(y - 15)

            B: (2y - 15)(y - 1)

            C: (2y - 3)(y - 5)

            D: (2y - 5)(y - 3)

The correct pair of binomial factors is the one with "-13y" as the middle term of its product.

    a) Which pair has "-13y" as the middle term of its product? Pair _____

    b) Therefore: $2y^2 - 13y + 15$ = (   )(   )

---

113.   Let's factor the trinomial: $5t^2 + 2t - 3$

    1) The only possible pair of letter terms is (5t and t).

    2) The possible pairs of number terms are (1 and -3) and (-1 and 3).

    3) The possible pairs of binomial factors are:

            A: (5t + 1)(t - 3)

            B: (5t - 3)(t + 1)

            C: (5t - 1)(t + 3)

            D: (5t + 3)(t - 1)

    a) Which pair of factors has a product whose middle term is "2t".
        Pair _____

    b) Therefore: $5t^2 + 2t - 3$ = (   )(   )

---

114.   Let's factor the trinomial: $7t^2 - 2t - 5$

    1) The only possible pair of letter terms is (7t and t).

    2) The possible pairs of number terms are (1 and -5) and (-1 and 5).

    3) The possible pairs of binomial factors are:

            A: (7t + 1)(t - 5)

            B: (7t - 5)(t + 1)

            C: (7t - 1)(t + 5)

            D: (7t + 5)(t - 1)

    a) Which pair of factors has a product whose middle term is "-2t"?
        Pair _____

    b) Therefore: $7t^2 - 2t - 5$ = (   )(   )

Answers column:

a) Pair C

b) (2y - 3)(y - 5)

a) Pair B

b) (5t - 3)(t + 1)

115. When the coefficient of the squared letter is a number other than "1", factoring is a process of "trial and error". Each possible pair of factors has to be checked by multiplying the binomial factors. That is:

$$3x^2 - 11x + 6 = (3x - 2)(x - 3) \text{ is correct, since:}$$

$$(3x - 2)(x - 3) = 3x^2 - 11x + 6$$

$$2y^2 - y - 1 = (2y + 1)(y - 1) \text{ is correct, since:}$$

$$(2y + 1)(y - 1) = 2y^2 - y - 1$$

Factor each trinomial below and check your results.

a) $5m^2 + 13m + 6 = $ _____

b) $7m^2 - 11m + 4 = $ _____

a) Pair D

b) $(7t + 5)(t - 1)$

---

116. Factor each trinomial below and check your results.

a) $11m^2 + 4m - 7 = $ _____

b) $2t^2 - 5t - 3 = $ _____

a) $(5m + 3)(m + 2)$

b) $(7m - 4)(m - 1)$

---

117. Let's factor the trinomial: $6x^2 + 19x + 8$.

1) The possible pairs of letter terms are (x and 6x) and (2x and 3x).

2) The possible pairs of number terms are (1 and 8) and (2 and 4).

3) The possible pairs of binomial factors are:

A: $(x + 1)(6x + 8)$ E: $(2x + 1)(3x + 8)$

B: $(x + 8)(6x + 1)$ F: $(2x + 8)(3x + 1)$

C: $(x + 2)(6x + 4)$ G: $(2x + 2)(3x + 4)$

D: $(x + 4)(6x + 2)$ H: $(2x + 4)(3x + 2)$

a) Which pair of factors has a product whose middle term is "19x"?
    Pair _____

b) Therefore: $6x^2 + 19x + 8 = ($ $)($ $)$

a) $(11m - 7)(m + 1)$

b) $(2t + 1)(t - 3)$

---

118. When the trinomial below is factored, the two binomials are identical.

$$9x^2 + 12x + 4 = (3x + 2)(3x + 2)$$

Factor these trinomials:

a) $16y^2 + 8y + 1 = $ _____

b) $4d^2 - 12d + 9 = $ _____

a) Pair E

b) $(2x + 1)(3x + 8)$

119. The trinomial below cannot be factored into binomials containing whole numbers.

$$2m^2 + 6m + 3$$

The possible pairs of factors for the trinomial are:

$$(2m + 3)(m + 1)$$
$$(2m + 1)(m + 3)$$

Neither pair has "6m" as the middle term of its product.

Factor if possible:

a) $3x^2 - x - 5 =$ _____

b) $4y^2 + 11y + 6 =$ _____

a) $(4y + 1)(4y + 1)$

b) $(2d - 3)(2d - 3)$

a) Not possible   b) $(y + 2)(4y + 3)$

SELF-TEST 15 (pp. 161-173)

Factor each binomial.

1. $25x^2 - 1$ 2. $100y^2 - 49$

Find each trinomial product.

3. $(t + 7)(t - 3)$ 4. $(a - 1)(a - 9)$

Factor each trinomial.

5. $r^2 + 7r + 12$ 7. $9d^2 + 6d + 1$

6. $w^2 - 9w + 20$ 8. $6x^2 - 11x - 10$

ANSWERS: 1. $(5x + 1)(5x - 1)$ 3. $t^2 + 4t - 21$ 5. $(r + 3)(r + 4)$ 7. $(3d + 1)(3d + 1)$
2. $(10y + 7)(10y - 7)$ 4. $a^2 - 10a + 9$ 6. $(w - 4)(w - 5)$ 8. $(3x + 2)(2x - 5)$

SUPPLEMENTARY PROBLEMS - CHAPTER 4

Assignment 13

Do these multiplications.

1. $(3y)(y)$ 2. $P(8P)$ 3. $(5x)(9x)$ 4. $(8r)(7r)$

Do these squarings.

5. $(2E)^2$ 6. $(5w)^2$ 7. $(11h)^2$ 8. $(30t)^2$

Do these multiplications.

9. $p(p + 1)$ 10. $x(7x - 6)$ 11. $2t(t - 1)$

12. $12y(2y + 7)$ 13. $5b(4b + 1)$ 14. $9R(6R - 7)$

Write the missing factor in each blank.

15. $60 = (12)(\quad)$ 16. $12h = (\quad)(2)$ 17. $32t^2 = (\quad)(4t)$ 18. $8d^2 = (d)(\quad)$

Factor each binomial completely.

19. $2t + 6$ 20. $5y + 5$ 21. $12r + 8$

22. $3h - 3$ 23. $16m - 4$ 24. $8x - 28$

25. $40d + 25$ 26. $24p - 18$ 27. $60w - 36$

Factor each binomial completely.

28. $s^2 + s$ 29. $2y^2 + 10y$ 30. $6r^2 + 3r$

31. $3x^2 - x$ 32. $4b^2 - 6b$ 33. $15w^2 - 10w$

34. $20r^2 + 12r$ 35. $8h^2 - 15h$ 36. $48t^2 + 80t$

Assignment 14

Do these multiplications.

1. $(x + 5)(x + 3)$ 2. $(r + 1)(r + 3)$ 3. $(m + 8)(m + 6)$

4. $(2y + 1)(y + 1)$ 5. $(d + 1)(4d + 7)$ 6. $(3t + 2)(2t + 3)$

7. $(5a + 6)(2a + 9)$ 8. $(3w + 5)(4w + 9)$ 9. $(8x + 1)(9x + 2)$

Do these multiplications.

10. $(t - 2)(t + 3)$ 11. $(b + 1)(b - 4)$ 12. $(y - 5)(y - 3)$

13. $(r + 1)(4r - 1)$ 14. $(5h + 2)(h - 2)$ 15. $(3p - 1)(4p - 5)$

16. $(2m - 7)(2m + 5)$ 17. $(4x + 9)(5x - 2)$ 18. $(7a - 2)(5a - 1)$

Do these multiplications involving the sum and difference of two terms.

19. $(a + 7)(a - 7)$ 20. $(V + 1)(V - 1)$ 21. $(3h + 2)(3h - 2)$

22. $(5t + 1)(5t - 1)$ 23. $(4y + 3)(4y - 3)$ 24. $(7x + 12)(7x - 12)$

Do these squarings.

25. $(R + 1)^2$ 26. $(w - 6)^2$ 27. $(2x + 5)^2$

28. $(6d - 1)^2$ 29. $(3 - t)^2$ 30. $(1 + 4y)^2$

Assignment 15

Factor each binomial.

1. $t^2 - 49$ 　　　　　 2. $9y^2 - 1$ 　　　　　 3. $4F^2 - 25$ 　　　　 4. $16x^2 - 81$

Find each trinomial product.

5. $(m + 5)(m + 1)$ 　　　　 6. $(w + 2)(w - 6)$ 　　　　 7. $(h - 7)(h - 9)$

8. $(P - 8)(P + 3)$ 　　　　 9. $(d - 4)(d - 5)$ 　　　 10. $(r + 10)(r - 9)$

Factor each trinomial.

11. $x^2 + 3x + 2$ 　　　　 12. $a^2 + 7a + 10$ 　　　 13. $y^2 + 6y + 9$

14. $t^2 - 4t + 3$ 　　　　 15. $R^2 - 6R + 8$ 　　　 16. $w^2 - w - 12$

17. $h^2 + 4h - 5$ 　　　　 18. $d^2 - 3d - 10$ 　　　 19. $m^2 + 4m - 21$

Factor each trinomial.

20. $2y^2 + 3y + 1$ 　　　　 21. $3x^2 + 7x + 2$ 　　　 22. $3p^2 + 11p + 6$

23. $2r^2 - 5r + 2$ 　　　　 24. $5t^2 - 8t + 3$ 　　　 25. $6w^2 - 11w + 4$

26. $4E^2 + 4E - 3$ 　　　　 27. $3d^2 + 2d - 8$ 　　　 28. $7x^2 - 4x - 3$

29. $8m^2 - 6m - 9$ 　　　　 30. $4b^2 - 4b + 1$ 　　　 31. $9w^2 + 12w + 4$

Chapter 5 QUADRATIC EQUATIONS

Equations like $x^2 - 25 = 0$, $3y^2 + 4y = 0$, and $t^2 - 5t + 6 = 0$ have two solutions. They are called "quadratic" equations. There are three types of quadratic equations - "pure", "incomplete", and "complete". In this chapter, we will discuss the methods for solving each type. Some formula evaluations and applied problems are included.

5-1 SQUARING SIGNED NUMBERS

In this section, we will discuss the meaning of "squaring" signed whole numbers, fractions, and decimal numbers. We will also show that whole numbers, fractions, and decimal numbers can be perfect squares.

1. To square any number, we multiply the number by itself. For example:

$$7^2 = (7)(7) = 49$$

$$\left(\frac{3}{4}\right)^2 = \left(\frac{3}{4}\right)\left(\frac{3}{4}\right) = \frac{9}{16}$$

$$(.5)^2 = (.5)(.5) = .25$$

Find the square of each number.

   a) $20^2 = $ _____   b) $\left(\frac{1}{2}\right)^2 = $ _____   c) $(.04)^2 = $ _____

2. The square of any negative number is a <u>positive</u> number. For example:

$$(-8)^2 = (-8)(-8) = 64$$

$$\left(-\frac{1}{5}\right)^2 = \left(-\frac{1}{5}\right)\left(-\frac{1}{5}\right) = \frac{1}{25}$$

$$(-.3)^2 = (-.3)(-.3) = .09$$

Find the square of each number.

   a) $(-1)^2 = $ _____   b) $\left(-\frac{5}{3}\right)^2 = $ _____   c) $(-1.1)^2 = $ _____

    a) 400

    b) $\frac{1}{4}$

    c) .0016

3. Any signed number has the same square as its opposite. For example:

The square of both 2 and -2 is 4.

The square of both $\frac{5}{6}$ and $-\frac{5}{6}$ is $\frac{25}{36}$.

The square of both 1.2 and -1.2 is _____.

    a) 1

    b) $\frac{25}{9}$

    c) 1.21

    1.44

176

4. The word names for expressions like $(-9)^2$ and $\left(\frac{3}{8}\right)^2$ are "-9 squared" and "$\frac{3}{8}$ squared". Therefore:

a) -10 squared = $(-10)^2$ = _____   b) $-\frac{1}{7}$ squared = $\left(-\frac{1}{7}\right)^2$ = _____

---

5. Any whole number that is the square of a signed whole number is called a "perfect square". That is:

Since $(-40)^2 = 1,600$, $1,600$ is a perfect square.

Since $125^2 = 15,625$, _____ is a perfect square.

a) 100

b) $\frac{1}{49}$

---

6. Any fraction that is the square of a signed fraction is called a "perfect square". That is:

Since $\left(\frac{7}{8}\right)^2 = \frac{49}{64}$, $\frac{49}{64}$ is a perfect square.

Since $\left(-\frac{1}{3}\right)^2 = \frac{1}{9}$, _____ is a perfect square.

15,625

---

7. Any decimal number that is the square of a signed decimal number is called a "perfect square". That is:

Since $(1.4)^2 = 1.96$, 1.96 is a perfect square.

Since $(-.8)^2 = .64$, _____ is a perfect square.

$\frac{1}{9}$

---

8. Following the example, complete the other evaluation.

When $x = 5$, $x^2 = 5^2 = 25$ When $x = -3$, $x^2 = (-3)^2 =$ _____

.64

---

9. Notice how we square before multiplying in the evaluations below.

When $x = 10$, $2x^2 = 2(10)^2 = 2(100) = 200$

When $x = -5$, $4x^2 = 4(-5)^2 = 4(\quad) =$ _____

9

---

10. Notice how we square before subtracting in the evaluations below.

When $x = 3$, $4x - x^2 = 4(3) - 3^2 = 12 - 9 = 3$

When $x = -2$, $4x - x^2 = 4(-2) - (-2)^2 =$ _____ $-$ _____ $=$ _____

4(25) = 100

(-8) − 4 = −12

5-2   FORMULA EVALUATIONS

In this section, we will discuss evaluations with formulas containing a squared letter.

11. In $\boxed{A = s^2}$, we can find A by simply squaring the value of "s". If $s = 8$, $A = s^2 = 8^2 = 64$ If $s = 30$, $A = s^2 = 30^2 = $ _____	
12. When the squared letter is one of the factors in a multiplication, we <u>square</u> before <u>multiplying</u>. An example is shown. Complete the other evaluation. In $\boxed{P = I^2R}$, find P when $I = 3$ and $R = 10$. $P = I^2R = (3^2)(10) = 9(10) = 90$ In $\boxed{A = 3.14r^2}$, find A when $r = 10$. $A = 3.14r^2 = $ _____	900
13. Following the example, complete the other evaluation. In $\boxed{s = \frac{1}{2}at^2}$, find "s" when $a = 12$ and $t = 4$. $s = \frac{1}{2}at^2 = \frac{1}{2}(12)(4^2) = \frac{1}{2}(12)(16) = 6(16) = 96$ In $\boxed{E = \frac{1}{2}mv^2}$, find E when $m = 8$ and $v = 10$. $E = \frac{1}{2}mv^2 = $ _____	$A = 314$, from: $3.14(100)$
14. When a squared letter is one of the terms of a fraction, <u>we simplify the terms</u> before <u>dividing</u>. An example is shown. Complete the other evaluation. In $\boxed{P = \dfrac{E^2}{R}}$, find P when $E = 9$ and $R = 10$. $P = \dfrac{E^2}{R} = \dfrac{9^2}{10} = \dfrac{81}{10} = 8.1$ In $\boxed{R = \dfrac{pL}{d^2}}$, find R when $p = 2$, $L = 50$, and $d = 5$. $R = \dfrac{pL}{d^2} = $ _____	$E = 400$, from: $\frac{1}{2}(8)(100)$
	$R = 4$, from: $\dfrac{100}{25}$

15. When a squared letter is one of the factors in a term of a fraction, we also <u>simplify</u> <u>the</u> <u>terms</u> <u>before</u> <u>dividing</u>. An example is shown. Complete the other evaluation.

In $\boxed{H = \dfrac{D^2 N}{2.5}}$, find H when $D = 2$ and $N = 20$.

$$H = \frac{D^2 N}{2.5} = \frac{(2^2)(20)}{2.5} = \frac{4(20)}{2.5} = \frac{80}{2.5} = 32$$

In $\boxed{P = \dfrac{rw}{2t^2}}$, find P when $r = 3$, $w = 100$, and $t = 5$.

$$P = \frac{rw}{2t^2} = \underline{\hspace{5cm}}$$

---

16. In the evaluation below, <u>we</u> <u>simplified</u> <u>the</u> <u>grouping</u> <u>before</u> <u>multiplying</u>. Complete the other evaluation.

In $\boxed{A = 0.785(D^2 - d^2)}$, find A when $D = 6$ and $d = 4$.

$$A = 0.785(D^2 - d^2)$$
$$= 0.785(6^2 - 4^2) = 0.785(36 - 16) = 0.785(20) = 15.7$$

In $\boxed{H = (1 - C^2)h}$, find H when $C = .5$ and $h = 100$.

$$H = (1 - C^2)h = \underline{\hspace{5cm}}$$

$P = 6$, from: $\dfrac{300}{50}$

---

17. In the evaluation below, we simplified the terms before adding. Complete the other evaluation.

In $\boxed{s = v_0 t + 16t^2}$, find "s" when $v_0 = 50$ and $t = 10$.

$$s = v_0 t + 16t^2$$
$$= 50(10) + 16(10^2) = 500 + 16(100) = 500 + 1{,}600 = 2{,}100$$

In $\boxed{I_A = I_C + md^2}$, find I_A when $I_C = 75$, $m = 4$, and $d = 5$.

$$I_A = I_C + md^2 = \underline{\hspace{5cm}}$$

$H = 75$, from:

$(.75)(100)$

$I_A = 175$, from:

$75 + 100$

18.   Don't confuse the subscript "2" with the "2" that is the symbol for squaring.

The subscript "2" is written <u>slightly</u> <u>below</u> <u>the</u> <u>letter</u>.
It usually means "second". That is:

v_2 means "the second velocity".

The symbol for squaring is written <u>slightly</u> <u>above</u> <u>the</u> <u>letter</u>. That is:

v^2 means "square the value of v".

Each formula below contains both the subscript "2" and the symbol for squaring. Following the example, complete the other evaluation.

In $\boxed{F = \dfrac{m_1 m_2}{rd^2}}$, find F when $m_1 = 8$, $m_2 = 10$, $r = 5$, and $d = 2$.

$$F = \frac{m_1 m_2}{rd^2} = \frac{(8)(10)}{(5)(2^2)} = \frac{(8)(10)}{(5)(4)} = \frac{80}{20} = 4$$

In $\boxed{E = \dfrac{kQ_1 Q_2}{d^2}}$, find E when $k = 4$, $Q_1 = 10$, $Q_2 = 5$, and $d = 2$.

$$E = \frac{kQ_1 Q_2}{d^2} = \underline{\hspace{4in}}$$

$E = 50$, from: $\dfrac{200}{4}$

5-3   THE SQUARE ROOTS OF POSITIVE NUMBERS

In this section, we will discuss what is meant by the "<u>square</u> <u>root</u>" of a positive number. We will show that any positive number has two square roots.

---

19.   A "<u>square</u> <u>root</u>" of a positive number N is "a number <u>whose</u> <u>square</u> <u>is</u> <u>N</u>".
That is:

Since $5^2 = 25$,   5 is a square root of 25 .

Since $8^2 = 64$,   8 is a square root of 64 .

Instead of saying "the square root of", we use the symbol $\sqrt{}$. That is:

Since $4^2 = 16$, $\sqrt{16} = 4$ Since $10^2 = 100$, $\sqrt{100} = \underline{\hspace{1in}}$

---

20.   For expressions like $\sqrt{49}$, the following terminology is used.

1) The entire expression $\sqrt{49}$ is called a <u>square</u> <u>root</u> <u>radical</u>.

2) The $\sqrt{}$ symbol is called a <u>radical</u> <u>sign</u> or simply a <u>radical</u>.

3) The number "49" under the radical sign is called the <u>radicand</u>.

Write the square root radical whose radicand is 16. $\underline{\hspace{1.5in}}$

10

$\sqrt{16}$

21.   A positive number also has a <u>negative</u> <u>square</u> <u>root</u>. That is:

        Since $(-6)^2 = 36$, -6 is a square root of 36 .

        Since $(-9)^2 = 81$, -9 is a square root of 81 .

Therefore, any positive number has two square roots, one positive and one negative. The two square roots are opposites. For example:

        Since $3^2 = 9$, 3 is the <u>positive</u> square root of 9 .

        Since $(-3)^2 = 9$, -3 is the <u>negative</u> square root of 9 .

Complete: a) The <u>positive</u> square root of 16 is _____ .

            b) The <u>negative</u> square root of 100 is _____ .

            c) The two square roots of 81 are _____ and _____ .

---

22.   The <u>positive</u> square root of a number is usually called its "<u>principal</u>" square root. Therefore:

        The <u>principal</u> (or positive) square root of 25 is 5 .

        The <u>negative</u> square root of 25 is -5 .

To indicate a <u>principal</u> square root, we write a "+" sign (or no sign) in front of the radical. To indicate a <u>negative</u> square root, we write a "-" sign in front of the radical. That is:

      $+\sqrt{16} = 4$ and $\sqrt{16} = 4$ $-\sqrt{16} = -4$

Complete these.

    a) $+\sqrt{81} =$ _____   b) $\sqrt{9} =$ _____   c) $-\sqrt{100} =$ _____

Answers:
a) 4
b) -10
c) 9 and -9

---

23.   The symbol "$\pm$" means "<u>both + and -</u>". It can be written either in front of a radical or a number.

      When written in front of a radical, "$\pm$" means both roots are desired. That is:

        $\pm\sqrt{25}$ means both $+\sqrt{25}$ and $-\sqrt{25}$.

      When written in front of a number, "$\pm$" means both signed values of that number. That is:

        ± 5 means both $+5$ and -5 .

      Therefore: $\pm\sqrt{25} = \pm 5$ means $+\sqrt{25} = +5$ and $-\sqrt{25} = -5$.

Using the "$\pm$" symbol, complete these:

    a) $\pm\sqrt{4} =$ _____   b) $\pm\sqrt{1} =$ _____   c) $\pm\sqrt{49} =$ _____

Answers:
a) 9
b) 3
c) -10

---

a) ± 2 b) ± 1 c) ± 7

24. Any whole number that is a perfect square has two whole numbers, one positive and one negative, as its square roots. For example:

Since both 30^2 and $(-30)^2 = 900$, $\pm\sqrt{900} = \pm30$.

However, the number "0" has only one square root. That is:

Since $0^2 = 0$, $\sqrt{0} =$ _____ .

---

25. Page 183 contains a table of square roots for whole numbers from 1 to 200. All of the square roots in the table are given to two decimal places, including the square roots of perfect squares. For example, the table has these entries:

Number	Square Root
9	3.00
144	12.00

Since "9" and "144" are perfect squares, the two 0's after the decimal point should be ignored. That is:

$$\sqrt{9} = 3 \qquad \sqrt{144} = 12$$

Except for perfect squares, the square roots of all other whole numbers from 1 to 200 are non-ending decimal numbers. In the table, they are rounded <u>to hundredths</u>. For example, the table has these entries:

Number	Square Root	
19	4.36	(from 4.358898...)
167	12.92	(from 12.922848...)

Though the entries above are not exact because they are rounded, we will treat them as if they are exact. That is:

We will say: $\sqrt{19} = 4.36$ $\qquad$ $\sqrt{167} =$ _____

---

26. When the square-root entry has a "0" in the hundredths place, that "0" can be ignored. For example:

$$\sqrt{41} = 6.40 \text{ or } 6.4 \qquad \sqrt{177} = 13.30 \text{ or } \rule{2cm}{0.4pt}$$

---

27. Though only principal (or positive) square roots are given in the table, it can be used to get both square roots. For example:

$$\pm\sqrt{196} = \pm14 \qquad \pm\sqrt{47} = \pm6.86 \qquad \pm\sqrt{171} = \rule{2cm}{0.4pt}$$

Answer column (right side):

0

12.92

13.3

±13.08

SQUARE ROOTS OF WHOLE NUMBERS FROM 1 TO 200

Number	Square root	Number	Square root	Number	Square root	Number	Square root
1	1.00	51	7.14	101	10.05	151	12.29
2	1.41	52	7.21	102	10.10	152	12.33
3	1.73	53	7.28	103	10.15	153	12.37
4	2.00	54	7.35	104	10.20	154	12.41
5	2.24	55	7.42	105	10.25	155	12.45
6	2.45	56	7.48	106	10.30	156	12.49
7	2.65	57	7.55	107	10.34	157	12.53
8	2.83	58	7.62	108	10.39	158	12.57
9	3.00	59	7.68	109	10.44	159	12.61
10	3.16	60	7.75	110	10.49	160	12.65
11	3.32	61	7.81	111	10.54	161	12.69
12	3.46	62	7.87	112	10.58	162	12.73
13	3.61	63	7.94	113	10.63	163	12.77
14	3.74	64	8.00	114	10.68	164	12.81
15	3.87	65	8.06	115	10.72	165	12.85
16	4.00	66	8.12	116	10.77	166	12.88
17	4.12	67	8.19	117	10.82	167	12.92
18	4.24	68	8.25	118	10.86	168	12.96
19	4.36	69	8.31	119	10.91	169	13.00
20	4.47	70	8.37	120	10.95	170	13.04
21	4.58	71	8.43	121	11.00	171	13.08
22	4.69	72	8.49	122	11.05	172	13.11
23	4.80	73	8.54	123	11.09	173	13.15
24	4.90	74	8.60	124	11.14	174	13.19
25	5.00	75	8.66	125	11.18	175	13.23
26	5.10	76	8.72	126	11.22	176	13.27
27	5.20	77	8.77	127	11.27	177	13.30
28	5.29	78	8.83	128	11.31	178	13.34
29	5.39	79	8.89	129	11.36	179	13.38
30	5.48	80	8.94	130	11.40	180	13.42
31	5.57	81	9.00	131	11.45	181	13.45
32	5.66	82	9.06	132	11.49	182	13.49
33	5.74	83	9.11	133	11.53	183	13.53
34	5.83	84	9.17	134	11.58	184	13.56
35	5.92	85	9.22	135	11.62	185	13.60
36	6.00	86	9.27	136	11.66	186	13.64
37	6.08	87	9.33	137	11.70	187	13.67
38	6.16	88	9.38	138	11.75	188	13.71
39	6.24	89	9.43	139	11.79	189	13.75
40	6.32	90	9.49	140	11.83	190	13.78
41	6.40	91	9.54	141	11.87	191	13.82
42	6.48	92	9.59	142	11.92	192	13.86
43	6.56	93	9.64	143	11.96	193	13.89
44	6.63	94	9.70	144	12.00	194	13.93
45	6.71	95	9.75	145	12.04	195	13.96
46	6.78	96	9.80	146	12.08	196	14.00
47	6.86	97	9.85	147	12.12	197	14.04
48	6.93	98	9.90	148	12.17	198	14.07
49	7.00	99	9.95	149	12.21	199	14.11
50	7.07	100	10.00	150	12.25	200	14.14

28. Any fraction that is a perfect square has two fractions as its square roots. That is:

Since $\left(\dfrac{5}{8}\right)^2$ and $\left(-\dfrac{5}{8}\right)^2 = \dfrac{25}{64}$, $\pm\sqrt{\dfrac{25}{64}} = \pm\dfrac{5}{8}$

Since $\left(\dfrac{1}{2}\right)^2$ and $\left(-\dfrac{1}{2}\right)^2 = \dfrac{1}{4}$, $\pm\sqrt{\dfrac{1}{4}} =$ _____

---

29. A fraction is a perfect square only if both of its terms are perfect squares. That is:

$\dfrac{81}{64}$ is a perfect square since both "81" and "64" are perfect squares.

$\dfrac{1}{36}$ is a perfect square since both "1" and "36" are perfect squares.

To find the square roots of perfect-square fractions, we find the square roots of both terms. Therefore:

$\pm\sqrt{\dfrac{81}{64}} = \pm\dfrac{9}{8}$ $\pm\sqrt{\dfrac{1}{36}} = \pm\dfrac{1}{6}$

Following the examples, complete these:

a) $\pm\sqrt{\dfrac{25}{49}} =$ _____   b) $\pm\sqrt{\dfrac{16}{9}} =$ _____   c) $\pm\sqrt{\dfrac{1}{25}} =$ _____

(answer in margin:) $\pm\dfrac{1}{2}$

---

30. Any decimal number that is a perfect square has two decimal numbers as its square roots. That is:

Since $(1.2)^2$ and $(-1.2)^2 = 1.44$, $\pm\sqrt{1.44} = \pm1.2$

Since $(.04)^2$ and $(-.04)^2 = .0016$, $\pm\sqrt{.0016} =$ _____

(answer in margin:) a) $\pm\dfrac{5}{7}$ b) $\pm\dfrac{4}{3}$ c) $\pm\dfrac{1}{5}$

---

31. To find the square roots of whole numbers greater than 200 or the square roots of decimal numbers, a hand-held calculator can be used. For example, a calculator was used to get the square roots below.

$$\sqrt{279} = 16.703293$$
$$\sqrt{41.8} = 6.465292$$
$$\sqrt{.069} = .26267851$$

Usually, square roots obtained from a calculator are rounded. For example, rounding the square roots above to two decimal places, we get:

$\sqrt{279} = 16.70$ $\sqrt{41.8} = 6.47$ $\sqrt{.069} =$ _____

(answer in margin:) $\pm.04$

(answer in margin:) .26

32.   Since the square of any number (positive or negative) is <u>positive</u>, there is no ordinary number that is the square root of a <u>negative</u> number. For example:

$$\sqrt{-16} \neq +4 \text{ , since } (+4)^2 = +16 \text{ .}$$

$$\sqrt{-16} \neq -4 \text{ , since } (-4)^2 = +16 \text{ .}$$

$$\sqrt{-\frac{9}{25}} \neq +\frac{3}{5} \text{ , since } \left(+\frac{3}{5}\right)^2 = +\frac{9}{25} \text{ .}$$

$$\sqrt{-\frac{9}{25}} \neq -\frac{3}{5} \text{ , since } \left(-\frac{3}{5}\right)^2 = +\frac{9}{25} \text{ .}$$

Though mathematicians have developed a method for handling the square roots of negative numbers, we will not discuss that method in this book.

5-4   THE THREE TYPES OF QUADRATIC EQUATIONS

Equations like $x^2 - 25 = 0$, $3y^2 + 4y = 0$, and $t^2 - 5t + 6 = 0$ have two solutions. They are called "quadratic" equations. There are three types of quadratic equations - "complete", "incomplete", and "pure". We will discuss the three types in this section.

33. $\boxed{x^2 - 36 = 0}$ is a "quadratic" equation. It has two solutions, 6 and -6. We checked 6 as its solution below. Check -6 as its solution.

$$x^2 - 36 = 0 \qquad\qquad x^2 - 36 = 0$$
$$(6)^2 - 36 = 0$$
$$36 - 36 = 0$$
$$0 = 0$$

34. $\boxed{y^2 - 5y = 0}$ is a "quadratic" equation. It has two solutions, 0 and 5. We checked "0" as its solution below. Check 5 as its solution.

$$y^2 - 5y = 0 \qquad\qquad y^2 - 5y = 0$$
$$(0)^2 - 5(0) = 0$$
$$0 - 0 = 0$$
$$0 = 0$$

(Answer to frame 33:)
$$(-6)^2 - 36 = 0$$
$$36 - 36 = 0$$
$$0 = 0$$

35. $\boxed{m^2 - 5m + 6 = 0}$ is a "quadratic" equation. It has two solutions, 2 and 3. We checked 2 as its solution below. Check 3 as its solution.

$$m^2 - 5m + 6 = 0 \qquad\qquad m^2 - 5m + 6 = 0$$
$$2^2 - 5(2) + 6 = 0$$
$$4 - 10 + 6 = 0$$
$$0 = 0$$

(Answer to frame 34:)
$$5^2 - 5(5) = 0$$
$$25 - 25 = 0$$
$$0 = 0$$

36. The three types of quadratic equations are called "complete", "incomplete", and "pure" quadratic equations. Their standard forms are summarized below. Note that the right side of each equation is "0".

$$3^2 - 5(3) + 6 = 0$$
$$9 - 15 + 6 = 0$$
$$0 = 0$$

Complete:	$x^2 + 3x - 10 = 0$
Incomplete:	$x^2 + 3x = 0$
Pure:	$x^2 - 10 = 0$

"Complete" quadratic equations have a trinomial on the left side. The trinomial contains a squared-letter term, a letter term, and a number term. Some examples are:

$$x^2 - 6x + 8 = 0 \qquad\qquad 3y^2 + 5y + 2 = 0$$

"Incomplete" quadratic equations have a binomial on the left side. The binomial contains a squared-letter term and a letter term. Some examples are:

$$m^2 - 3m = 0 \qquad\qquad 4d^2 + 20d = 0$$

"Pure" quadratic equations contain a binomial on the left side. The binomial contains a squared-letter term and a number term. Some examples are:

$$t^2 - 64 = 0 \qquad\qquad 16p^2 - 81 = 0$$

For each quadratic equation, state whether it is "complete", "incomplete", or "pure".

a) $m^2 - 25 = 0$ _____

b) $t^2 - 6t + 5 = 0$ _____

c) $p^2 - 8p = 0$ _____

---

37. For each quadratic equation, state whether it is "complete", "incomplete", or "pure".

a) $6x^2 + 12x = 0$ _____

b) $4V^2 - 100 = 0$ _____

c) $3y^2 + 14y + 8 = 0$ _____

a) pure

b) complete

c) incomplete

---

a) incomplete

b) pure

c) complete

5-5   SOLVING PURE QUADRATIC EQUATIONS

In this section, we will discuss a method for solving "pure" quadratic equations.

---

38.   In the standard form of a pure quadratic equation, both the squared-letter term and the number term are on the same side. For example:

$$x^2 - 49 = 0 \qquad\qquad y^2 - 4 = 0$$

To solve the equations above, we ordinarily use the addition axiom to get the squared-letter term on one side and the number term on the other side. We get:

$$x^2 = 49 \qquad\qquad y^2 = 4$$

The two solutions of each equation above are <u>the two square roots of the number</u>. That is:

For $x^2 = 49$, $x = \pm\sqrt{49} = 7$ and -7

For $y^2 = 4$, $y = \pm\sqrt{4} = $ _____ and _____

2 and -2

---

39.   To solve the pure quadratic below, we used the addition axiom to get t^2 on one side and 100 on the other side. Solve the other equation.

$$t^2 - 100 = 0 \qquad\qquad m^2 - 1 = 0$$
$$t^2 = 100$$
$$t = \pm\sqrt{100}$$
$$t = 10 \text{ and } -10$$

m = 1 and -1

---

40.   In each equation below, the number is a perfect-square fraction. The two solutions are <u>the two square roots of the fraction</u>. That is:

For $x^2 = \dfrac{36}{25}$, $x = \pm\sqrt{\dfrac{36}{25}} = \dfrac{6}{5}$ and $-\dfrac{6}{5}$

For $y^2 = \dfrac{1}{4}$, $y = \pm\sqrt{\dfrac{1}{4}} = $ _____ and _____

$\dfrac{1}{2}$ and $-\dfrac{1}{2}$

---

41.   To solve the equation below, we used the multiplication axiom to isolate t^2 and then found both square roots of $\dfrac{9}{16}$. Solve the other equation.

$$16t^2 = 9 \qquad\qquad 49d^2 = 100$$
$$t^2 = \dfrac{9}{16}$$
$$t = \pm\sqrt{\dfrac{9}{16}}$$
$$t = \dfrac{3}{4} \text{ and } -\dfrac{3}{4}$$

42.   To solve the equation below, we used both axioms to isolate x^2 and then found both square roots of $\frac{81}{64}$. Solve the other equation.

$$64x^2 - 81 = 0 \qquad\qquad 9b^2 - 1 = 0$$

$$64x^2 = 81$$

$$x^2 = \frac{81}{64}$$

$$x = \frac{9}{8} \text{ and } -\frac{9}{8}$$

| $d = \frac{10}{7}$ and $-\frac{10}{7}$ |

---

43.   We used the multiplication axiom to isolate x^2 below. Notice that we multiplied both sides by 2. Solve the other equation.

$$\frac{1}{2}x^2 = 8 \qquad\qquad 12 = \frac{1}{3}y^2$$

$$2\left(\frac{1}{2}x^2\right) = 2(8)$$

$$x^2 = 16$$

$$x = 4 \text{ and } -4$$

| $b = \frac{1}{3}$ and $-\frac{1}{3}$ |

---

44.   We used both axioms to solve the equation below. Solve the other equation.

$$\frac{1}{2}t^2 - 32 = 0 \qquad\qquad \frac{1}{4}m^2 - 25 = 0$$

$$\frac{1}{2}t^2 = 32$$

$$t^2 = 64$$

$$t = 8 \text{ and } -8$$

| $y = 6$ and -6 |

---

45.   To clear the fraction below, we multiplied both sides by 3. Solve the other equation.

$$\frac{x^2}{3} = 12 \qquad\qquad 25 = \frac{y^2}{4}$$

$$3\left(\frac{x^2}{3}\right) = 3(12)$$

$$x^2 = 36$$

$$x = 6 \text{ and } -6$$

| $m = 10$ and -10 |

| $y = 10$ and -10 |

46. Following the example, solve the other equation.

$$\frac{4x^2}{5} = 5 \qquad\qquad 16 = \frac{9y^2}{4}$$

$$\cancel{5}\left(\frac{4x^2}{\cancel{5}}\right) = 5(5)$$

$$4x^2 = 25$$

$$x^2 = \frac{25}{4}$$

$$x = \frac{5}{2} \text{ and } -\frac{5}{2}$$

---

47. When a squared-letter term is divided by itself, the quotient is "1". That is:

$$\frac{x^2}{x^2} = 1 \qquad\qquad \frac{5y^2}{5y^2} = 1$$

The fact above is used in the multiplications below.

$$x^2\left(\frac{7}{x^2}\right) = \frac{(x^2)(7)}{x^2} = \left(\frac{x^2}{x^2}\right)(7) = 1(7) = 7$$

$$5y^2\left(\frac{3}{5y^2}\right) = \frac{(5y^2)(3)}{5y^2} = \left(\frac{5y^2}{5y^2}\right)(3) = 1(3) = 3$$

We can use cancelling to perform the same multiplications. That is:

$$\cancel{x^2}\left(\frac{7}{\cancel{x^2}}\right) = 7 \qquad \cancel{5y^2}\left(\frac{3}{\cancel{5y^2}}\right) = 3 \qquad 2d^2\left(\frac{1}{2d^2}\right) = \underline{}$$

Answer (top right): $y = \frac{8}{3} \text{ and } -\frac{8}{3}$

---

48. To clear the fraction below, we multiplied both sides by $4d^2$. Solve the other equation.

$$\frac{25}{4d^2} = 9 \qquad\qquad 81 = \frac{49}{t^2}$$

$$\cancel{4d^2}\left(\frac{25}{\cancel{4d^2}}\right) = 4d^2(9)$$

$$25 = 36d^2$$

$$d^2 = \frac{25}{36}$$

$$d = \frac{5}{6} \text{ and } -\frac{5}{6}$$

Answer (right): $\cancel{2d^2}\left(\frac{1}{\cancel{2d^2}}\right) = 1$

Answer (bottom right): $t = \frac{7}{9} \text{ and } -\frac{7}{9}$

49.   To clear the fractions in the proportion below, we multiplied both sides by $18d^2$. Solve the other equation.

$$\frac{25}{18} = \frac{2}{d^2} \qquad\qquad\qquad \frac{4}{3} = \frac{3}{y^2}$$

$$\cancel{18d^2}^{d^2}\left(\frac{25}{\cancel{18}}\right) = \cancel{18d^2}^{18}\left(\frac{2}{\cancel{d^2}}\right)$$

$$25d^2 = 36$$

$$d^2 = \frac{36}{25}$$

$$d = \frac{6}{5} \text{ and } -\frac{6}{5}$$

---

50.   When the number is not a perfect square, we use the square-root table. That is:

$$\text{For } x^2 = 69, \quad x = \pm\sqrt{69} = \pm 8.31$$

Use the square-root table to solve these.

a) $t^2 - 94 = 0$ b) $\frac{1}{4}m^2 = 30$

$y = \frac{3}{2}$ and $-\frac{3}{2}$

---

51.   In solving the equation below, we obtained a fraction that is not a perfect square. Therefore, we performed a division and then used the square-root table. Solve the other equation.

$$2p^2 = 30 \qquad\qquad\qquad 5d^2 - 975 = 0$$

$$p^2 = \frac{30}{2}$$

$$p^2 = 15$$

$$p = \pm 3.87$$

a) $t = \pm\sqrt{94} = \pm 9.7$

b) $m = \pm\sqrt{120} = \pm 10.95$

---

52.   Solve each equation.

a) $\frac{180}{3t^2} = 2$ b) $\frac{5}{2} = \frac{x^2}{20}$

$d = \pm\sqrt{195} = \pm 13.96$

---

a) $t = \pm\sqrt{30} = \pm 5.48$

b) $x = \pm\sqrt{50} = \pm 7.07$

53.   Following the example, solve the other proportion.

$$\frac{4x}{5} = \frac{8}{x} \qquad\qquad \frac{5}{y} = \frac{y}{9}$$

$$\overset{x}{\cancel{5x}}\left(\frac{4x}{\cancel{5}}\right) = \overset{5}{\cancel{5x}}\left(\frac{8}{\cancel{x}}\right)$$

$$4x^2 = 40$$

$$x^2 = 10$$

$$x = \pm\sqrt{10} = \pm 3.16$$

---

54.   To solve $9 - x^2 = 0$, we can begin by adding either -9 or x^2 to both sides. Both methods are shown.

$$9 - x^2 = 0 \qquad\qquad\qquad 9 - x^2 = 0$$

$$(-9) + 9 - x^2 = 0 + (-9) \qquad 9 - x^2 + x^2 = 0 + x^2$$

$$-1x^2 = -9 \qquad\qquad\qquad\qquad 9 = x^2$$

$$x^2 = \frac{-9}{-1} = 9 \qquad\qquad\qquad x = \pm 3$$

$$x = \pm 3$$

Use either method to solve these.

a) $49 - y^2 = 0$ \qquad\qquad b) $100 - 4t^2 = 0$

$y = \pm\sqrt{45} = \pm 6.71$

---

55.   In solving each equation below, we obtained a negative number on the right side.

$$x^2 + 25 = 0 \qquad\qquad 9y^2 + 4 = 0$$

$$x^2 = -25 \qquad\qquad\qquad 9y^2 = -4$$

$$x = \pm\sqrt{-25} \qquad\qquad\quad y^2 = -\frac{4}{9}$$

$$y = \pm\sqrt{-\frac{4}{9}}$$

We have already seen that negative numbers do not have ordinary numbers as their square roots. Therefore, though each equation above has two solutions, the solutions are not ordinary numbers. We will not discuss solutions of that type in this book.

a) $y = \pm 7$

b) $t = \pm 5$

SELF-TEST 16 (pages 176-192)

1. In $\boxed{G = K(p^2 - t^2)}$, find G
    when $K = 3$, $p = 6$, and $t = 2$.

2. In $\boxed{w = \dfrac{mv^2}{2r}}$, find "w"
    when $m = 40$, $v = 5$, and $r = 10$.

3. In $\boxed{y = 3x - x^2}$, find "y" when $x = -1$.

Find the two solutions of each quadratic equation.

4. $25y^2 - 1 = 0$

5. $81 - 36x^2 = 0$

6. $10 = \dfrac{5t^2}{8}$

7. $\dfrac{9}{2w^2} = 8$

8. $\dfrac{1}{4} = \dfrac{r^2}{12}$

9. $\dfrac{16}{x} = \dfrac{x}{10}$

ANSWERS:

1. $G = 96$

2. $w = 50$

3. $y = -4$

4. $y = \dfrac{1}{5}$ and $-\dfrac{1}{5}$

5. $x = \dfrac{3}{2}$ and $-\dfrac{3}{2}$

6. $t = 4$ and -4

7. $w = \dfrac{3}{4}$ and $-\dfrac{3}{4}$

8. $r = \pm 1.73$

9. $x = \pm 12.65$

5-6   THE PRINCIPLE OF ZERO PRODUCTS

All incomplete and some complete quadratics can be solved by the factoring method. The basic principle used in the factoring method is the "principle of zero products". We will discuss that principle in this section.

---

56.   If one factor in a multiplication is "0", the product is "0". For example:

$$(5)(0) = 0 \qquad\qquad (0)(9) = 0$$

Therefore, if the product in a multiplication is "0", one of the factors must be "0". That is:

If $ab = 0$, either "a" or "b" must be "0".

If $xy = 0$, either "x" or "y" must be _____ .

57.   The equation below involves a multiplication. The factors are $(x - 2)$
and $(x - 3)$.

$$(x - 2)(x - 3) = 0$$

Since the product is "0", the equation is true when either $(x - 2)$ or $(x - 3)$
is "0". Therefore, we can find the two solutions of the equation by solving
the two equations below as we have done.

$x - 2 = 0$	$x - 3 = 0$
$x = 2$	$x = 3$

We checked 2 as one solution of the original equation. Check 3 as the second
solution.

$(x - 2)(x - 3) = 0$	$(x - 2)(x - 3) = 0$
$(2 - 2)(2 - 3) = 0$	
$(0)(-1) = 0$	
$0 = 0$	

0

---

58.   The equation below is true if either $(y + 4)$ or $(y - 5)$ is "0".

$$(y + 4)(y - 5) = 0$$

Therefore, to find the two solutions, we solved the equations below.

$y + 4 = 0$	$y - 5 = 0$
$y = -4$	$y = 5$

The two solutions of the original equation are _____ and _____ .

$(3 - 2)(3 - 3) = 0$

$(1)(0) = 0$

$0 = 0$

---

59.   We solved the equation below by setting both factors equal to "0". Solve
the other equation.

$(t + 2)(t + 8) = 0$	$(m - 1)(m + 10) = 0$
$t + 2 = 0$ | $t + 8 = 0$	
$t = -2$ | $t = -8$	

−4 and 5

---

60.   The equation below is true if either $(3x - 2)$ or $(2x + 5)$ is "0".

$$(3x - 2)(2x + 5) = 0$$

Therefore to find the two solutions, we solved the equations below.

$3x - 2 = 0$	$2x + 5 = 0$
$3x = 2$	$2x = -5$
$x = \dfrac{2}{3}$	$x = -\dfrac{5}{2}$

The two solutions of the original equation are _____ and _____

$m = 1$ and −10

61.   We solved the equation below by setting both factors equal to "0".
       Solve the other equation.

$$(3p + 1)(4p - 7) = 0 \qquad\qquad (5t - 9)(6t - 1) = 0$$

$$3p + 1 = 0 \quad | \quad 4p - 7 = 0$$
$$3p = -1 \quad | \quad 4p = 7$$
$$p = -\frac{1}{3} \quad | \quad p = \frac{7}{4}$$

$\dfrac{2}{3}$ and $-\dfrac{5}{2}$

---

62.   The equation below is true if either (x) or $(x - 3)$ is "0".

$$x(x - 3) = 0$$

Therefore to find the two solutions, we solved the equations below.

$$x = 0 \qquad\qquad x - 3 = 0$$
$$x = 3$$

The two solutions of the original equation are _____ and _____.

$t = \dfrac{9}{5}$ and $\dfrac{1}{6}$

---

63.   In the last frame, we saw that "0" is one of the solutions of $x(x - 3) = 0$.
       The "0" solution is frequently overlooked. Following the example, solve
       the other equation.

$$m(m + 6) = 0 \qquad\qquad y(y - 12) = 0$$

$$m = 0 \quad | \quad m + 6 = 0$$
$$\qquad | \quad m = -6$$

0 and 3

---

64.   We solved the equation below by setting both factors equal to "0". Solve
       the other equation.

$$b(2b - 1) = 0 \qquad\qquad x(5x + 6) = 0$$

$$b = 0 \quad | \quad 2b - 1 = 0$$
$$\qquad | \quad 2b = 1$$
$$\qquad | \quad b = \frac{1}{2}$$

$y = 0$ and 12

---

65.   We solved one equation below. Solve the other equation.

$$4y(y + 3) = 0 \qquad\qquad 7t(4t - 3) = 0$$

$$4y = 0 \quad | \quad y + 3 = 0$$
$$y = \frac{0}{4} \quad | \quad \mathbf{y = -3}$$
$$y = 0 \quad |$$

$x = 0$ and $-\dfrac{6}{5}$

$t = 0$ and $\dfrac{3}{4}$

66.   The principle of "setting the factors equal to zero" to solve an equation applies <u>only</u> <u>when</u> <u>the</u> <u>product</u> <u>is</u> "0". Therefore, it <u>could</u> <u>not</u> <u>be</u> <u>used</u> with either equation below.

$$(x - 3)(x - 9) \;=\; \overset{\downarrow}{5} \qquad\qquad y(y + 8) \;=\; \overset{\downarrow}{9}$$

In which equations below does the principle apply? _____

   a) $(3x - 1)(2x + 3) \;=\; 7$ b) $d(d - 1) \;=\; 0$ c) $p(p + 4) \;=\; 12$

Only (b)

5-7   SOLVING INCOMPLETE QUADRATICS BY THE FACTORING METHOD

In this section, we will discuss the factoring method for solving incomplete quadratic equations.

67.   To solve the incomplete quadratic at the right, we factored the left side and then set both factors equal to "0".

$$x^2 - 5x \;=\; 0$$
$$x(x - 5) \;=\; 0$$

The two solutions are "0" and "5". We checked "0" as a solution of the original equation below. Check "5" as its solution.

$$x = 0 \;\Big|\; x - 5 \;=\; 0$$
$$x \;=\; 5$$

$$x^2 - 5x \;=\; 0 \qquad\qquad x^2 - 5x \;=\; 0$$
$$(0)^2 - 5(0) \;=\; 0$$
$$0 - 0 \;=\; 0$$
$$0 \;=\; 0$$

68.   Use the steps below to solve the incomplete quadratic at the right.

$$y^2 + 8y \;=\; 0$$

   a) Factor the left side.

   b) Solve the new equation by setting both factors equal to "0".

   c) The two solutions of the original equation are _____ and _____.

$$(5)^2 - 5(5) \;=\; 0$$
$$25 - 25 \;=\; 0$$
$$0 \;=\; 0$$

69.   We used the same steps to solve one equation below. Solve the other equation.

$$m^2 + m \;=\; 0 \qquad\qquad t^2 - t \;=\; 0$$
$$m(m + 1) \;=\; 0$$
$$m = 0 \;\Big|\; m + 1 \;=\; 0$$
$$m \;=\; -1$$

a) $y(y + 8) = 0$

b) $y = 0 \;\Big|\; y + 8 = 0$
   $ y = -8$

c) 0 and -8

70. We used the factoring method to solve one equation below. Solve the other equation.

$$2x^2 - 3x = 0$$

$$x(2x - 3) = 0$$

$$x = 0 \quad | \quad 2x - 3 = 0$$

$$2x = 3$$

$$x = \frac{3}{2}$$

$$7y^2 + 5y = 0$$

t = 0 and 1, from:

$$t(t - 1) = 0$$

$$t = 0 \quad | \quad t - 1 = 0$$

$$t = 1$$

71. When using the factoring method, we can sometimes factor out a number as well as the letter. An example is shown. Solve the other equation.

$$2t^2 - 6t = 0$$

$$2t(t - 3) = 0$$

$$2t = 0 \quad | \quad t - 3 = 0$$

$$t = 0 \quad | \quad t = 3$$

$$5p^2 + 10p = 0$$

y = 0 and $-\dfrac{5}{7}$, from:

$$y(7y + 5) = 0$$

$$y = 0 \quad | \quad 7y + 5 = 0$$

$$7y = -5$$

$$y = -\frac{5}{7}$$

72. When using the factoring method, we usually factor completely. Solve the other equation.

$$8x^2 + 12x = 0$$

$$4x(2x + 3) = 0$$

$$4x = 0 \quad | \quad 2x + 3 = 0$$

$$x = 0 \quad | \quad 2x = -3$$

$$x = -\frac{3}{2}$$

$$12y^2 - 6y = 0$$

p = 0 and -2, from:

$$5p(p + 2) = 0$$

$$5p = 0 \quad | \quad p + 2 = 0$$

$$p = 0 \quad | \quad p = -2$$

73. Though the equation below is not in standard form, it is an incomplete quadratic because it contains a squared-letter term and a letter term.

$$2R^2 = 8R$$

To get it in standard form, we used the addition axiom below. Notice that we converted the addition of (-8R) to a subtraction.

$$2R^2 + (-8R) = 8R + (-8R)$$

$$2R^2 - 8R = 0$$

Using the same steps, write each equation below in standard form.

a) $x^2 = 10x$

b) $7m^2 = -4m$

y = 0 and $\dfrac{1}{2}$, from:

$$6y(2y - 1) = 0$$

74.   To solve the equation below, we put it in standard form first and then used the factoring method. Solve the other equation.

$$t^2 = 2t \qquad\qquad 7F^2 = 3F$$

$$t^2 - 2t = 0$$

$$t(t - 2) = 0$$

$$t = 0 \quad | \quad t - 2 = 0$$

$$\qquad\qquad\; t = 2$$

a) $x^2 - 10x = 0$

b) $7m^2 + 4m = 0$

---

75.   Though the incomplete quadratic below is not in standard form, we solved it without rearranging. Solve the other equation.

$$7x - x^2 = 0 \qquad\qquad 2y - 5y^2 = 0$$

$$x(7 - x) = 0$$

$$x = 0 \quad | \quad 7 - x = 0$$

$$\qquad\qquad\; x = 7$$

$F = 0$ and $\dfrac{3}{7}$, from:

$$7F^2 - 3F = 0$$

$y = 0$ and $\dfrac{2}{5}$, from:

$$y(2 - 5y) = 0$$

5-8   THE FACTORING METHOD AND COMPLETE QUADRATICS OF THE FORM: $x^2 + bx + c = 0$

In this section, we will discuss the factoring method for solving complete quadratic equations in which the coefficient of the squared letter is "1".

---

76.   To solve the complete quadratic at the right, we factored the left side and then set both factors equal to "0".

The two solutions are "-1" and "-2". We checked "-1" as a solution of the original equation below. Check "-2" as a solution.

$$x^2 + 3x + 2 = 0 \qquad\qquad x^2 + 3x + 2 = 0$$

$$(-1)^2 + 3(-1) + 2 = 0$$

$$1 + (-3) + 2 = 0$$

$$3 + (-3) = 0$$

$$0 = 0$$

$$x^2 + 3x + 2 = 0$$

$$(x + 1)(x + 2) = 0$$

$$x + 1 = 0 \quad | \quad x + 2 = 0$$

$$x = -1 \quad | \quad\;\; x = -2$$

$$(-2)^2 + 3(-2) + 2 = 0$$

$$4 + (-6) + 2 = 0$$

$$6 + (-6) = 0$$

$$0 = 0$$

77. Use the steps below to solve the complete quadratic at the right.

$$y^2 - 5y + 6 = 0$$

   a) Factor the left side.

   b) Solve the new equation by setting both factors equal to "0".

   c) The two solutions of the original equation are _____ and _____ .

---

78. We used the factoring method to solve one equation below. Solve the other equation.

$$t^2 + 3t - 10 = 0 \qquad\qquad m^2 - 8m - 9 = 0$$

$$(t - 2)(t + 5) = 0$$

$$t - 2 = 0 \quad\big|\quad t + 5 = 0$$

$$t = 2 \quad\big|\quad t = -5$$

a) $(y - 2)(y - 3) = 0$

b) $y - 2 = 0 \ \big| \ y - 3 = 0$
      $y = 2 \ \big| \ \ \ \ \ y = 3$

c) 2 and 3

---

79. We solved the equation at the left below. Notice that both solutions are -3. A solution of that type is called a <u>double root</u>. Solve the other equation.

$$x^2 + 6x + 9 = 0 \qquad\qquad y^2 - 4y + 4 = 0$$

$$(x + 3)(x + 3) = 0$$

$$x + 3 = 0 \quad\big|\quad x + 3 = 0$$

$$x = -3 \quad\big|\quad x = -3$$

m = -1 and 9, from:

  $(m + 1)(m - 9) = 0$

$m + 1 = 0 \ \big| \ m - 9 = 0$

  $m = -1 \ \big| \ \ \ m = 9$

---

80. Though the equation below is not in standard form, it is a complete quadratic equation because it contains a squared-letter term, a letter term, and a number term.

$$m^2 = 7m - 8$$

To put it in standard form, we used the addition axiom below. Notice that we converted the addition of (-7m) to a subtraction.

$$m^2 + (-7m) + 8 = 7m + (-7m) - 8 + 8$$

$$m^2 - 7m + 8 = 0$$

Using the same steps, write each equation in standard form.

   a) $t^2 + 6 = 5t$ b) $d^2 = d + 2$

Both solutions are 2.

---

a) $t^2 - 5t + 6 = 0$

b) $d^2 - d - 2 = 0$

81. The equation below contains an instance of the distributive principle. We put it in standard form. Write the other equation in standard form.

$$7 + x(x + 1) = 2$$
$$7 + x^2 + x = 2$$
$$x^2 + x + 5 = 0$$

$$(p + 3)(p - 3) = 4p$$

---

82. Some fractional equations are really complete quadratic equations. For example, we cleared the fraction and then put the equation below in standard form. Write the other equation in standard form.

$$x + \frac{2}{x} = 3$$

$$x\left(x + \frac{2}{x}\right) = x(3)$$

$$x(x) + \cancel{x}\left(\frac{2}{\cancel{x}}\right) = 3x$$

$$x^2 + 2 = 3x$$

$$x^2 - 3x + 2 = 0$$

$$F = \frac{7}{F} + 1$$

$p^2 - 4p - 9 = 0$

---

83. We usually write the standard form of complete quadratic so that the "x^2" term is positive. To convert the equation below to standard form, we multiplied both sides by -1 to get rid of the "-" in front of x^2. Doing so is the same as changing the sign of each term.

$$-x^2 + 5x - 6 = 0$$
$$-1(-x^2 + 5x - 6) = -1(0)$$
$$x^2 - 5x + 6 = 0$$

Following the example, write the other equation in standard form.

$$\frac{5}{t} = \frac{t}{7} + 2$$

$$7t\left(\frac{5}{t}\right) = 7t\left(\frac{t}{7} + 2\right)$$

$$7\cancel{t}\left(\frac{5}{\cancel{t}}\right) = \cancel{7}t\left(\frac{t}{\cancel{7}}\right) + 7t(2)$$

$$35 = t^2 + 14t$$

$$t^2 + 14t - 35 = 0$$

$$\frac{5}{m} + 4 = \frac{m}{3}$$

$F^2 - F - 7 = 0$

$m^2 - 12m - 15 = 0$

84.   When a complete quadratic is not in standard form, the first step is putting it in standard form. An example is shown. Solve the other quadratic.

$$y^2 = y + 6$$ $$p^2 + 15 = 8p$$

$$y^2 - y - 6 = 0$$

$$(y + 2)(y - 3) = 0$$

$$y + 2 = 0 \quad | \quad y - 3 = 0$$

$$y = -2 \quad | \quad y = 3$$

---

85.   Using the same steps, solve each equation.

   a) $x(x - 1) = 2$ b) $4 = y + \dfrac{3}{y}$

   p = 3 and 5, from:

   $$p^2 - 8p + 15 = 0$$

---

86.   In the equation below, each term has the common factor "10". We can simplify by factoring out the common numerical factor and then multiplying both sides by its reciprocal. We get:

$$10x^2 + 50x + 40 = 0$$

$$10(x^2 + 5x + 4) = 0$$

$$\frac{1}{10}(10)(x^2 + 5x + 4) = \frac{1}{10}(0)$$

$$1 \quad (x^2 + 5x + 4) = 0$$

$$x^2 + 5x + 4 = 0$$

To solve the equation below, we began by factoring out the largest common factor. Use the same method to solve the other equation.

$$3x^2 + 12x + 9 = 0$$ $$4d^2 - 16d - 20 = 0$$

$$3(x^2 + 4x + 3) = 0$$

$$x^2 + 4x + 3 = 0$$

$$(x + 1)(x + 3) = 0$$

$$x + 1 = 0 \quad | \quad x + 3 = 0$$

$$x = -1 \quad | \quad x = -3$$

a) x = -1 and 2, from:

$$x^2 - x - 2 = 0$$

b) y = 1 and 3, from:

$$y^2 - 4y + 3 = 0$$

---

d = -1 and 5, from:

$$d^2 - 4d - 5 = 0$$

5-9   THE FACTORING METHOD AND COMPLETE QUADRATICS OF THE FORM: $ax^2 + bx + c = 0$

In this section, we will discuss the factoring method for solving complete quadratic equations in which the coefficient of the squared letter is a number other than "1".

---

87.   The same method is used to solve a complete quadratic when the coefficient of the squared letter is a number other than "1". That is, we factor the left side and then set both factors equal to "0". An example is shown. Solve the other equation.

$$3x^2 + 5x + 2 = 0 \qquad\qquad 2y^2 - 7y + 6 = 0$$

$$(3x + 2)(x + 1) = 0$$

$3x + 2 = 0$	$x + 1 = 0$
$3x = -2$	$x = -1$
$x = -\dfrac{2}{3}$	

---

88.   Following the example, solve the other equation.

$$2p^2 - 5p - 12 = 0 \qquad\qquad 5d^2 + 3d - 2 = 0$$

$$(2p + 3)(p - 4) = 0$$

$2p + 3 = 0$	$p - 4 = 0$
$2p = -3$	$p = 4$
$p = -\dfrac{3}{2}$	

$y = \dfrac{3}{2}$ and 2, from:

$$(2y - 3)(y - 2) = 0$$

$2y - 3 = 0$	$y - 2 = 0$
$2y = 3$	$y = 2$
$y = \dfrac{3}{2}$	

---

89.   To solve the equation below, we began by putting it in standard form. Solve the other equation.

$$7x^2 = 3x + 4 \qquad\qquad 3m^2 + 2 = 7m$$

$$7x^2 - 3x - 4 = 0$$

$$(7x + 4)(x - 1) = 0$$

$7x + 4 = 0$	$x - 1 = 0$
$7x = -4$	$x = 1$
$x = -\dfrac{4}{7}$	

$d = \dfrac{2}{5}$ and -1, from:

$$(5d - 2)(d + 1) = 0$$

$5d - 2 = 0$	$d + 1 = 0$
$5d = 2$	$d = -1$
$d = \dfrac{2}{5}$	

$m = \dfrac{1}{3}$ and 2, from:

$$3m^2 - 7m + 2 = 0$$

$$(3m - 1)(m - 2) = 0$$

90. We usually write the standard form of complete quadratics so that the coefficient of "x^2" is <u>positive</u>. To convert the equation below to standard form, we multiplied both sides by -1 to get rid of the negative coefficient "-3". Doing so is <u>the</u> <u>same</u> <u>as</u> <u>changing</u> <u>the</u> <u>sign</u> <u>of</u> <u>each</u> <u>term</u>.

$$-3x^2 - 5x + 4 = 0$$

$$-1(-3x^2 - 5x + 4) = -1(0)$$

$$3x^2 + 5x - 4 = 0$$

Solve each equation by putting it in standard form first.

a) $4y(y - 1) = 3$

b) $\dfrac{1}{x} + \dfrac{1}{3} = \dfrac{2x}{3}$

91. To solve the equation below, we began by factoring out the common numerical factor. Use the same method to solve the other equation.

$$30t^2 - 80t + 50 = 0 \qquad\qquad 4d^2 + 2d - 30 = 0$$

$$10(3t^2 - 8t + 5) = 0$$

$$3t^2 - 8t + 5 = 0$$

$$(3t - 5)(t - 1) = 0$$

$3t - 5 = 0$	$t - 1 = 0$
$3t = 5$	$t = 1$
$t = \dfrac{5}{3}$	

a) $y = -\dfrac{1}{2}$ and $\dfrac{3}{2}$, from:

$$4y^2 - 4y - 3 = 0$$

b) $x = \dfrac{3}{2}$ and -1, from:

$$2x^2 - x - 3 = 0$$

d = $\dfrac{5}{2}$ and -3, from:

$$2d^2 + d - 15 = 0$$

SELF-TEST 17 (pages 192-203)

Solve each equation.

1. $2x^2 + x = 0$

2. $4y^2 - 6y = 0$

3. $w^2 = 3w$

4. $t^2 + t - 6 = 0$

5. $\dfrac{x}{2} - \dfrac{4}{x} = 1$

6. $2r^2 - 5r - 3 = 0$

7. $\dfrac{10}{p} = 15p + 25$

ANSWERS:

1. $x = 0$ and $x = -\dfrac{1}{2}$

2. $y = 0$ and $y = \dfrac{3}{2}$

3. $w = 0$ and $w = 3$

4. $t = 2$ and $t = -3$

5. $x = -2$ and $x = 4$

6. $r = 3$ and $r = -\dfrac{1}{2}$

7. $p = -2$ and $p = \dfrac{1}{3}$

5-10   LIMITATIONS OF THE FACTORING METHOD

Some complete quadratic equations are not ordinarily solved by the factoring method either because they cannot be factored into binomials containing whole numbers or because they are too difficult to factor. We will give examples of both types in this section.

92.   There are some trinomials that cannot be factored into binomials in which the numbers are whole numbers. Two examples are given below.

For $x^2 + 4x + 6$, the possible pairs of factors are:

$$(x + 1)(x + 6)$$
$$(x + 2)(x + 3)$$

Neither pair has "4x" as the middle term of its product.

For $2m^2 + 6m + 3$, the possible pairs of factors are:

$$(2m + 3)(m + 1)$$
$$(2m + 1)(m + 3)$$

Neither pair has "6m" as the middle term of its product.

When a complete quadratic contains a trinomial that cannot be factored into binomials containing whole numbers, we do not solve it by the factoring method.

Which of the following quadratics would not be solved by the factoring method? _____

a) $y^2 + 9y + 14 = 0$ c) $5m^2 - 6m - 2 = 0$

b) $t^2 - 11t + 16 = 0$ d) $3x^2 - 8x + 5 = 0$

---

93.   When the numbers in the quadratic are large, we do not ordinarily use the factoring method to solve it because the factoring is too difficult. An example is shown. Complete the other solution.

$$50y^2 - 495y + 144 = 0 \qquad\qquad y^2 - 44y + 288 = 0$$

$$(5y - 48)(10y - 3) = 0 \qquad\qquad (y - 8)(y - 36) = 0$$

$5y - 48 = 0$	$10y - 3 = 0$
$5y = 48$	$10y = 3$
$y = \dfrac{48}{5}$	$y = \dfrac{3}{10}$

Only (b) and (c).

(a) can be factored:

$(y + 2)(y + 7) = 0$

(d) can be factored:

$(3x - 5)(x - 1) = 0$

$y = 8$ and 36

94.   Even though the numbers in the quadratic below are quite small, the number of possible pairs of factors is quite large as you can see. Therefore, the equation is difficult to solve by the factoring method.

$$6x^2 + 5x - 14 = 0$$

$(6x + 1)(x - 14) = 0$	$(3x + 1)(2x - 14) = 0$
$(6x - 1)(x + 14) = 0$	$(3x - 1)(2x + 14) = 0$
$(6x + 14)(x - 1) = 0$	$(3x + 14)(2x - 1) = 0$
$(6x - 14)(x + 1) = 0$	$(3x - 14)(2x + 1) = 0$
$(6x + 2)(x - 7) = 0$	$(3x + 2)(2x - 7) = 0$
$(6x - 2)(x + 7) = 0$	$(3x - 2)(2x + 7) = 0$
$(6x + 7)(x - 2) = 0$	$(3x + 7)(2x - 2) = 0$
$(6x - 7)(x + 2) = 0$	$(3x - 7)(2x + 2) = 0$

a) By checking the factorings above, find the correct factoring for the equation. _____

b) Therefore, the solutions of the equation are _____ and _____.

95.   The quadratic equations below contain decimal coefficients and numbers.

$$1.56x^2 + 2.47x + 3.14 = 0$$
$$31.9y^2 - 64.8y + 14.2 = 0$$

Neither trinomial can be factored into binomials containing whole numbers. Therefore, would we try to solve the quadratics above by the factoring method? _____

a) $(6x - 7)(x + 2) = 0$

b) $x = \dfrac{7}{6}$ and -2

No

5-11 IDENTIFYING "a", "b", AND "c" IN THE STANDARD FORM OF QUADRATIC EQUATIONS

When a quadratic equation cannot be easily solved by the factoring method, we use the quadratic formula to solve it. In order to use the quadratic formula, we must be able to identify "a", "b", and "c" in the standard form of quadratic equations. We will discuss the method in this section.

96.   Any quadratic equation can be written in the following standard form:

$$ax^2 + bx + c = 0$$

In the standard form, "a" is the coefficient of the squared letter, "b" is the coefficient of the letter, and "c" is the number term. Two examples are shown.

$$2x^2 + 5x + 3 = 0 \qquad 5y^2 - 4y - 1 = 0$$

a = 2	a = 5
b = 5	b = -4
c = 3	c = -1

Identify "a", "b", and "c" in each equation below.

a) $3m^2 + 4m - 9 = 0$ a = _____, b = _____, c = _____

b) $7t^2 - 2t + 8 = 0$ a = _____, b = _____, c = _____

97. When "a" or "b" is either "1" or "-1", the "1" or "-1" is not ordinarily shown. For example:

$$x^2 + x - 6 = 0 \qquad\qquad y^2 - y + 12 = 0$$

It is helpful to write each "1" or "-1" before identifying "a", "b", and "c". We get:

$$1x^2 + 1x - 6 = 0 \qquad\qquad 1y^2 - 1y + 12 = 0$$

a = 1   a = 1
b = 1   b = -1
c = -6   c = 12

Identify "a", "b", and "c" in each equation below.

a) $t^2 + t - 30 = 0$ a = _____, b = _____, c = _____

b) $d^2 - d - 2 = 0$ a = _____, b = _____, c = _____

a) a = 3
   b = 4
   c = -9

b) a = 7
   b = -2
   c = 8

---

98. When "a" is negative, we usually multiply by -1 to make it positive before identifying "a", "b", and "c". An example is shown.

$$-5x^2 + 7x - 3 = 0$$
$$-1(-5x^2 + 7x - 3) = -1(0)$$
$$5x^2 - 7x + 3 = 0$$
$$a = 5, \quad b = -7, \quad c = 3$$

Write each equation below so that "a" is positive and then identify "a", "b", and "c".

a) $-10y^2 - 3y + 7 = 0$ b) $-t^2 + 9t + 5 = 0$

_____   _____

a = ___, b = ___, c = ___   a = ___, b = ___, c = ___

a) a = 1
   b = 1
   c = -30

b) a = 1
   b = -1
   c = -2

---

99. The equation below is not in standard form because there are some non-zero terms on the right side.

$$2d^2 = d + 3$$

To identify "a", "b", and "c", we must put it in standard form first. We get:

$$2d^2 - d - 3 = 0$$
$$a = 2, \quad b = -1, \quad c = -3$$

a) $10y^2 + 3y - 7 = 0$
   a = 10, b = 3, c = -7

b) $t^2 - 9t - 5 = 0$
   a = 1, b = -9, c = -5

Continued on following page.

99.   Continued

Put each equation in standard form and then identify "a", "b", and "c".

a) $(x + 1)(x - 5) = 3$

b) $\dfrac{7}{y} = \dfrac{3y}{2} + 1$

_____   _____

a = ____ , b = ____ , c = ____   a = ____ , b = ____ , c = ____

---

100.   In the standard form of pure quadratics, the letter term is not usually shown because its coefficient is "0". For example:

$$x^2 - 16 = 0 \qquad\qquad 2y^2 - 80 = 0$$

However, when identifying "a", "b", and "c", it is helpful to insert the letter term with its "0" coefficient. We get:

$$x^2 + 0x - 16 = 0 \qquad\qquad 2y^2 + 0y - 80 = 0$$

a) For the <u>left</u> equation: a = _____ , b = _____ , c = _____

b) For the <u>right</u> equation: a = _____ , b = _____ , c = _____

a) $x^2 - 4x - 8 = 0$

    a = 1, b = -4, c = -8

b) $3y^2 + 2y - 14 = 0$

    a = 3, b = 2, c = -14

---

101.   In the standard form of incomplete quadratics, the number term is not usually shown because it is "0". For example:

$$x^2 - 6x = 0 \qquad\qquad 3y^2 + 12y = 0$$

However, when identifying "a", "b", and "c", it is helpful to insert "0" as the number term. We get:

$$x^2 - 6x + 0 = 0 \qquad\qquad 3y^2 + 12y + 0 = 0$$
$$a = 1, b = -6, c = 0 \qquad a = 3, b = 12, c = 0$$

Identify "a", "b", and "c" for each equation below.

a) $m^2 - 1 = 0$ a = _____ , b = _____ , c = _____

b) $5p^2 - p = 0$ a = _____ , b = _____ , c = _____

a) a = 1
    b = 0
    c = -16

b) a = 2
    b = 0
    c = -80

---

a) a = 1, b = 0, c = -1, from:

    $m^2 + 0m - 1 = 0$

b) a = 5, b = -1, c = 0, from:

    $5p^2 - p + 0 = 0$

5-12   THE QUADRATIC FORMULA

Though the quadratic formula is a general method that can be used to solve any type of quadratic equation, it is ordinarily used only to solve complete quadratics that are not easily solved by the factoring method. We will discuss the quadratic formula in this section.

---

102.   The quadratic formula is given below. It contains "a", "b", and "c" from the standard form of quadratic equations. Since any quadratic equation has two solutions, there are two forms of the formula. The only difference between them is the "+" and "-" in front of the radical in the numerator (see the arrows).

$$\text{First solution} = \frac{(-b) \overset{\downarrow}{+} \sqrt{b^2 - 4ac}}{2a}$$

$$\text{Second solution} = \frac{(-b) \overset{\downarrow}{-} \sqrt{b^2 - 4ac}}{2a}$$

If we use the factoring method to solve $x^2 + 3x - 10 = 0$ we get 2 and -5 as its solutions from $(x - 2)(x + 5) = 0$. Let's use the quadratic formula to solve the same equation.

$$x^2 + 3x - 10 = 0$$

$$a = 1, \quad b = 3, \quad c = -10$$

First Solution

$$x = \frac{(-b) + \sqrt{b^2 - 4ac}}{2a}$$

$$= \frac{(-3) + \sqrt{(-3)^2 - 4(1)(-10)}}{2(1)}$$

$$= \frac{(-3) + \sqrt{9 - (-40)}}{2}$$

$$= \frac{(-3) + \sqrt{49}}{2}$$

$$= \frac{(-3) + 7}{2}$$

$$= \frac{4}{2} = 2$$

Second Solution

$$x = \frac{(-b) - \sqrt{b^2 - 4ac}}{2a}$$

$$= \frac{(-3) - \sqrt{(-3)^2 - 4(1)(-10)}}{2(1)}$$

$$= \frac{(-3) - \sqrt{9 - (-40)}}{2}$$

$$= \frac{(-3) - \sqrt{49}}{2}$$

$$= \frac{(-3) - 7}{2}$$

$$= \frac{-10}{2} = -5$$

Did we get the same solutions that would be obtained by the factoring method? _____

Yes

103.   The formulas for the two solutions are identical except for the "+" and "-" in front of the radical. Therefore, we usually write one formula with a "±" symbol. The "±" means "<u>add</u> the radical for the first solution" and "<u>subtract</u> the radical for the second solution".

$$\text{Two solutions } = \frac{(-b) \pm \sqrt{b^2 - 4ac}}{2a}$$

Let's use the single formula to solve the equation below.

$$y^2 + 6y - 16 = 0$$

$$a = 1, \quad b = 6, \quad c = -16$$

$$\text{Two solutions } = \frac{(-6) \pm \sqrt{(6)^2 - 4(1)(-16)}}{2(1)}$$

$$= \frac{(-6) \pm \sqrt{36 - (-64)}}{2}$$

$$= \frac{(-6) \pm \sqrt{100}}{2}$$

a) The first solution $= \dfrac{(-6) + 10}{2} = $ _____

b) The second solution $= \dfrac{(-6) - 10}{2} = $ _____

---

104.   There are a few points you should notice about plugging values into the quadratic formula.

$$\text{Two solutions } = \frac{(-b) \pm \sqrt{b^2 - 4ac}}{2a}$$

1) (-b) means "<u>the opposite of b</u>".   If $b = 4$, $-b = -4$
                                              If $b = -5$, $-b = +5$

2) b^2 (under the radical) is <u>always a positive number</u>.

   If $b = 4$, $b^2 = (4)^2 = +16$
   If $b = -5$, $b^2 = (-5)^2 = +25$

3) Watch the signs with $b^2 - 4ac$ under the radical.

   "4ac" is a set of <u>three</u> factors. Perform the multiplication <u>before</u> handling the subtraction. That is:

   If $a = 1$ $b^2 - 4ac = (-10)^2 - [4(1)(-6)]$
       $b = -10$ $= 100 - [-24]$
       $c = -6$ $= 100 + [+24] = 124$

   If $a = -3$ $b^2 - 4ac = (12)^2 - [4(-3)(-9)]$
       $b = 12$ $= 144 - [+108]$
       $c = -9$ $= 144 + [-108]$
                   $= 36$

For the equation: $2x^2 - 5x - 3 = 0$

a) $-b = $ _____   b) $b^2 = $ _____   c) $b^2 - 4ac = $ _____

---

a)   2, from: $\dfrac{4}{2}$

b)   -8, from: $\dfrac{-16}{2}$

105.   For the equation: $7m^2 - 6m - 1 = 0$

a) $-b =$ _____   b) $b^2 =$ _____   c) $b^2 - 4ac =$ _____

a) +5

b) 25

c) 49

---

106.   <u>The</u> <u>quadratic</u> <u>formula</u> <u>should</u> <u>be</u> <u>memorized</u>. Therefore, study the formula as it is given in frame 103 and then write it from memory below.

Two solutions =

a) +6

b) 36

c) 64

---

107.   Let's use the formula to solve the equation below.

$$3t^2 - 2t - 5 = 0$$
$$a = 3, \quad b = -2, \quad c = -5$$

Two solutions $= \dfrac{(2) \pm \sqrt{(-2)^2 - 4(3)(-5)}}{2(3)}$

a) First solution = _____   b) Second solution = _____

$\dfrac{-b \pm \sqrt{b^2 - 4ac}}{2a}$

---

108.   Though we ordinarily solve incomplete quadratics by the factoring method, we can use the quadratic formula to solve them. An example is shown.

$$3x^2 + 4x = 0$$
$$a = 3, \quad b = 4, \quad c = 0$$

Two solutions $= \dfrac{(-4) \pm \sqrt{(4)^2 - 4(3)(0)}}{2(3)}$

$= \dfrac{(-4) \pm \sqrt{16 - 0}}{6}$

$= \dfrac{(-4) \pm \sqrt{16}}{6}$

a) First solution is _____   b) Second solution is _____

a) $\dfrac{5}{3}$, from: $\dfrac{10}{6}$

b) -1, from: $\dfrac{-6}{6}$

---

a) 0, from: $\dfrac{0}{6}$

b) $-\dfrac{4}{3}$, from: $\dfrac{-8}{6}$

109. Though we ordinarily solve pure quadratics by isolating the squared letter, we can use the quadratic formula to solve them. An example is shown.

$$y^2 - 9 = 0$$

$$a = 1, \quad b = 0, \quad c = -9$$

$$\text{Two solutions} = \frac{0 \pm \sqrt{(0)^2 - 4(1)(-9)}}{2(1)}$$

$$= \frac{0 \pm \sqrt{0 - (-36)}}{2}$$

$$= \frac{0 \pm \sqrt{36}}{2}$$

a) First solution is _____   b) Second solution is _____

---

110. Because we cannot factor the equation below into binomials containing whole numbers, we used the formula to solve it.

$$y^2 + 9y + 7 = 0$$

$$a = 1, \quad b = 9, \quad c = 7$$

$$\text{Two solutions} = \frac{(-9) \pm \sqrt{(9)^2 - 4(1)(7)}}{2(1)}$$

$$= \frac{(-9) \pm \sqrt{81 - 28}}{2}$$

$$= \frac{(-9) \pm \sqrt{53}}{2}$$

When the radicand is not a perfect square (like 53), the solutions are sometimes left in the radical form above. However, we can also use the square-root table to **find** $\sqrt{53}$ and then report the solutions as decimal numbers. We get:

$$\text{The first solution} = \frac{(-9) + 7.28}{2} = \frac{-1.72}{2} = -0.86$$

$$\text{The second solution} = \frac{(-9) - 7.28}{2} = \text{_____} = \text{_____}$$

a) 3, from: $\dfrac{6}{2}$

b) -3, from: $\dfrac{-6}{2}$

$$\frac{-16.28}{2} = -8.14$$

111.   To simplify the calculations needed when using the quadratic formula, we should factor out obvious common numerical factors first. For example, we factored out 100 below.

$$500x^2 - 100x + 900 = 0$$
$$100(5x^2 - x + 9) = 0$$
$$5x^2 - x + 9 = 0$$

Let's solve this quadratic equation: $20y(y + 4) - 70 = 0$

a) Write the equation in standard form.

b) Now use the quadratic formula to solve it.

112.   In the following equation, "a", "b", and "c" are decimal numbers.

$$1.37x^2 + 0.85x + 2.14 = 0$$

We can eliminate the decimals by multiplying both sides by 100. We get:

$$100(1.37x^2 + 0.85x + 2.14) = 100(0)$$
$$100(1.37x^2) + 100(0.85x) + 100(2.14) = 0$$
$$137x^2 + 85x + 214 = 0$$

By multiplying both sides by either 10 or 100, eliminate the decimals in these.

a) $2.1y^2 - 4.6y + 1.8 = 0$ b) $4.99F^2 - 6.66F - 2.58 = 0$

a) $20y^2 + 80y - 70 = 0$

or

$2y^2 + 8y - 7 = 0$

b) $y = 0.74$ and -4.74

or

$y = \dfrac{-8 \pm \sqrt{120}}{4}$

113.   After the decimals are eliminated, we can sometimes simplify the resulting equation by factoring out a common numerical factor. For example, we could factor out a 5 below.

$$1.5y^2 - 3.5y + 2.5 = 0$$
$$15y^2 - 35y + 25 = 0$$
$$3y^2 - 7y + 5 = 0$$

Eliminate the decimals in these and then simplify further if possible.

a) $1.3x^2 + 4.7x - 1.5 = 0$ b) $1.2t^2 - 1.6t + 2.4 = 0$

a) $21y^2 - 46y + 18 = 0$

b) $499F^2 - 666F - 258 = 0$

114.   Let's solve this equation: $\dfrac{1.5}{D} + 3.5D = 7.5$

   a) Write the equation in standard form.

   b) Now use the quadratic formula to find the two solutions.

a) $13x^2 + 47x - 15 = 0$
    (Not possible to simplify further)

b) $3t^2 - 4t + 6 = 0$
    (4 was a common factor)

---

115.   When the radical reduces to a negative number, the equation does not have ordinary numbers as its solutions. For example:

$$x^2 - 2x + 7 = 0$$

Two solutions $= \dfrac{2 \pm \sqrt{(-2)^2 - 4(1)(7)}}{2(1)}$

$= \dfrac{2 \pm \sqrt{4 - 28}}{2}$

$= \dfrac{2 \pm \sqrt{-24}}{2}$

In this book, we will not solve equations like the one above.

a) $3.5D^2 - 7.5D + 1.5 = 0$
    $35D^2 - 75D + 15 = 0$
    $7D^2 - 15D + 3 = 0$

b) $D = 1.92$ and 0.22

    or

    $D = \dfrac{15 \pm \sqrt{141}}{14}$

---

116.   The following strategy can be used to solve quadratic equations.

Pure quadratics   - Isolate the squared letter and then find both square roots of the number on the other side.

Incomplete quadratics - Use the factoring method.

Complete quadratics   - Try the factoring method first. If the factoring is too difficult or if you can't find factors in which the numbers are whole numbers, switch immediately to using the formula. Don't waste a lot of time trying to factor.

5-13   FORMULA EVALUATIONS REQUIRING EQUATION-SOLVING

In this section, we will discuss some formula evaluations that require solving a quadratic equation.

---

117.   To do the formula evaluation below, we must solve a pure quadratic equation.

Find the side of a square if its area is 144 square centimeters.

$$\boxed{A = s^2}$$

$$144 = s^2$$

$$s = \pm\sqrt{144} = \pm12 \text{ centimeters}$$

Of the two answers (+12 centimeters and -12 centimeters), only one makes sense. Which one? _____

---

118.   Here is another formula problem that leads to a pure quadratic equation.

| +12 centimeters

The distance "d" in feet that an object falls from rest in "t" seconds is given by the formula:

$$\boxed{d = 16t^2}$$

How many seconds does it take an object to fall 64 feet?

$$64 = 16t^2$$

$$t^2 = \frac{64}{16} = 4$$

$$t = \pm2 \text{ seconds}$$

Of the two answers (+2 seconds and -2 seconds), only one makes sense. Which one? _____

---

119.   When solving for a letter that is squared in a formula, we report <u>only the</u> <u>positive root</u> because ordinarily only the positive root makes sense. An example is given. Complete the other evaluation.

| +2 seconds

In the formula below, find I when $P = 72$ and $R = 2$.

In the formula below, find V when $h = 1$.

$$\boxed{P = I^2R}$$
$$72 = I^2(2)$$

$$I^2 = \frac{72}{2} = 36$$

$$I = \sqrt{36} = 6$$

$$\boxed{h = \frac{V^2}{64}}$$

| V = 8

120. Following the example, complete the other evaluation. Use the square-root table. <u>Remember</u> <u>to</u> <u>report</u> <u>only</u> <u>the</u> <u>positive</u> <u>root</u>.

In the formula below, find "t" when $s = 100$ and $a = 10$.

$$\boxed{s \;=\; \frac{1}{2}at^2}$$

$$100 \;=\; \frac{1}{2}(10)\,(t^2)$$

$$100 \;=\; 5t^2$$

$$t^2 \;=\; \frac{100}{5} \;=\; 20$$

$$t \;=\; \sqrt{20} \;=\; 4.47$$

In the formula below, find "v" when $E = 600$ and $m = 20$.

$$\boxed{E \;=\; \frac{1}{2}mv^2}$$

---

121. In the formula below, find "t" when $s = 100$ and $g = 5$.

$$\boxed{s \;=\; \frac{gt^2}{2}}$$

$$100 \;=\; \frac{5t^2}{2}$$

$$2(100) \;=\; \cancel{2}\!\left(\frac{5t^2}{\cancel{2}}\right)$$

$$200 \;=\; 5t^2$$

$$t^2 \;=\; \frac{200}{5} \;=\; 40$$

$$t \;=\; \sqrt{40} \;=\; 6.32$$

In the formula below, find "d" when $F = 3$, $m_1 = 6$, $m_2 = 10$, and $r = 2$.

$$\boxed{F \;=\; \frac{m_1 m_2}{rd^2}}$$

$v = 7.75$ from: $\sqrt{60}$

---

122. In the formula below, find "d_2" when $I_1 = 1$, $I_2 = 4$, and $d_1 = 10$.

$$\boxed{\frac{I_1}{I_2} \;=\; \frac{(d_2)^2}{(d_1)^2}}$$

$$\frac{1}{4} \;=\; \frac{(d_2)^2}{(10)^2}$$

$$\frac{1}{4} \;=\; \frac{(d_2)^2}{100}$$

$$\overset{25}{\cancel{100}}\!\left(\frac{1}{\cancel{4}}\right) \;=\; \cancel{100}\left[\frac{(d_2)^2}{\cancel{100}}\right]$$

$$25 \;=\; (d_2)^2$$

$$d_2 \;=\; \sqrt{25} \;=\; 5$$

In the formula below, find "r_o" when $Q_o = 2$, $Q_p = 3$, and $r_p = 10$.

$$\boxed{\frac{Q_o}{Q_p} \;=\; \frac{(r_p)^2}{(r_o)^2}}$$

$d = 3.16$ from: $\sqrt{10}$

$r_o = 12.25$, from: $\sqrt{150}$

123.   The formula below shows the relationship between distance traveled (s), initial velocity (v_o), and time traveled (t) for an object that is thrown vertically downward.

$$s = v_o t + 16t^2$$

If the initial velocity is 48 feet per second, we can find the amount of time needed to travel 64 feet by solving the complete quadratic below.

$$64 = 48t + 16t^2$$

a) Write the equation in the simplest standard form.

b) Find the two solutions of the equation.

c) Only one of the solutions makes sense. Which one? _____

---

a) $t^2 + 3t - 4 = 0$ b) t = 1 and -4   c) t = 1 second

5-14   APPLIED PROBLEMS

In this section, we will discuss some applied problems that require setting up and solving quadratic equations.

124.   To solve the problem below, we have to solve a pure quadratic equation.

The length of a rectangle is 3 times its width. If the area of the rectangle is 48 square feet, what are its dimensions?

Let: x = width of the rectangle
           3x = length of the rectangle

Then: $A = LW$

          48 = (3x)(x)

          $48 = 3x^2$

          $x^2 = 16$

          x = ±4 feet

For the problem, -4 feet does not make sense as a value for width. Therefore:

a) the width of the rectangle is _____ feet.

b) the length of the rectangle is _____ feet.

125.   To solve the problem below, we have to solve a complete quadratic equation.

a)   4 feet

b)   12 feet

The length of a rectangle is 3 meters more than its width. If the area of the rectangle is 54 square meters, what are its dimensions?

Let: x = width of the rectangle
x + 3 = length of the rectangle

Then: $\boxed{A = LW}$

$54 = (x + 3)(x)$

$54 = x^2 + 3x$

$x^2 + 3x - 54 = 0$

$(x - 6)(x + 9) = 0$

$x = +6 \text{ or } -9 \text{ meters}$

For the problem, -9 meters does not make sense as a value for width. Therefore:

a) the width of the rectangle is _____ meters.

b) the length of the rectangle is _____ meters.

126.   The Pythagorean Theorem says this: In a right triangle, the sum of the squares of the legs equals the hypotenuse squared. That is:

a) 6 meters

b) 9 meters

$$\boxed{(\ell_1)^2 + (\ell_2)^2 = h^2}$$

We can use the Pythagorean Theorem to solve the problem below.

In a right triangle, one leg is 7 inches longer than the other. If the length of the hypotenuse is 13 inches, how long is each leg?

Let: ℓ_1 = the shortest leg
$\ell_2 = \ell_1 + 7$

Then: $(\ell_1)^2 + (\ell_2)^2 = h^2$

$(\ell_1)^2 + (\ell_1 + 7)^2 = 13^2$

$(\ell_1)^2 + (\ell_1)^2 + 14\ell_1 + 49 = 169$

$2(\ell_1)^2 + 14\ell_1 - 120 = 0$

$(\ell_1)^2 + 7\ell_1 - 60 = 0$

Use the factoring method to complete the solution.

$\ell_1 = $ _____ $\ell_2 = $ _____

127.   The formula for the total resistance of an electric circuit consisting of two resistors in parallel is given below.

$$\frac{1}{R_t} = \frac{1}{R_1} + \frac{1}{R_2}$$

We can use the above formula to solve the problem below.

We must **connect** two resistors in parallel so that the total resistance is 20 ohms. One of the two resistors must be 10 ohms greater than the other. Find the size of the two resistors.

Let: R_1 = the smaller resistor
            R_2 = $R_1 + 10$

Then: $\dfrac{1}{R_t} = \dfrac{1}{R_1} + \dfrac{1}{R_2}$

            $\dfrac{1}{20} = \dfrac{1}{R_1} + \dfrac{1}{R_1 + 10}$

Clearing the fraction and writing the equation in standard form, we get:

$(R_1)^2 - 30R_1 - 200 = 0$

Using the quadratic formula, we get these two solutions:

R_1 = 35.6 ohms and –5.6 ohms

a) Only one value makes sense for R_1. Therefore R_1 = _____

b) R_2 = $R_1 + 10$ = _____

ℓ_1 = 5 inches

ℓ_2 = 12 inches

from:

$(\ell_1 - 5)(\ell_1 + 12) = 0$

---

a) 35.6 ohms

b) 45.6 ohms

<u>SELF</u>-<u>TEST</u> <u>18</u> (pages <u>204</u>-<u>219</u>)

Solve each equation.

1. $8x^2 + 2x - 3 = 0$

2. $20y + 70 = 10y^2$

3. $\dfrac{1}{R} - R = 1$

4.   In $\boxed{P = \dfrac{E^2}{R}}$, find E

    when $P = 20$ and $R = 8$.

5. A rectangle's length and width are 5 meters and 3 meters, respectively. It is desired to increase the length and the width <u>by</u> <u>the</u> <u>same</u> <u>amount</u> to obtain a larger rectangle whose area is 25 square meters. Find the amount of the increase.

    <u>Note</u>: If "x" is the amount of the increase, then $(x + 5)(x + 3) = 25$.

<u>ANSWERS</u>:

1. x = 0.5
    x = -0.75

2. y = 3.83
    y = -1.83

3. R = 0.62
    R = -1.62

4. E = 12.65

5. 1.10 meters

SUPPLEMENTARY PROBLEMS - CHAPTER 5

Assignment 16

Do these squarings.

1. $\left(-\dfrac{1}{4}\right)^2$ 2. $(-50)^2$ 3. $\left(-\dfrac{3}{8}\right)^2$ 4. $(-0.2)^2$

Do these evaluations.

5.   When $x = -3$, $5x^2 = $ _____   6.   When $x = -1$, $3x - 2x^2 = $ _____

Do these formula evaluations.

7.   In $\boxed{d = w^2}$, find "d"
     when $w = 15$.

8.   In $\boxed{M = \dfrac{mL^2}{3}}$, find M
     when $m = 4$ and $L = 6$.

9.   In $\boxed{r = \dfrac{v^2}{a}}$, find "r"
     when $v = 10$ and $a = 5$.

10.   In $\boxed{E = \dfrac{I}{s^2}}$, find E
     when $I = 24$ and $s = 2$.

11.   In $\boxed{s = \dfrac{1}{2}at^2}$, find "s"
     when $a = 32$ and $t = 3$.

12.   In $\boxed{K = \dfrac{D^2 - d^2}{t}}$, find K
     when $D = 8$, $d = 4$, and $t = 12$.

13.   In $\boxed{I = \dfrac{h(a^2 + b^2)}{4}}$, find I
     when $h = 8$, $a = 4$, and $b = 3$.

14.   In $\boxed{w = pt + ct^2}$, find "w"
     when $p = 10$, $t = 5$, and $c = 6$.

Find these square roots.

15. $-\sqrt{49}$ 16. $\pm\sqrt{\dfrac{1}{25}}$ 17. $\pm\sqrt{15}$ 18. $\pm\sqrt{193}$

Find the two solutions of each quadratic equation.

19. $x^2 - 1 = 0$

20. $1 - 4t^2 = 0$

21. $9y^2 = 100$

22. $\dfrac{4w^2}{25} = 1$

23. $\dfrac{25p^2}{2} = 8$

24. $27 = \dfrac{4}{3d^2}$

25. $\dfrac{2y^2}{3} = \dfrac{3}{8}$

26. $\dfrac{9}{5} = \dfrac{5}{4d^2}$

27. $\dfrac{1}{8} = \dfrac{2}{r^2}$

28. $10 - m^2 = 0$

29. $\dfrac{3y^2}{4} = 60$

30. $\dfrac{24}{2t^2} = 4$

31. $\dfrac{3x}{20} = \dfrac{6}{x}$

32. $\dfrac{100}{r} = \dfrac{2r}{3}$

33. $\dfrac{d}{2} = \dfrac{3}{d}$

Assignment 17

Find the two solutions of each equation.

1. $(t + 1)(t - 7) = 0$ 2. $(2w - 1)(3w - 5) = 0$ 3. $8p(5p + 2) = 0$

Solve each equation by factoring.

4. $x^2 + x = 0$ 5. $4y^2 - 3y = 0$ 6. $2d^2 - d = 0$

7. $r^2 + 6r = 0$ 8. $5m - 15m^2 = 0$ 9. $12t^2 + 8t = 0$

10. $2h + 10h^2 = 0$ 11. $20w^2 = 5w$ 12. $12R = 8R^2$

Solve each equation by factoring.

13. $x^2 + 4x + 3 = 0$ 14. $t^2 - 3t - 10 = 0$ 15. $k^2 + 6k - 7 = 0$

16. $y^2 + 10y + 25 = 0$ 17. $r^2 - 2r + 1 = 0$ 18. $x(x - 2) = 15$

19. $P(P + 1) = 6$ 20. $5F - 4 = F^2$ 21. $\dfrac{12}{h} = h + 1$

22. $\dfrac{2}{m} + 1 = m$ 23. $\dfrac{y}{2} = \dfrac{4}{y} + 1$ 24. $20x^2 + 40x - 60 = 0$

Solve each equation by factoring.

25. $2t^2 + 3t - 2 = 0$ 26. $3d^2 - 2d - 1 = 0$ 27. $2y^2 - 7y + 3 = 0$

28. $3x^2 + 5x + 2 = 0$ 29. $w(2w - 7) = 4$ 30. $9 - 7x = 6(x^2 + 1)$

31. $3h^2 = 2(h + 4)$ 32. $P(10P + 1) = 2$ 33. $\dfrac{3}{m} + 2m = \dfrac{11}{2}$

34. $\dfrac{9}{V} = V + \dfrac{9}{2}$ 35. $10y^2 - 55y + 75 = 0$ 36. $150x^2 - 10x - 60 = 0$

Assignment 18

Solve each equation using the quadratic formula.

1. $x^2 - 2x - 8 = 0$ 2. $6t^2 + t - 2 = 0$ 3. $4w^2 - w - 3 = 0$

4. $2R^2 + 3R - 20 = 0$ 5. $y^2 + 4y + 1 = 0$ 6. $3d^2 - 5d - 1 = 0$

7. $5p^2 - 8p + 2 = 0$ 8. $4F^2 + F - 7 = 0$ 9. $h^2 + 2h - 1 = 0$

Continued on following page.

Put each equation in standard form and solve it using the quadratic formula.

10. $5x - 2 = 3x^2$ 11. $11y + 3(y^2 + 2) = 0$ 12. $t(2t - 9) = 5$

13. $\dfrac{5r}{2} + r^2 = 6$ 14. $\dfrac{1}{N} = N - 1$ 15. $\dfrac{5}{4w} - \dfrac{3}{2} = w$

Before solving each equation, simplify it by factoring out a common numerical factor or by eliminating the decimals, or by doing both.

16. $40x^2 - 80x - 120 = 0$ 17. $1.2p^2 - 4.2p + 3.6 = 0$ 18. $0.18y^2 + 0.15y + 0.03 = 0$

Do these formula evaluations. Report the positive solution only.

19.   In $\boxed{P = I^2R}$, find I
       when $P = 20$ and $R = 5$.

20.   In $\boxed{A = 3.14r^2}$, find "r"
       when $A = 6.28$.

21.   In $\boxed{F = \dfrac{mv^2}{r}}$, find "v"
       when $F = 72$, $m = 8$, and $r = 16$.

22.   In $\boxed{Z^2 = R^2 + X^2}$, find R
       when $Z = 8$ and $X = 6$.

23.   In $\boxed{W = \dfrac{a_1 a_2}{ks^2}}$, find "s" when
       $W = 5$, $a_1 = 10$, $a_2 = 12$, and $k = 3$.

24.   In $\boxed{P = ht^2 - dt}$, find "t"
       when $P = 10$, $h = 2$, and $d = 1$.

Solve these applied problems.

25.   If the area of a rectangle is 12 square meters and if its length is 4 meters more than its width, find its dimensions.

26.   If the hypotenuse of a right triangle is 20 centimeters and if one leg is twice as long as the other leg, find the lengths of the two legs.

27.   If each side of a square is increased 3 inches, the area of the resulting square is 49 square inches. Find the length of each side of the original square.

28.   The area of a right triangle is one-half the product of the two legs. If its area is 10 square centimeters and if one leg is 5 centimeters longer than the other leg, find the lengths of the two legs.

29.   If a ball is projected vertically upward with a velocity of 160 feet per second, in "t" seconds it will reach a height of "h" feet according to the equation: $h = 160t - 16t^2$
       In how many seconds will it reach a height of 400 feet?

30.   The sum of a positive number N and its square is 40 . Find N.

Chapter 6 GRAPHING

In this chapter, we will discuss the procedures for reading and constructing graphs. The coordinate system is reviewed. Both "xy" equations and formulas are graphed. Linear and non-linear graphs are included.

6-1 SOLUTION-TABLES FOR LINEAR "xy" EQUATIONS

An equation containing both "x" and "y" is called an "xy" equation. An "xy" equation whose graph is a straight line is called a "linear" equation. To graph a linear "xy" equation, we begin by making up a table that contains some of its solutions. We will discuss <u>solutions</u> and <u>solution-tables</u> for linear "xy" equations in this section.

1. When an equation contains both "x" and "y", it is "true" when some <u>pairs of values</u> are substituted for the letters and "false" when other <u>pairs of values</u> are substituted for the letters. For example:

   $y = 2x$ is <u>true</u> for (x = 4, y = 8) but <u>false</u> for (x = 1, y = 3).

$y = 2x$	$y = 2x$
$8 = 2(4)$	$3 \neq 2(1)$
$8 = 8$	$3 \neq 2$

   $2x - y = 7$ is <u>true</u> for (x = 5, y = 3) but <u>false</u> for (x = 7, y = 5).

$2x - y = 7$	$2x - y = 7$
$2(5) - 3 = 7$	$2(7) - 5 \neq 7$
$10 - 3 = 7$	$14 - 5 \neq 7$
$7 = 7$	$9 \neq 7$

   a) Is $y = 3x - 2$ true for (x = 4, y = 1)? _____

   b) Is $x + y = 12$ true for (x = 5, y = 7)? _____

2. When a pair of values makes an "xy" equation true, we say that the pair satisfies the equation. For example:

   (x = 2, y = 10) satisfies $y = 3x + 4$, since $10 = 3(2) + 4$

   a) Does (x = 3, y = 15) satisfy $y = 5x$? _____

   b) Does (x = 4, y = 6) satisfy $y = 2x - 5$? _____

   a) No, since:
      $1 \neq 3(4) - 2$

   b) Yes, since:
      $5 + 7 = 12$

223

3. When a pair of values satisfies an "xy" equation, the pair is called a "<u>solution</u>" of the equation.   For example:

    (x = 4, y = 3) is a solution of 2x - y = 5, since 2(4) - 3 = 5.

    a) Is (x = 10, y = 4) a solution of x - y = 3? _____

    b) Is (x = 2, y = 1) a solution of 3x + 2y = 8? _____

    a) Yes, since:
        15 = 5(3)

    b) No, since:
        $6 \neq 2(4) - 5$

4. The solution of an "xy" equation can include one or two negative values. For example:

    (x = -2, y = -8) is a solution of y = 4x, since -8 = 4(-2).

    a) Is (x = -4, y = 1) a solution of y = x + 5? _____

    b) Is (x = -3, y = -6) a solution of 2x - y = 4? _____

    a) No, since:
        $10 - 4 \neq 3$

    b) Yes, since:
        3(2) + 2(1) = 8

5. The solution of an "xy" equation can contain one or two fractions. For example:

    $\left(x = \dfrac{5}{4}, \ y = -\dfrac{1}{4}\right)$ is a solution of x + y = 1, since $\dfrac{5}{4} + \left(-\dfrac{1}{4}\right) = 1$.

    a) Is $\left(x = \dfrac{2}{3}, \ y = 8\right)$ a solution of y = 6x? _____

    b) Is $\left(x = \dfrac{3}{2}, \ y = -\dfrac{1}{2}\right)$ a solution of x - y = 2? _____

    a) Yes, since:
        1 = (-4) + 5

    b) No, since:
        $2(-3) - (-6) \neq 4$

6. The solution of an "xy" equation can contain one or two decimal numbers. For example:

    (x = 2.5, y = 1.5) is a solution of x - y = 1, since 2.5 - 1.5 = 1

    a) Is (x = 3.4, y = 2.6) a solution of x + y = 5? _____

    b) Is (x = 2.5, y = 10) a solution of y = 4x? _____

    a) No, since:
        $8 \neq 6\left(\dfrac{2}{3}\right)$

    b) Yes, since:
        $\dfrac{3}{2} - \left(-\dfrac{1}{2}\right) = 2$

7. <u>Any "xy" equation</u> has many solutions. For example, some solutions for $\boxed{y = 2x}$ are listed below.

x = 5, y = 10	x = -1, y = -2
x = 1.5, y = 3	x = -2.5, y = -5
$x = \dfrac{1}{2}$, y = 1	$x = -\dfrac{7}{2}$, y = -7
x = 0, y = 0	x = -40, y = -80

    How many more solutions are there for the same equation? _____

    a) No, since:
        $3.4 + 2.6 \neq 5$

    b) Yes, since:
        10 = 4(2.5)

    An infinite number

8. To find some solutions for the equation below, we substituted 3, 1, 0, -1, and -3 for "x" and found the corresponding values for "y". We then summarized the five solutions in a table called a <u>solution-table</u>.

$y = 5x$

$y = 5x$

If $x = 3$,	$y = 5(3) = 15$.
If $x = 1$,	$y = 5(1) = 5$.
If $x = 0$,	$y = 5(0) = 0$.
If $x = -1$,	$y = 5(-1) = -5$.
If $x = -3$,	$y = 5(-3) = -15$.

x	y
3	15
1	5
0	0
-1	-5
-3	-15

Are all of the possible solutions for $y = 5x$ listed in the table? _____

---

9. When making up a solution-table to graph a linear "xy" equation, we begin by picking some values of "x" to substitute. The following guidelines can be used.

    1) Pick some positive values, "0", and some negative values.

    2) Usually pick smaller whole-number values to make the evaluations easier.

Complete the solution-table for each equation below.

> No. There are an infinite number of possible solutions.

a) $y = x + 1$

x	y
2	
1	
0	
-1	
-2	

b) $y = 3 - x$

x	y
4	
2	
0	
-2	
-4	

---

10. In a solution-table, the values of "x" should be listed <u>in order</u> as they are below. Complete the tables.

a) $y = 3x + 4$

x	y
5	
3	
0	
-3	
-5	

b) $y = 5 - 2x$

x	y
4	
2	
0	
-2	
-4	

a)
x	y
2	3
1	2
0	1
-1	0
-2	-1

b)
x	y
4	-1
2	1
0	3
-2	5
-4	7

---

a)
x	y
5	19
3	13
0	4
-3	-5
-5	-11

b)
x	y
4	-3
2	1
0	5
-2	9
-4	13

11.   To evaluate for "y" below when x = 2, we simplified the numerator first and then divided.

$$y = \frac{10 - 3x}{2} \qquad y = \frac{10 - 3(2)}{2} = \frac{10 - 6}{2} = \frac{4}{2} = 2$$

Using the same procedure, complete the following tables.

a) $y = \dfrac{12 - x}{4}$ b) $y = \dfrac{8 - 3x}{2}$

x	y
8	
4	
0	
-4	
-8	

x	y
4	
2	
0	
-2	
-4	

a)
x	y
8	1
4	2
0	3
-4	4
-8	5

b)
x	y
4	-2
2	1
0	4
-2	7
-4	10

6-2   SOLVING FOR "y" AND SOLUTION-TABLES

It is easier to make up a solution-table for an "xy" equation when "y" is solved-for. However, in some "xy" equations, "y" is not solved-for. In this section, we will show how to rearrange "xy" equations to solve for "y" and then use such rearrangements to make up solution-tables.

12.   In the equation below, "y" is not solved-for. Therefore, when we substituted a value for "x" in the equation as it stands, we had to solve an equation to find the corresponding value for "y". Complete the other evaluation.

Find "y" when "x" is 3.   Find "y" when "x" is -1.

$$x + y = 10 \qquad\qquad\qquad 2x + y = 5$$
$$3 + y = 10$$
$$(-3) + 3 + y = 10 + (-3)$$
$$y = 7$$

y = 7, from:
(-2) + y = 5

13.   Following the example, complete the other evaluation by solving an equation.

Find "y" when "x" is 4.   Find "y" when "x" is 3.

$$x - y = 1$$ $$3x - 2y = 5$$

$$4 - y = 1$$

$$(-4) + 4 - 1y = 1 + (-4)$$

$$-1y = -3$$

$$y = \frac{-3}{-1} = 3$$

---

14.   When making up a solution-table, several pairs of values are needed. It is easier to find those pairs if "y" is solved-for. Therefore, we ordinarily solve for "y" before making up a table. We will discuss various steps needed to solve for "y" in the following frames.

To solve for "y" below, we added "x" to both sides. Solve for "y" in the other equation.

$$y - x = 1$$ $$y - 2x = 5$$

$$y - x + x = 1 + x$$

$$y = 1 + x$$

$$\text{or} \quad y = x + 1$$

y = 2, from:
9 - 2y = 5

---

15.   Any addition of a negative quantity can be written as a subtraction. For example:

$$5 + (-x) \quad \text{can be written} \quad 5 - x$$

$$7 + (-3x) \quad \text{can be written} \quad 7 - 3x$$

When solving for "y" below, we wrote the addition of a negative term as a subtraction in the final step. Solve for "y" in the other equation.

$$x + y = 10$$ $$5x + y = 12$$

$$(-x) + x + y = 10 + (-x)$$

$$y = 10 + (-x)$$

$$y = 10 - x$$

y = 5 + 2x
or
y = 2x + 5

---

y = 12 - 5x

16. To solve for "y" below, we divided the whole right side of the equation by 2.

$$2y = 3x + 5$$

$$y = \frac{3x + 5}{2}$$

We used the same process in the final step to solve for "y" below. Solve for "y" in the other equation.

$$4y - 3x = 7$$

$$4y - 3x + 3x = 3x + 7$$

$$4y = 3x + 7$$

$$y = \frac{3x + 7}{4}$$

$$3y - x = 10$$

---

17. Following the example, solve for "y" in the other equation.

$$2x + 5y = 10$$

$$(-2x) + 2x + 5y = 10 + (-2x)$$

$$5y = 10 - 2x$$

$$y = \frac{10 - 2x}{5}$$

$$x + 4y = 20$$

$$y = \frac{x + 10}{3}$$

---

18. In an earlier chapter, we gave the following definition for the opposite of a subtraction.

The opposite of (a – b) is [(–a) + b].

However, [(–a) + b] can be written [b + (–a)] or (b – a). Therefore, we can get the opposite of a subtraction by simply interchanging the terms. That is:

The opposite of (a – b) is (b – a).

To confirm that (a – b) and (b – a) are opposites, complete their addition below to show that their sum is "0".

$$(a - b) + (b - a) = a - b + b - a$$
$$= \underline{a - a} + \underline{b - b}$$
$$= \underline{\quad} + \underline{\quad} = \underline{\quad}$$

$$y = \frac{20 - x}{4}$$

---

19. Using the definition from the last frame, we can write the opposite of any subtraction by simply interchanging the two terms. That is:

The opposite of 2x – 5 is 5 – 2x.

The opposite of 10 – x is x – 10.

$$0 + 0 = 0$$

Write the opposite of each subtraction.

a) x – 9 _____   b) 7 – 3x _____

20.   If we replace each side of an equation with its opposite, the new equation is equivalent to the original one.   For example:

$$-5x = 15$$
and
$$5x = -15$$

are equivalent because the solution of each is "-3".

Replacing each side of an equation by its opposite is called the "<u>oppositing principle for equations</u>".   That principle was used to solve for "y" below.

$$-y = 10 - 2x$$
$$y = 2x - 10$$

Solve for "y" in each equation by replacing each side with its opposite.

a) $-y = 7 - 5x$

y = _____

b) $-y = 6 - x$

y = _____

a) 9 - x

b) 3x - 7

---

21.   When solving for "y" below, we used the <u>oppositing principle</u> in the final step to get rid of the "-" in front of "y".   Solve for "y" in the other equation.

$$2x - y = 15$$
$$(-2x) + 2x - y = 15 + (-2x)$$
$$-y = 15 - 2x$$
$$y = 2x - 15$$

$$x - y = 3$$

a) y = 5x - 7

b) y = x - 6

---

22.   We solved for "y" in each equation below.   Using that form of the equation, complete the solution-table.

a) $\boxed{2x + y = 10}$

or

$\boxed{y = 10 - 2x}$

x	y
3	
1	
0	
-1	
-3	

b) $\boxed{x - y = 3}$

or

$\boxed{y = x - 3}$

x	y
4	
2	
0	
-2	
-4	

y = x - 3

---

a) x	y	b) x	y
3	4	4	1
1	8	2	-1
0	10	0	-3
-1	12	-2	-5
-3	16	-4	-7

23.   We solved for "y" in each equation below. Using that form of the equation, complete the solution-table.

a)   2x - y = 9

or

y = 2x - 9

x	y
5	
3	
0	
-3	
-5	

b)   3x + 2y = 10

or

$y = \dfrac{10 - 3x}{2}$

x	y
4	
2	
0	
-2	
-4	

24.   Complete the solution-table for each equation.

a)   x + y = 12

x	y
6	
3	
0	
-3	
-6	

b)   5x - y = 10

x	y
4	
2	
0	
-2	
-4	

a)
x	y
5	1
3	-3
0	-9
-3	-15
-5	-19

b)
x	y
4	-1
2	2
0	5
-2	8
-4	11

a)
x	y
6	6
3	9
0	12
-3	15
-6	18

b)
x	y
4	10
2	0
0	-10
-2	-20
-4	-30

6-3   READING POINTS ON THE COORDINATE SYSTEM

In this section, we will introduce the coordinate system that is used to graph "xy" equations. Then we will discuss the method for reading the coordinates of points on the coordinate system.

25.   The coordinate system is shown at the right. It consists of a horizontal and a vertical number line.

The horizontal number line is called the horizontal axis.

The vertical number line is called the vertical axis.

The two number lines together are called coordinate axes.

The two coordinate axes intersect at "0" on each. That point is called the _____ .

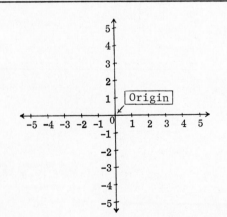

26. In the coordinate system at the right, we have included some additional horizontal and vertical lines that form a "gridwork". The gridwork makes it easier to use the coordinate system.

Each point in the coordinate system represents a pair of values, one related to the horizontal axis and one related to the vertical axis.

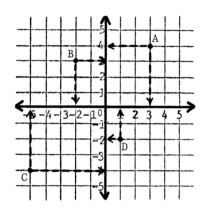

origin

To find the pair of values for point A:

1) We drew an arrow down to the horizontal axis. The arrow points to "3".

2) We drew an arrow over to the vertical axis. The arrow points to "4".

3) Therefore, point A represents 3 on the horizontal axis and 4 on the vertical axis.

Similarly: Point B represents _____ on the horizontal axis and _____ on the vertical axis.

Point C represents _____ on the horizontal axis and _____ on the vertical axis.

Point D represents _____ on the horizontal axis and _____ on the vertical axis.

27. Mathematicians use the two axes to represent the values of "x" and "y", the letters ordinarily used in two-letter equations. They put:

> x-values on the horizontal axis
>
> y-values on the vertical axis

They label the axes "x" and "y" as we have done at the right.

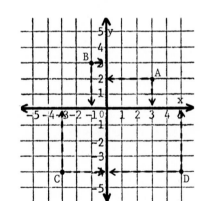

B) -2 ... 3

C) -5 ... -4

D) 1 ... -2

On the coordinate system shown:

Point A represents: x = 3, y = 2.

Point B represents: x = _____, y = _____

Point C represents: x = _____, y = _____

Point D represents: x = _____, y = _____

28. Since the axes are used to represent x-values and y-values, they are called the "x-axis" and the "y-axis".

a) The "x-axis" is another name for the _____ (horizontal, vertical) axis.

b) The "y-axis" is another name for the _____ (horizontal, vertical) axis.

B   x = -1, y = 3

C   x = -3, y = -4

D   x = 5, y = -4

29.   Point A at the right represents: x = 4, y = 2.

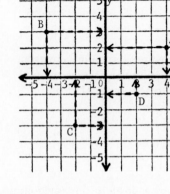

   "4" and "2" are called the coordinates
      of point A.

   "4" is called the x-coordinate.

   "2" is called the y-coordinate.

The coordinates of point A are usually
written as we have done below. That is,
they are enclosed in parentheses and
separated by a comma with the
x-coordinate written first.

         (4 , 2)

| a) horizontal |
| b) vertical |

Write the coordinates of points B, C, and D below. Be sure to write the x-coordinate first.

      B (,   )   C (,   )   D (,   )

---

30.   When writing a pair of coordinates in parentheses, the order in which they are
written makes a difference. That is:

            (4, 7)   means: x = 4, y = 7
            (7, 4)   means: x = 7, y = 4

Since the order in which they are written makes a difference, a pair of coordinates
is called an "ordered pair".

   a) In the ordered pair (-1, -2), the x-coordinate is _____.

   b) In the ordered pair (-5, -3), the y-coordinate is _____.

| B (-4, 3) |
| C (-2, -3) |
| D (2, -1) |

---

31.   Only part of the scale is shown
on each axis at the right.
Therefore, you have to fill in
some missing numbers to read the
the scales.   For example:

The coordinates of A are (8, 6).
Write the coordinates of B, C,
and D below.

      B _____
      C _____
      D _____

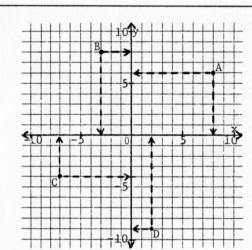

| a) -1 |
| b) -3 |

| B (-3, 8) |
| C (-7, -4) |
| D (2, -9) |

32. We do not always use the same
     scale on each axis. For example,
     we counted "by 2's" on the y-axis
     at the right.

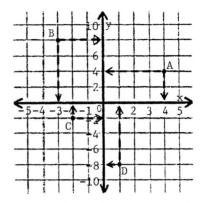

    The coordinates of point A are (4, 4).

    The coordinates of point B are (,).

    The coordinates of point C are (,).

    The coordinates of point D are (,).

33. When scaling the axes at the
     right, we counted "by 5's" on
     the x-axis and "by 10's" on the
     y-axis. Write the coordinates
     of these points:

    A _____

    B _____

    C _____

    D _____

B (-3, 8)

C (-2, -2)

D (1, -8)

A (10, 30)   B (-5, 20)   C (-25, -30)   D (15, -50)

6-4   PLOTTING POINTS ON THE COORDINATE SYSTEM

If we know the coordinates of a point, we can locate or "plot" the point on the coordinate system. We will
discuss the method in this section.

34. The coordinate axes divide the coordinate system into four parts. These four parts are called
     quadrants. We have labeled the four quadrants in Figure 1 below. Notice that they are numbered
     in a counter-clockwise direction, beginning with the upper right quadrant.

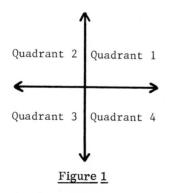

Figure 1

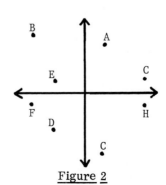

Figure 2

On the coordinate system in Figure 2 above, there are two points labeled in each quadrant.

   a) Points B and E lie in Quadrant _____.   b) Points C and H lie in Quadrant _____.

35.   To locate or "plot" a point on the coordinate system, we simply reverse the procedure for reading the coordinates of a point. For example, we have plotted points A and B at the right by drawing arrows from the axes.

   A (2, 4)

   B (-4, -3)

Plot and label the four points below on the same coordinate system.

   C (3, 1)   E (-2, 5)
   D (3, -2)   F (-2, -4)

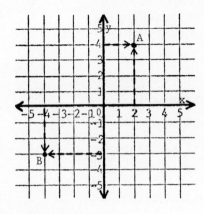

a) Quadrant 2

b) Quadrant 4

---

36.   The correct plotting of the points from the last frame is shown at the right.

   Point C is in quadrant _____.

   Point D is in quadrant _____.

   Point E is in quadrant _____.

   Point F is in quadrant _____.

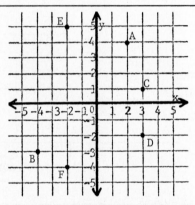

See next frame for answer.

---

37.   When the scales on the axes are different, be careful when plotting points. For example, we plotted points A and B on the coordinate system at the right.

   A (-10, 20)

   B (2, -30)

Plot and label the four points below on the same coordinate system.

   C (-6, -10)   E (2, 50)
   D (-4, 40)   F (10, -20)

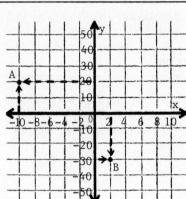

C (quadrant 1)

D (quadrant 4)

E (quadrant 2)

F (quadrant 3)

See next frame for answer.

38. The correct plotting of the points from the last frame is shown at the right.

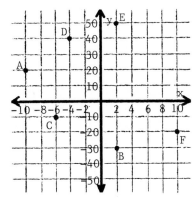

C is in quadrant _____.

D is in quadrant _____.

E is in quadrant _____.

F is in quadrant _____.

---

39. Only part of the scale is shown on each axis at the right. We plotted the two points below.

A (3, 7)

B (-8, -4)

Plot and label the four points below on the same coordinate system.

C (9, 2)   E (-2, -6)
D (-6, 5)   F (4, -9)

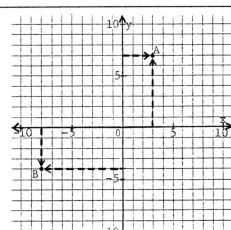

C (quadrant 3)

D (quadrant 2)

E (quadrant 1)

F (quadrant 4)

---

40. The correct plotting of the points from the last frame is shown at the right.

Two other names are used for the coordinates of a point.
That is:

The x-coordinate is called the "abscissa".

The y-coordinate is called the "ordinate".

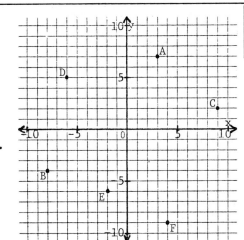

See next frame for answer.

Since the coordinates of A are (3, 7):   its abscissa is 3.
                                          its ordinate is 7.

Since the coordinates of B are (-8, -4):   a) its abscissa is _____.

                                          b) its ordinate is _____.

---

a) -8

b) -4

41.   Write the coordinates of a point:

    a) if its abscissa is 5 and its ordinate is 3.   (,   )

    b) if its ordinate is -1 and its abscissa is -4.   (,   )

---

42.   a) Another name for the x-coordinate is the _____ .

    b) Another name for the y-coordinate is the _____ .

a) (5, 3)

b) (-4, -1)

---

a) abscissa   b) ordinate

> Note: The labels "horizontal", "x", and "abscissa" go together.
>
> The labels "vertical", "y", and "ordinate" go together.
>
> To help you remember what goes together, notice that alphabetically:
>
>   horizontal comes before vertical
>
>   x comes before y
>
>   abscissa comes before ordinate

6-5   READING AND PLOTTING POINTS ON THE AXES

In this section, we will read and plot points on the coordinate axes.

43.   We have plotted various points on the coordinate system below. Points I, J, K, and L lie on the horizontal axis.

a) Points A, B, C, and D have the same y-coordinate (or ordinate). It is ____ .

b) Points E, F, G, and H have the same y-coordinate (or ordinate). It is ____ .

c) Points I, J, K, and L lie on the horizontal axis. They have the same y-coordinate (or ordinate). It is ____ .

d) The y-coordinate (or ordinate) of any point on the horizontal axis is ____ .

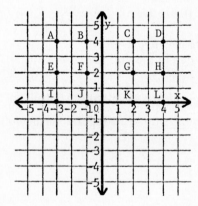

a) 4

b) 2

c) 0

d) 0

44. Eight points have been plotted on the coordinate system at the right. Write the coordinates of each point:

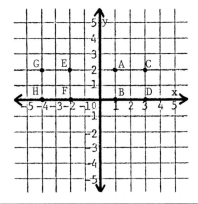

A _____   E _____

B _____   F _____

C _____   G _____

D _____   H _____

45. We have plotted various points on the coordinate system below. Some of the points lie on the vertical axis.

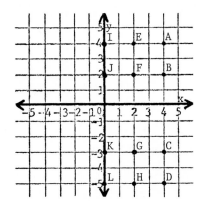

a) Points A, B, C, and D have the same x-coordinate (or abscissa). It is _____.

b) Points E, F, G, and H have the same x-coordinate (or abscissa). It is _____.

c) Points I, J, K, and L lie on the vertical axis. Each point has the same x-coordinate (or abscissa). It is _____.

d) The x-coordinate (or abscissa) of any point on the vertical axis is _____.

A (1, 2)
B (1, 0)

C (3, 2)
D (3, 0)

E (-2, 2)
F (-2, 0)

G (-4, 2)
H (-4, 0)

46. Various points have been plotted on the coordinate system at the right. Write the coordinates of each point.

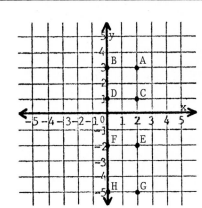

A _____   E _____

B _____   F _____

C _____   G _____

D _____   H _____

a) 4

b) 2

c) 0

d) 0

A (2, 3)   C (2, 1)   E (2, -2)   G (2, -5)

B (0, 3)   D (0, 1)   F (0, -2)   H (0, -5)

47. Four points have been plotted
    on the coordinate system at the
    right. Write the coordinates
    of each point:

    A _____

    B _____

    C _____

    D _____

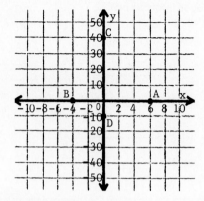

---

48. Plot the following points on
    the coordinate system at
    the right:

    A (4, 0)

    B (0, 4)

    C (-3, 0)

    D (0, -3)

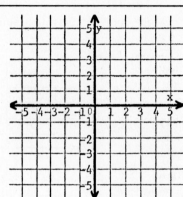

A (6, 0)

B (-4, 0)

C (0, 40)

D (0, -10)

---

49. The correct plotting of the points
    from the last frame is shown at
    the right.

    As you can see from the graph,
    the point where the two axes
    intersect is called the "origin".
    Since the origin lies on both
    axes:

    a) its x-coordinate is _____.

    b) its y-coordinate is _____.

    c) its coordinates are (,   )

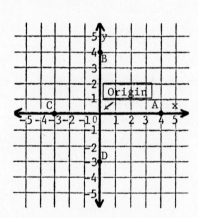

See next frame
for answer.

---

a) 0

b) 0

c) (0, 0)

SELF-TEST 19 (pages 223-239)

Complete the solution-table for each equation.

1. | x + y = 2 |

x	y
6	
3	
0	
-2	
-5	

2. | 3x - y = 1 |

x	y
4	
1	
0	
-1	
-3	

3. | 2y - x = 8 |

x	y
8	
4	
0	
-2	
-6	

4. Write the coordinates of these points:

A _____   D _____

B _____   E _____

C _____   F _____

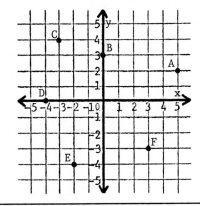

5. The point (4, -1) lies in quadrant _____.

6. Write the coordinates of the point whose ordinate is 5 and whose abscissa is -2. _____

ANSWERS:

1.
x	y
6	-4
3	-1
0	2
-2	4
-5	7

2.
x	y
4	11
1	2
0	-1
-1	-4
-3	-10

3.
x	y
8	8
4	6
0	4
-2	3
-6	1

4. A (5, 2)

B (0, 3)

C (-3, 4)

D (-4, 0)

E (-2, -4)

F (3, -3)

5. quadrant 4   6. (-2, 5)

6-6   GRAPHING LINEAR "xy" EQUATIONS

A linear "xy" equation is an equation whose graph is a straight line.   To graph a linear "xy" equation on the coordinate system, three steps are used:

      1) Make up a solution-table.

      2) Plot the points representing those solutions.

      3) Draw a straight line through the plotted points.

In this section, we will discuss the method for graphing linear "xy" equations.

---

50.   The three steps needed to graph the
       following equation are discussed below.

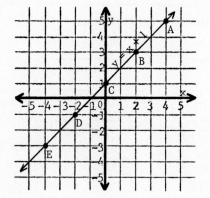

$$y = x + 1$$

    Step 1: Make up a solution-table.

	x	y
A	4	5
B	2	3
C	0	1
D	-2	-1
E	-4	-3

    Step 2: Plot the points in the solution-table.

    Step 3: Draw a straight line through the plotted points.
              Label the graph "y = x + 1".

The straight line shown is the graph of $y = x + 1$.

    a) Does the straight line (or graph) pass through the origin? _____

    b) The straight line (or graph) passes through what three quadrants? _____

---

51.   When making up a solution-table for graphing, we always use "0" and a few
       positive and negative values for x. Even though different values are used for
       x, we get the same graph. For example, we used two different tables to graph
       y = 2x below and on the following page. Notice that the graphed lines are identical.

a) No

b) 1, 2, 3

$$y = 2x$$

	x	y
A	2	4
B	1	2
C	0	0
D	-1	-2
E	-2	-4

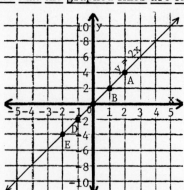

Continued on following page.

51.   Continued

y = 2x

	x	y
A	5	10
B	3	6
C	0	0
D	-3	-6
E	-5	-10

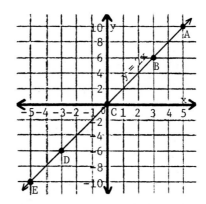

a) Does the graphed line pass through the origin? _____

b) The graphed line passes through what two quadrants? _____

52.   At the right, we graphed the equation:

x + y = 2

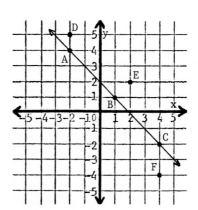

Any point <u>on the graphed line</u> represents one solution of the equation. Therefore, its coordinates satisfy the equation.

Any point <u>off the graphed line</u> does not represent a solution of the equation. Therefore, its coordinates <u>do not</u> <u>satisfy</u> the equation.

Let's examine the points <u>on</u> the line first.

a) Write the coordinates of the three points <u>on</u> the graphed line:

        A (,   )   B (,   )   C (,   )

b) Do the coordinates of these points satisfy the equation x + y = 2? _____

Now let's examine the points <u>off</u> the line.

c) Write the coordinates of the three points <u>off</u> the graphed line:

        D (,   )   E (,   )   F (,   )

d) Do the coordinates of these points satisfy the equation x + y = 2? _____

53.   On the two coordinate systems below, we graphed the equation x + y = 1 from the table of solutions shown:

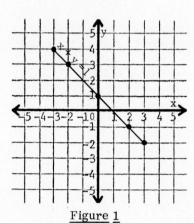

x + y = 1	
x	y
3	-2
2	-1
0	1
-2	3
-3	4

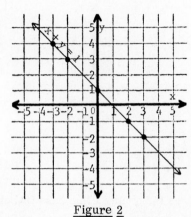

Figure 1   Figure 2

Note: In Figure 1 we stopped the graphed line at the last plotted point on each end.

In Figure 2 we extended the graphed line to the edge of the coordinate system shown and put arrowheads at each end of it.

Figure 2 is the correct graphing of x + y = 1 for these reasons:

1) The line should be extended to the edge of the coordinate system to show that there are other solutions, like (5, -4) and (-4, 5), beyond -3 and 3 on that part of the line.

2) Arrowheads should be put at each end of the graphed line to show that there are solutions, like (10, -9) and (-20, 21), beyond the edge of that part of the coordinate system.

---

54.   Using the tables provided, graph each equation. Be sure to extend each line to the edge of the coordinate system and put arrowheads on each end of the line.

y = 2x + 1	
x	y
2	5
1	3
0	1
-1	-1
-2	-3

x + y = 2	
x	y
4	-2
2	0
0	2
-2	4
-4	6

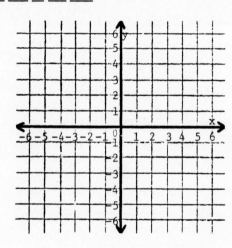

See next frame
for answer.

55. The correct graphs of the equations from the last frame are shown at the right.

   a) Does either graphed line pass through the origin? _____

   b) The graph of $x + y = 2$ lies in what three quadrants? _____

   c) The graph of $y = 2x + 1$ lies in what three quadrants? _____

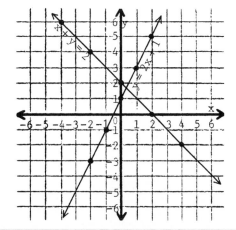

56. Complete each solution-table and then graph both equations on the coordinate system provided.

$y = 3x$	
x	y
3	
1	
0	
-1	
-3	

$2x + y = 1$	
x	y
4	
2	
0	
-2	
-4	

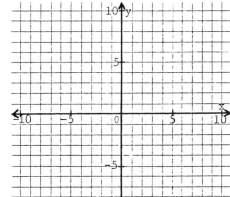

a) No

b) 1, 2, 4

c) 1, 2, 3

57. Using some positive values, "0", and some negative values for x, make up a solution-table and then graph each equation.

$2x - y = 5$	
x	y

$x + 2y = 4$	
x	y

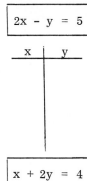

x	y	x	y
3	9	4	-7
1	3	2	-3
0	0	0	1
-1	-3	-2	5
-3	-9	-4	9

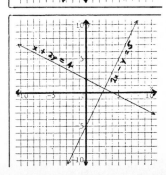

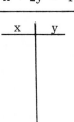

6-7   SOLUTION-TABLES FOR NON-LINEAR "xy" EQUATIONS

An "xy" equation is non-linear if its graph is a curved line. To graph a non-linear equation, we also begin by making up a solution-table. We will discuss solution-tables for non-linear "xy" equations in this section.

---

58.   To find solutions for $\boxed{y = x^2}$, we square various values of <u>x</u>. For example:

If $x = 3$, $y = 3^2 = 9$

a) If $x = \frac{3}{2}$, $y =$ _____

If $x = -\frac{1}{2}$, $y = \left(-\frac{1}{2}\right)^2 = \frac{1}{4}$

b) If $x = -4$, $y =$ _____

---

59.   We found two solutions for $\boxed{y = 2x^2}$ below.

If $x = \frac{3}{2}$, $y = 2\left(\frac{3}{2}\right)^2 = 2\left(\frac{9}{4}\right) = \frac{9}{2}$ or $4\frac{1}{2}$

If $x = -5$, $y = 2(-5)^2 = 2(25) = 50$

Following the examples, find these solutions for the same equation.

a) If $x = 1$, $y =$ _____

b) If $x = -10$, $y =$ _____

a) $\frac{9}{4}$

b) 16

---

60.   Complete the solution-tables for these equations:

a) $\boxed{y = x^2}$

x	y
3	
1	
$\frac{1}{2}$	
0	
$-\frac{1}{2}$	
-1	
-3	

b) $\boxed{y = 4x^2}$

x	y
5	
$\frac{3}{2}$	
1	
0	
-1	
$-\frac{3}{2}$	
-5	

a) 2, from 2(1)

b) 200, from 2(100)

a)
x	y
3	9
1	1
$\frac{1}{2}$	$\frac{1}{4}$
0	0
$-\frac{1}{2}$	$\frac{1}{4}$
-1	1
-3	9

b)
x	y
5	100
$\frac{3}{2}$	9
1	4
0	0
-1	4
$-\frac{3}{2}$	9
-5	100

61. We found two solutions for $\boxed{y = x^2 - 2x}$ below.

$$\text{If } x = 4, \ y = 4^2 - 2(4) = 16 - 8 = 8$$

$$\text{If } x = -\frac{1}{2}, \ y = \left(-\frac{1}{2}\right)^2 - 2\left(-\frac{1}{2}\right) = \frac{1}{4} + 1 = 1\frac{1}{4}$$

Following the examples, find these solutions for the same equation.

a) If $x = \frac{5}{2}$, $y =$ _____

b) If $x = -2$, $y =$ _____

62. We found two solutions for $\boxed{y = 4x - x^2}$ below.

$$\text{If } x = \frac{1}{2}, \ y = 4\left(\frac{1}{2}\right) - \left(\frac{1}{2}\right)^2 = 2 - \frac{1}{4} = \frac{7}{4} \text{ or } 1\frac{3}{4}$$

$$\text{If } x = -1, \ y = 4(-1) - (-1)^2 = (-4) - (1) = -5$$

Following the examples, find these solutions for the same equations.

a) If $x = 3$, $y =$ _____

b) If $x = -\frac{1}{2}$, $y =$ _____

a) $\frac{5}{4}$ or $1\frac{1}{4}$, from $\frac{25}{4} - 5$

b) 8, from $4 + 4$

63. Complete the solution-tables for these equations.

a) $\boxed{y = x^2 - 4x}$

x	y
5	
4	
3	
2	
1	
0	
$-\frac{1}{2}$	
-1	

b) $\boxed{y = 2x - x^2}$

x	y
3	
$\frac{5}{2}$	
2	
1	
0	
$-\frac{1}{2}$	
-1	
-2	

a) 3, from: $12 - 9$

b) $-\frac{9}{4}$ or $-2\frac{1}{4}$, from: $(-2) - \left(\frac{1}{4}\right)$

a)
x	y
5	5
4	0
3	-3
2	-4
1	-3
0	0
$-\frac{1}{2}$	$2\frac{1}{4}$
-1	5

b)
x	y
3	-3
$\frac{5}{2}$	$-1\frac{1}{4}$
2	0
1	1
0	0
$-\frac{1}{2}$	$-1\frac{1}{4}$
-1	-3
-2	-8

64.   To find two solutions for $\boxed{xy = 40}$ below, we substituted and then solved an equation.

Find $\underline{y}$ when $\underline{x}$ is 8.   Find $\underline{y}$ when $\underline{x}$ is –10.

$$xy = 40$$ $$xy = 40$$
$$8y = 40$$ $$-10y = 40$$
$$y = 5$$ $$y = -4$$

However, before finding solutions for equations like those above, we ordinarily solve for "y". To do so, we divide 40 by "x". We get:

$$xy = 40$$

$$y = \frac{40}{x}$$

Using the new form of the equation, let's find the same two solutions.

If $x = 8$, $y = \frac{40}{8} = 5$ If $x = -10$, $y = \frac{40}{-10} = $ _____

---

65.   Following the example, solve for "y" in the other equations.

$$xy = 24$$ a) $xy = 10$ b) $xy = 100$

$$y = \frac{24}{x}$$ $y = $ _____ $y = $ _____

	–4

---

66.   If we substitute "0" for $\underline{x}$ in the equation below, we get a division by "0". Since division by "0" is impossible (or undefined), there is no solution for the equation when $\underline{x} = 0$.

For $\boxed{y = \dfrac{24}{x}}$, when $x = 0$, $y = \dfrac{24}{0}$ (impossible)

a) $y = \dfrac{10}{x}$

b) $y = \dfrac{100}{x}$

Remembering the fact above, complete the solution-tables for these equations.

a) $\boxed{xy = 48}$

or

$\boxed{y = \dfrac{48}{x}}$

x	y
8	
6	
4	
2	
0	
-2	
-4	
-6	
-8	

b) $\boxed{xy = 100}$

or

$\boxed{y = \dfrac{100}{x}}$

x	y
20	
10	
5	
2	
0	
-2	
-5	
-10	
-20	

67. To cube a number, we use the number as a factor <u>three</u> times. A small "3" is used as the symbol for cubing. That is:

$$5^3 = (5)(5)(5) = 25(5) = 125$$

Following the example, cube each number below.

   a) $2^3 = (2)(2)(2) = $ _____

   b) $4^3 = (4)(4)(4) = $ _____

a) x	y	b) x	y
8	6	20	5
6	8	10	10
4	12	5	20
2	24	2	50
0	No solution	0	No solution
-2	-24	-2	-50
-4	-12	-5	-20
-6	-8	-10	-10
-8	-6	-20	-5

68. When a negative number is cubed, its cube is <u>negative</u>. For example:

$$(-3)^3 = (-3)(-3)(-3) = (9)(-3) = -27$$

Following the example, cube each number below.

   a) $(-1)^3 = (-1)(-1)(-1) = $ _____   b) $(-10)^3 = (-10)(-10)(-10) = $ _____

a) 8

b) 64

69. To find solutions for $\boxed{y = x^3}$, we cube values of <u>x</u>. That is:

   If $x = 1$, $y = 1^3 = 1$ a) If $x = -1$, $y = (-1)^3 = $ _____

   If $x = 0$, $y = 0^3 = 0$ b) If $x = \frac{1}{2}$, $y = \left(\frac{1}{2}\right)^3 = $ _____

a) -1, from (1)(-1)

b) -1,000, from (100)(-10)

a) -1   b) $\frac{1}{8}$

6-8   ESTIMATING THE COORDINATES AND POSITION OF POINTS

When reading graphs or graphing equations, we frequently have to read or plot points whose coordinates lie between the given values on the axes. In such cases, we have to estimate coordinates or the positions of points on the coordinate system. We will discuss both types of estimation in this section.

70. When reading a point on a graph, we have to estimate its coordinates if they lie between the given values on the axes. For example, we counted by 1's on the x-axis below. Therefore, A is approximately $1\frac{1}{2}$ or $\frac{3}{2}$ and B is approximately $-\frac{1}{2}$.

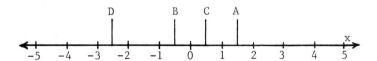

Estimate the values of the following points:

   C is approximately _____   D is approximately _____

(C) $\frac{1}{2}$ (D) $-2\frac{1}{2}$ or $-\frac{5}{2}$

71. On the x-axis below, we counted by 2's. Therefore, A is approximately 5 and B is approximately
     -1$\frac{1}{2}$ or -1.5.

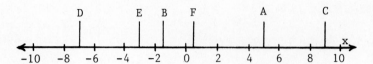

Estimate the value of the following points:

C is approximately _____   E is approximately _____

D is approximately _____   F is approximately _____

---

72. On the x-axis below, we counted by 10's. Therefore, A is approximately
     15 and B is approximately -32.

(C) 9	(E) -3
(D) -7	(F) $\frac{1}{2}$ or 0.5

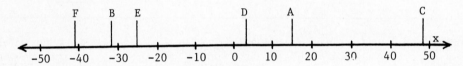

Estimate the values of the following points:

C is approximately _____   E is approximately _____

D is approximately _____   F is approximately _____

---

73. On the y-axis at the right, we
     counted by 20's. Therefore,
     A is approximately 70 and
     B is approximately -25.

(C) 48	(E) -25
(D) 3	(F) -41

     Estimate the values of the
     following points:

     C is approximately _____

     D is approximately _____

     E is approximately _____

     F is approximately _____

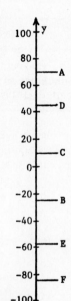

74.   On the y-axis at the right, we counted by 5's. Therefore, A is approximately 8 and B is approximately -11.

Estimate the values of the following points:

C is approximately _____

D is approximately _____

E is approximately _____

F is approximately _____

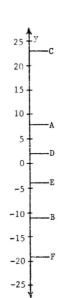

(C) 10

(D) 45

(E) -57

(F) -84

75.   On the y-axis at the right, we counted by 50's. Therefore, A is approximately 120 and B is approximately -90.

Estimate the values of the following points:

C is approximately _____

D is approximately _____

E is approximately _____

F is approximately _____

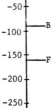

(C) 23   (E) -4

(D) 2   (F) -19

76.   Several points are plotted on the coordinate system at the right. Estimate the coordinates of each point.

A (,   )

B (,   )

C (,   )

D (,   )

E (,   )

F (,   )

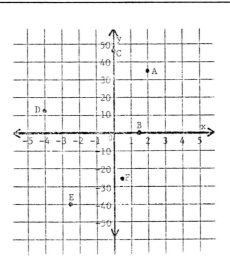

(C) 30   (E) -25

(D) 240   (F) -160

77. Estimate the coordinates of each point on the graph at the right.

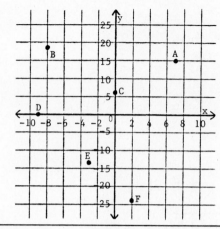

A (,)

B (,)

C (,)

D (,)

E (,)

F (,)

A (2, 35)

B ($1\frac{1}{2}$, 0)

C (0, 47)

D (-4, 12)

E (-$2\frac{1}{2}$, -40)

F ($\frac{1}{2}$, -25)

78. Estimate the coordinates of each point on the graph at the right.

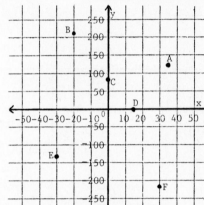

A (,)

B (,)

C (,)

D (,)

E (,)

F (,)

A (7, 15)

B (-8, 18)

C (0, 6)

D (-9, 0)

E (-3, -13)

F (2, -24)

79. When plotting a point, we have to estimate its position if its coordinates lie between the given values on the axes. For example, we estimated the position of $2\frac{1}{2}$ or $\frac{5}{2}$ on the x-axis below. Estimate the positions of these numbers on the same axis.

$$\frac{1}{2}, \quad 4\frac{1}{2}, \quad -\frac{1}{2}, \quad -\frac{3}{2}, \quad -3\frac{1}{2}$$

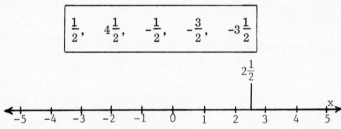

A (35, 125)

B (-20, 210)

C (0, 80)

D (15, 0)

E (-30, -130)

F (30, -220)

80. We estimated the positions of 3 and $7\frac{1}{2}$ on the x-axis below. Estimate the positions of these numbers on the same axis.

$$5, \quad \frac{1}{2}, \quad -3, \quad -9, \quad -1\frac{1}{2}$$

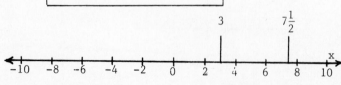

Your x-axis should look approximately like this:

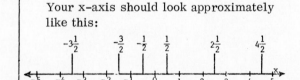

81. We estimated the position of 17 on each x-axis below. Estimate the positions of these numbers <u>on</u> <u>both</u> <u>axes</u>.

$$6, \quad 24, \quad -7, \quad -13, \quad -21$$

Your axis should look approximately like this:

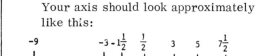

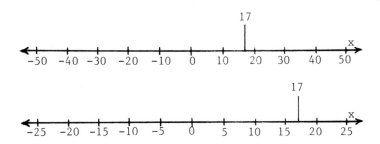

82. We estimated the position of 30 on each y-axis below. Estimate the position of these numbers <u>on</u> <u>both</u> <u>axes</u>.

$$95, \quad 12, \quad -25, \quad -55, \quad -87$$

Your answers should look approximately like this:

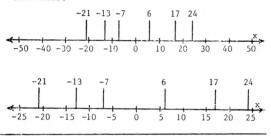

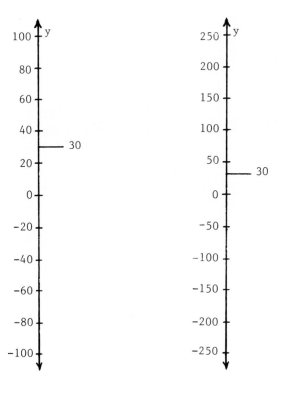

Answers to Frame 82

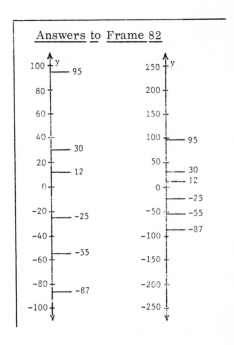

83.   Plot and label the following points on the coordinate system below.

A (4, 25)

B (0, 42)

C ($\frac{1}{2}$, 8)

D (−3, 35)

E (−5, −33)

F ($\frac{3}{2}$, −20)

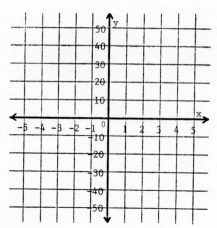

Answer to Frame 83:

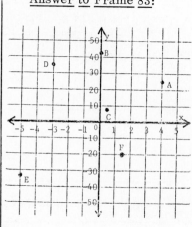

84.   Plot and label the following points on the coordinate system below.

A (8, 70)

B (3, 44)

C (−6, 35)

D (−7, 0)

E (−4, −95)

F (5, −37)

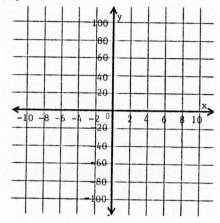

Answer to Frame 84:

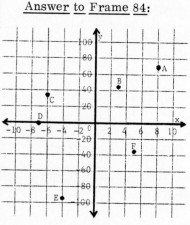

85.   Plot and label the following points on the coordinate system below.

A (35, 125)

B (0, 80)

C (−45, 212)

D (−25, −188)

E (−40, −227)

F (17, −95)

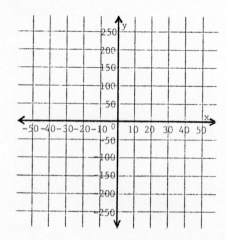

Answer to Frame 85:

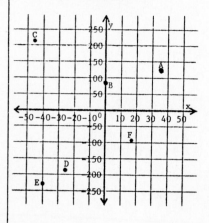

6-9   GRAPHING NON-LINEAR "xy" EQUATIONS

A non-linear "xy" equation is one whose graph is a curved line rather than a straight line. We will discuss the procedure for graphing non-linear "xy" equations in this section.

86.   The graph of a non-linear "xy" equation is a curved line rather than a straight line. The sketches of the graphs of three non-linear "xy" equations are shown below. Notice the different shapes of the three curves.

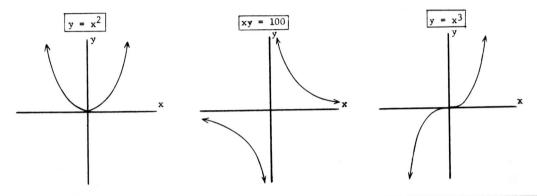

87.   Using the solution-table below, we graphed $y = x^2$. Its graph is a curved line called a "<u>parabola</u>".

$$\boxed{y = x^2}$$

	x	y			x	y
A	0	0		A	0	0
B	1	1		G	-1	1
C	2	4		H	-2	4
D	3	9		I	-3	9
E	4	16		J	-4	16
F	5	25		K	-5	25

When the graph of an equation is a curve, the graph still represents <u>all</u> the solutions of the equation. Therefore:

1.   The coordinates of any point <u>on the graph</u> <u>satisfy</u> the equation. For example, point D (3, 9) is <u>on the graph.</u> It coordinates satisfy the equation.

$$y = x^2$$
$$9 = 3^2$$
$$9 = 9$$

2.   The coordinates of any point <u>off the graph</u> <u>do not satisfy</u> the equation. For example, point P (4, 5) is <u>off the graph.</u> Show that its coordinates <u>do not</u> <u>satisfy</u> the equation.

$$y = x^2$$

88.   Using the solution-table below, we graphed $y = 2x^2$ two different ways. At <u>the left</u>, we drew the graph by connecting the points with straight lines. At <u>the right</u>, we drew the graph by connecting the points with a smooth curve.

$5 \neq 4^2$

$5 \neq 16$

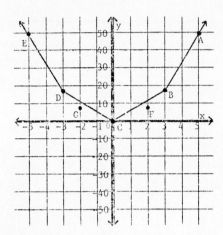

	$y = 2x^2$	
	x	y
A	5	50
B	3	18
C	0	0
D	-3	18
E	-5	50

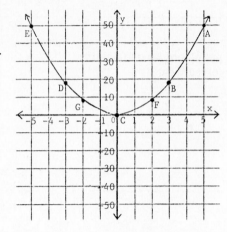

Two other points that satisfy the equation are F (2, 8) and G (-2, 8). They should lie on the graph. Therefore:

The <u>left</u> graph is <u>incorrect</u> because F and G <u>do</u> <u>not</u> <u>lie</u> on it.

The <u>right</u> graph is <u>correct</u> because F and G <u>do</u> <u>lie</u> on it.

> WHEN GRAPHING A NON-LINEAR EQUATION, BE SURE TO CONNECT THE PLOTTED POINTS WITH A SMOOTH CURVE.

---

89.   The following procedure can be used to graph a non-linear "xy" equation.

1. Plot enough points so that the outline of the graph is clear. Then draw a smooth curve through the plotted points.

2. Always substitute "0" for <u>x</u> to see whether the graph passes through the origin.

Using the solution-table below, graph $y = 4x^2$. <u>Be</u> <u>sure</u> <u>to</u> <u>draw</u> <u>a</u> <u>smooth</u> <u>curve</u> <u>through</u> <u>the</u> <u>plotted</u> <u>points</u>.

$y = 4x^2$			
x	y	x	y
0	0	0	0
1	4	-1	4
2	16	-2	16
3	36	-3	36
4	64	-4	64
5	100	-5	100

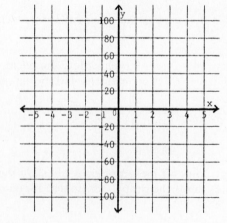

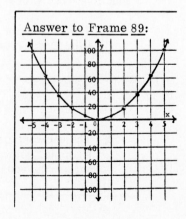

Answer to Frame 89:

90.   A solution-table and the graph for $y = x^2 - 2x$ are shown below. Its graph is also called a "parabola".

$y = x^2 - 2x$			
x	y	x	y
1	-1	$\frac{1}{2}$	$-\frac{3}{4}$
$1\frac{1}{2}$	$-\frac{3}{4}$	0	0
2	0	-1	3
3	3	-2	8
4	8		

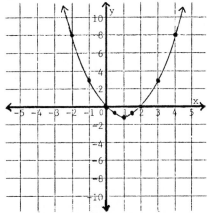

Complete the solution-table for $y = 4x - x^2$ below and then graph the equation.

x	y	x	y
-1		$2\frac{1}{2}$	
0		3	
1		4	
$1\frac{1}{2}$		5	
2			

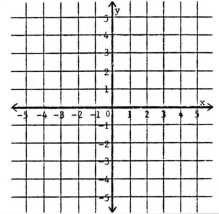

Answer to Frame 90:

x	y	x	y
-1	-5	$2\frac{1}{2}$	$3\frac{3}{4}$
0	0	3	3
1	3	4	0
$1\frac{1}{2}$	$3\frac{3}{4}$	5	-5
2	4		

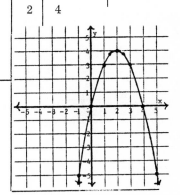

91.   Complete the solution-table for $y = x^2 - 2x - 3$ below and then graph the equation. Its graph is also a "parabola".

x	y	x	y
1		$\frac{1}{2}$	
$1\frac{1}{2}$		0	
2		-1	
3		-2	
4			

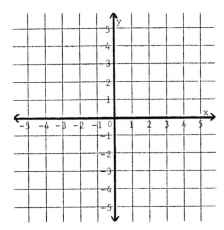

Answer to Frame 91:

x	y	x	y
1	-4	$\frac{1}{2}$	$-3\frac{3}{4}$
$1\frac{1}{2}$	$-3\frac{3}{4}$	0	-3
2	-3	-1	0
3	0	-2	5
4	5		

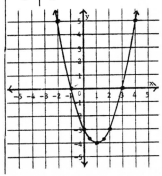

92.   We used the solution-table below to graph xy = 24. The graph is called a "hyperbola". Notice these points about it:

1. When $\underline{x} = 0$, $y = \frac{24}{0}$ which is an impossible or undefined operation. Therefore, there is no solution for x = 0 and no corresponding point on the graph.

2. The graph has two parts, one in Quadrant 1 and one in Quadrant 3. The two parts are not connected.

3. Though the curves approach the axes, they do not touch the axes.

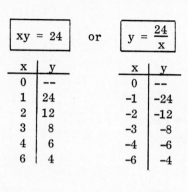

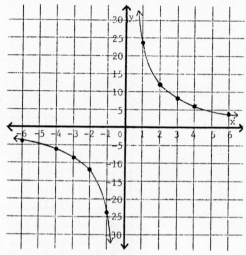

Complete the solution-table for xy = 48 and then graph the equation. Its graph is also a "hyperbola".

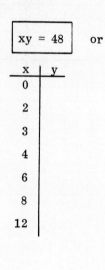

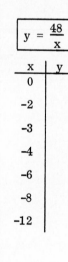

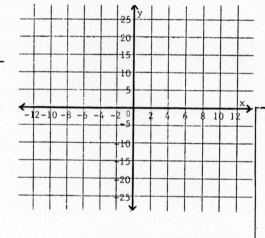

Answer to Frame 92:

x	y	x	y
0	--	0	--
2	24	-2	-24
3	16	-3	-16
4	12	-4	-12
6	8	-6	-8
8	6	-8	-6
12	4	-12	-4

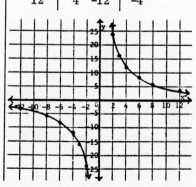

SELF-TEST 20 (pages 240-257)

Graph each equation.

1. $\boxed{x - y = 2}$ 2. $\boxed{y = x^2 - 4x}$

x	y
-2	
0	
2	
4	

x	y
-1	
0	
1	
2	
3	
4	
5	

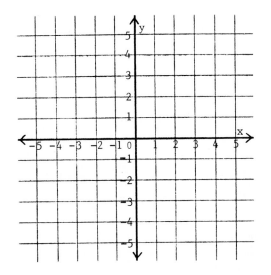

3. Graph: $\boxed{xy = 60}$

x	y	x	y
0			
1			
2			
3			
4			
6			
10			

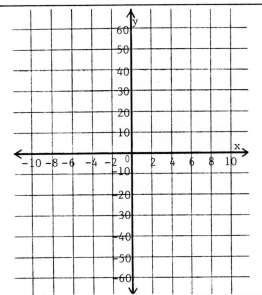

ANSWERS:

1-2.

x	y	x	y
-2	-4	-1	5
0	-2	0	0
2	0	1	-3
4	2	2	-4
		3	-3
		4	0
		5	5

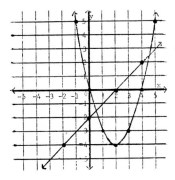

3.

x	y	x	y
0	---	0	---
1	60	-1	-60
2	30	-2	-30
3	20	-3	-20
4	15	-4	-15
6	10	-6	-10
10	6	-10	-6

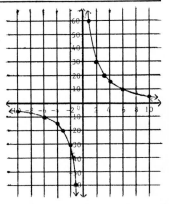

6-10   GRAPHING TWO-VARIABLE FORMULAS

In this section, we will discuss what is meant by "variables" in a formula.   Then we will discuss the procedure for graphing formulas that contain only two variables.

---

93.   In the equation: $y = 3x + 5$

"x" and "y" are called <u>variables</u> because we can satisfy the equation by substituting <u>various</u> pairs of values for them.

"3" and "5" are called <u>constants</u> because they do not change for the various pairs of values for "x" and "y".

a) In $xy = 50$, "x" and "y" are called _____ .

b) In $y = 4x^2 + 1$, "4" and "1" are called _____ .

---

94.   Formulas also contain variables and constants.   For example:

In $F = \frac{9}{5}C + 32$:   "F" and "C" are <u>variables</u>.

$\frac{9}{5}$ and 32 are <u>constants</u>.

a) In $PV = 200$, the variables are _____ and _____ .

b) In $s = 16t^2$, the constant is _____ .

a) variables

b) constants

---

95.   When graphing an "xy" equation:

1) We <u>always</u> put the x-scale on the horizontal axis.

2) We <u>always</u> put the y-scale on the vertical axis.

a) P and V

b) 16

When graphing a two-variable formula, there is no rigid rule about which variable goes on which axis.   The choice is somewhat arbitrary.   However, there are accepted conventions with certain formulas.   Those conventions have to be learned for the individual formulas.

With the formula $E = 100I$:

a) If you are told to "graph I as <u>x</u>", you should put the I-scale on the

_____ (horizontal/vertical) axis.

b) If you are told to "graph I as <u>y</u>", you should put the I-scale on the

_____ (horizontal/vertical) axis.

---

96.   The <u>abscissa</u> is another name for the x-coordinate.   The <u>ordinate</u> is another name for the y-coordinate.

With the formula $F = \frac{9}{5}C + 32$:

a) If you are told to "plot C as the abscissa", you should put the C-scale

on the _____ (horizontal/vertical) axis.

b) If you are told to "plot C as the ordinate", you should put the C-scale

on the _____ (horizontal/vertical) axis.

a) horizontal

b) vertical

97.   The formula below shows the relationship between degrees-Celsius and
       degrees-Fahrenheit. Using the solution-table, graph the formula. Notice
       that we plotted C as <u>x</u> or as the abscissa.

a) horizontal

b) vertical

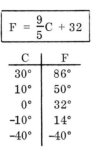

$$F = \frac{9}{5}C + 32$$

C	F
30°	86°
10°	50°
0°	32°
-10°	14°
-40°	-40°

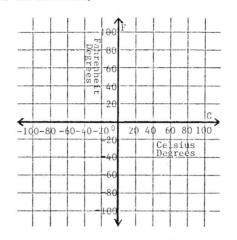

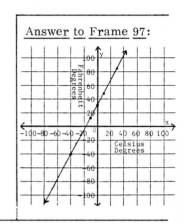

Answer to Frame 97:

98.   Though there are negative values for both degrees-Fahrenheit and degrees-Celsius, negative values
       are not generally used for most variables in formulas. For example, we generally do not use
       negative values for distance, time, pressure, power, and so on. Therefore, when graphing most
       formulas, only positive values appear on the graph. <u>Since only positive values are plotted, the
       graphs of most formulas appear only in quadrant 1.</u> An example is discussed below.

       The formula below shows the relationship between distance and time for an object traveling
       at a constant velocity of 100 kilometers per hour.

       $$d = 100t$$ where "d" is distance (in kilometers)
                        and "t" is time (in hours)

       Using the solution-table below, we graphed the formula.

t	d
0	0
1	100
2	200
3	300
4	400
5	500

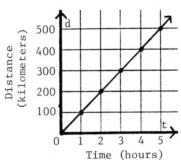

       Notice these two points about the graph:

       1) Only quadrant 1 was used.

       2) We plotted "t" as <u>x</u> or as the abscissa.

99.   The formula below is related to the concept of "load line" in transistor electronics.

$$E + 8I = 24$$

where E is voltage (in volts)
and I is current (in amperes)

Complete the solution-table below and graph the formula with E plotted as <u>x</u>.   Its graph is
a straight line.

E	I
0	
8	
16	
24	

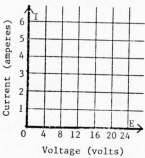

Voltage (volts)

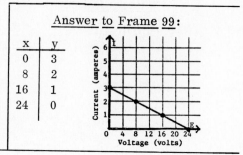

Answer to Frame 99:

x	y
0	3
8	2
16	1
24	0

Voltage (volts)

100.   The formula below shows the distance a dropped object will fall from rest in a given period
of time.

$$s = 5t^2$$

where "s" is distance (in meters)

and "t" is time (in seconds)

Complete the solution-table below and graph the formula.   Its graph is
half a parabola because only quadrant 1 is used.

t	s
0	
1	
2	
3	
4	
5	

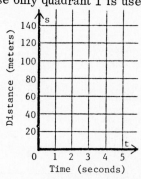

Time (seconds)

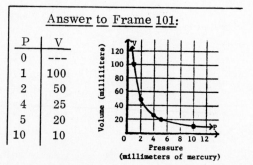

Answer to Frame 100:

t	s
0	0
1	5
2	20
3	45
4	80
5	125

Time (seconds)

101.   The formula below is a specific instance of Boyle's Law which gives the relationship between
the pressure and volume of a gas.

$$PV = 100$$

where P is pressure (in millimeters of mercury)
and V is volume (in milliliters)

Complete the solution-table below and graph the formula.   Its graph is
half a hyperbola because only quadrant 1 is used.

P	V
0	---
1	
2	
4	
5	
10	

Pressure
(millimeters of mercury)

Answer to Frame 101:

P	V
0	---
1	100
2	50
4	25
5	20
10	10

Pressure
(millimeters of mercury)

6-11   READING THE GRAPHS OF FORMULAS

In this section, we will discuss the method for reading the graphs of formulas.   Then we will show how graphs can be used to get approximate answers to applied problems.

---

102.   The graph at the right represents the distance a car will travel in a given period of time at a constant velocity of 75 kilometers per hour.   To read most points on the graph, we have to estimate values on the axes.   For example:

Point A represents this fact: The car travels approximately 110 kilometers in approximately 1.5 hours.

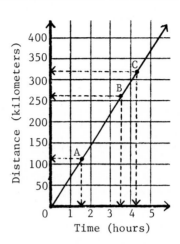

a)   Point B represents this fact: The car travels approximately _____ kilometers in approximately _____ hours.

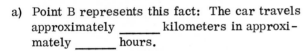

b)   Point C represents this fact: The car travels approximately _____ kilometers in approximately _____ hours.

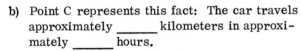

---

103.   The graph at the right shows the relationship between degrees-Celsius and degrees-Fahrenheit.

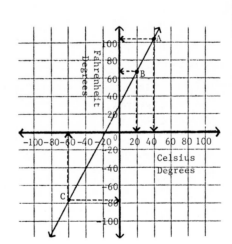

a)   Point A represents this fact: A Celsius temperature of 40° is approximately equal to a Fahrenheit temperature of _____°.

b)   Point B represents this fact: A Celsius temperature of 20° is approximately equal to a Fahrenheit temperature of _____°.

c)   Point C represents this fact: A Celsius temperature of –60° is approximately equal to a Fahrenheit temperature of _____°.

a) 260 km ... 3.5 hours

b) 320 km ... 4.3 hours

---

a) 105°

b) 68°

c) –75°

104. The graph at the right represents the distance an object will fall from rest in a given period of time. We can use the graph to solve the following problem:

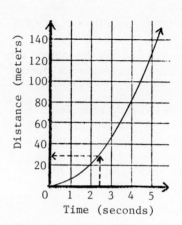

Approximately how far will an object fall in 2.5 seconds?

To solve the problem, we use these steps:

1) Draw a vertical arrow up from 2.5 on the time-axis to the graphed line.

2) Then draw a horizontal line from that point of intersection to the distance-axis.

3) Since the horizontal arrow points to approximately 30 on the distance-axis, the object drops approximately 30 meters in 2.5 seconds.

Using the same method, complete these:

a) Approximately how far will an object fall in 4.5 seconds?   _____ meters

b) Approximately how far will an object fall in 1.5 seconds?   _____ meters

105. The graph below is a specific example of Hooke's Law. That is, it shows how far a spring will stretch when various amounts of force are applied to it. Use it to complete these:

a) If a force of 150 grams is applied, the spring will stretch approximately _____ centimeters.

b) If a force of 75 grams is applied, the spring will stretch approximately _____ centimeters.

c) If a force of 225 grams is applied, the spring will stretch approximately _____ centimeters.

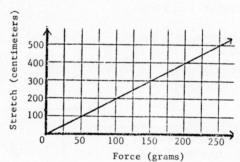

a) 100 meters

b) 10 meters

a) 300 centimeters

b) 150 centimeters

c) 450 centimeters

106. The graph at the right shows the relationship between degrees-Fahrenheit and degrees-Celsius, with degrees-Fahrenheit plotted on the x-axis. Using the arrows, find the Celsius temperature that is approximately equal to these Fahrenheit temperatures.

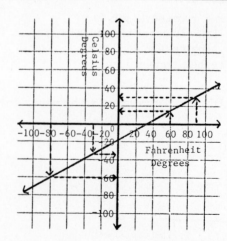

a) 60° Fahrenheit

_____ Celsius

b) 90° Fahrenheit

_____ Celsius

c) -30° Fahrenheit

_____ Celsius

d) -80° Fahrenheit

_____ Celsius

107. The graph shown below is a specific example of Ohm's Law. That is, it shows the relationship between voltage and current in an electric circuit with a constant resistance. We can use the graph to solve the following problem.

a) 15°
b) 30°
c) -35°
d) -60°

Approximately how much voltage is needed to get a current of 10 amperes?

To solve the problem, we use these steps:

1) Draw a horizontal arrow from 10 on the current-axis to the graphed line.

2) Then draw a vertical arrow from that point of intersection to the voltage-axis.

3) Since the vertical arrow points to 50 on the voltage-axis, we need approximately 50 volts to get a current of 10 amperes.

Using the same steps, complete these:

a) Approximately how much voltage is needed to get a current of 4 amperes? _____ volts

b) Approximately how much voltage is needed to get a current of 13 amperes? _____ volts

108. The graph below shows the relationship between the pressure and volume of a gas at a constant temperature.

a) 20 volts
b) 65 volts

Use the arrows to get approximate answers for these:

a) When the volume is 100 milliliters, the pressure is _____ milliliters of mercury.

b) When the volume is 50 milliliters, the pressure is _____ millimeters of mercury.

c) When the volume is 25 milliliters, the pressure is _____ millimeters of mercury.

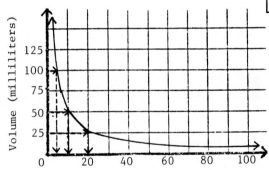

109. A Fahrenheit-Celsius graph is shown at the right. Using the arrows, find the approximate Fahrenheit temperature that corresponds to each of the following Celsius temperatures.

a) 5
b) 10
c) 20

a) 25° Celsius

_____ Fahrenheit

b) -30° Celsius

_____ Fahrenheit

c) -75° Celsius

_____ Fahrenheit

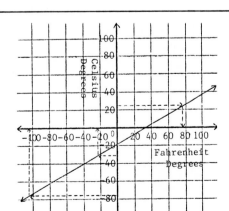

110. The graph below shows how far two different springs will stretch when various amounts of force are applied to them.

   a) If a force of 300 grams is applied, how far is each spring stretched?

       Spring #1 _____

       Spring #2 _____

   b) To stretch each spring 100 centimeters, how much force must be applied to it?

       Spring #1 _____

       Spring #2 _____

   c) Which spring stretches more easily? _____

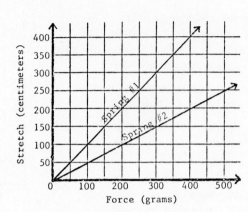

a) 77° (76° to 78°)

b) −22° (−21° to −23°)

c) −103° (−102° to −104°)

111. The graph below relates voltage and current for two electrical circuits with different resistances.

   a) If a voltage of 50 volts is applied, how much current do we get in each circuit?

       Circuit #1 _____

       Circuit #2 _____

   b) To get a current of 6 amperes, how much voltage must be applied to each circuit?

       Circuit #1 _____

       Circuit #2 _____

   c) The more resistance there is in a circuit, the less current we get for the amount of voltage applied. Which circuit has more resistance? _____

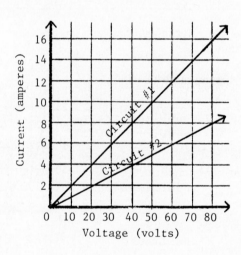

a) #1 300 centimeters
   #2 150 centimeters

b) #1 100 grams
   #2 200 grams

c) Spring #1

112. When a graph is based on a formula, we obtain only an <u>approximate</u> solution to a problem by reading the graph. To get an <u>exact</u> solution, we must substitute in the formula and evaluate. An example is discussed on the next page.

a) #1 10 amperes
   #2 5 amperes

b) #1 30 volts
   #2 60 volts

c) Circuit #2

Continued on following page.

112.   Continued

The graph below shows the relationship between distance and time when traveling at a constant velocity of 120 kilometers per hour.   The graph is based on the formula: $\boxed{d = 120t}$

Use the graph to get <u>approximate</u> answers:

    a)   In 3.4 hours, the distance traveled
        is approximately _____ kilometers.

    b)   The object travels 534 kilometers in
        approximately _____ hours.

Now substitute in $d = 120t$ to get <u>exact</u> answers:

    c)   In 3.4 hours, the distance traveled
        is exactly _____ kilometers.

    d)   The object travels 534 kilometers in
        exactly _____ hours.

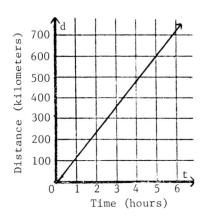

---

113.   The formula for the distance an object falls from rest in a given amount of time is:

    $\boxed{s = 5t^2}$ where "s" is distance (in meters)
                      and "t" is time (in seconds)

    a)   410 kilometers
    b)   4.5 hours
    c)   408 kilometers
    d)   4.45 hours

The graph of the formula is shown.

Use the graph to get <u>approximate</u> answers:

    a)   In 2.5 seconds, an object falls
        approximately _____ meters.

    b)   To fall 90 meters, it takes
        approximately _____ seconds.

Now substitute in $s = 5t^2$ to get <u>exact</u> answers:

    c)   In 2.5 seconds, an object falls
        exactly _____ meters.

    d)   To fall 90 meters, it takes
        exactly _____ seconds.
        (Round to hundredths.)

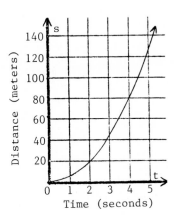

---

    a)   30 meters
    b)   4.2 seconds
    c)   31.25 meters
    d)   4.24 seconds

6-12   GRAPHING THREE-VARIABLE FORMULAS

Most formulas contain more than two variables. In this section, we will show a method for graphing formulas that contain <u>three</u> variables.

114.   In formulas, almost all letters are <u>variables</u> and all numbers are <u>constants</u>. For example:

In   P = EI: the <u>variables</u> are P, E, and I.
                   there are no <u>constants</u>.

In   A = $\frac{1}{2}$bh: a) the variables are _____

                   b) the constant is _____

115.   An occasional letter in a formula is really a constant. For example:

In   C = πd, π stands for the number 3.1416... Therefore,
               π is a constant.

   Note: In 3.1416..., the three periods mean that more decimal
         places could be added indefinitely.

In   s = $\frac{1}{2}$gt^2, "g" is acceleration due to gravity. At sea level:

   g = 32.17... if distance (s) is measured in feet.

   g = 9.80... if distance (s) is measured in meters.

What do the three periods after 32.17... and 9.80... mean? _____

a) A, b, h

b) $\frac{1}{2}$

116.   The formula below is called "Ohm's Law". It shows the relationship between voltage (E), current (I), and resistance (R) in an electric circuit.

                   E = IR

Since the coordinate system contains <u>only two</u> axes, we can only graph the relationship between <u>two</u> variables at one time. For example, if we put the E-scale on one axis and the I-scale on the other axis, there is no axis left for the R-scale.

However, we can graph the formula above <u>if we set one of the three variables equal to a constant</u>. For example, if we set R equal to 70 ohms, we get:

       E = I(70)   or   E = 70I

We can graph E = 70I because that formula contains only _____ variables.

That more decimal places could be added indefinitely.

two

117.   When setting R equal to a constant to graph E = IR, we can set it equal to various constants. For example:

If we set R equal to 30 ohms, we get: E = 30I

If we set R equal to 60 ohms, we get: E = 60I

By setting R equal to 30 ohms, 60 ohms, and 150 ohms, we get the following <u>family</u> <u>of</u> <u>formulas</u>.

E = 30I   E = 60I   E = 150I

Each formula in the family can be plotted on the same graph.   To do so, we must make up a solution-table for each.

E = 30I		E = 60I		E = 150I	
E	I	E	I	E	I
0	0	0	0	0	0
30	1	60	1	150	1
60	2	120	2	300	2
150	5	300	5	450	3

Using the above values, we graphed each formula below.   Notice that we labeled each graphed line with its R-value.

Approximately how much voltage is needed to get 2 amperes of current in each circuit?

R = 30   _____ volts

R = 60   _____ volts

R = 150   _____ volts

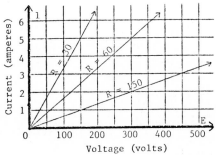

118.   There are three variables in d = vt

where "d" is distance (in meters)
    "v" is velocity (in meters per second)
    "t" is time (in seconds)

R = 30	<u>60</u> volts
R = 60	<u>120</u> volts
R = 150	<u>300</u> volts

To graph the relationship between "v" and "t", we must set "d" equal to a constant or constants. If we set "d" equal to 80 meters, 160 meters, and 320 meters, we get the following <u>family</u> <u>of</u> <u>formulas</u>.

vt = 80   vt = 160   vt = 320

When the three formulas are graphed with "t" plotted as the abscissa, we get the <u>family</u> <u>of</u> <u>curves</u> shown on the graph at the top of the next page.

Continued on following page.

118.   Continued

a) Approximately what velocity is needed to travel 80 meters in 6 seconds? _____

b) Approximately what velocity is needed to travel 160 meters in 11 seconds? _____

c) Approximately what velocity is needed to travel 320 meters in 7 seconds? _____

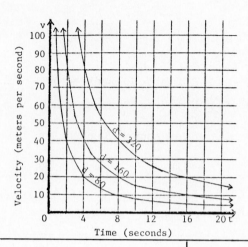

---

119.   There are three variables in $\boxed{E = \frac{1}{2}mv^2}$

where "E" is kinetic energy (in ergs)

"m" is mass (in grams)

"v" is velocity (in centimeters per second)

a) 13 meters per second

b) 15 meters per second

c) 45 meters per second

To graph the relationship between E and "v", we must set "m" equal to a constant or constants. If we set "m" equal to 100 grams, 300 grams, and 500 grams, we get the following family of formulas.

$$E = 50v^2 \qquad E = 150v^2 \qquad E = 250v^2$$

The corresponding family of curves is given on the graph below.

a) With a mass of 100 grams, approximately how much kinetic energy is produced with a velocity of 4.5 centimeters per second? _____

b) With a mass of 300 grams, approximately how much kinetic energy is produced with a velocity of 3.5 centimeters per second? _____

c) With a mass of 500 grams, what velocity is required to produce a kinetic energy of 2,000 ergs? _____

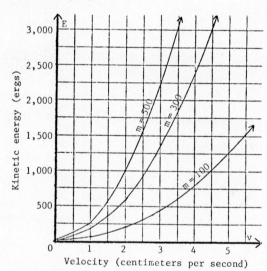

---

a) 1,000 ergs

b) 1,800 ergs

c) 2.8 centimeters per second

SELF-TEST 21 (pages 258-269)

1. For the electrical power formula $\boxed{EI = 12}$,
   complete the solution-table and construct the
   graph.

E	I
0	
1	
2	
4	
6	
8	
16	
24	

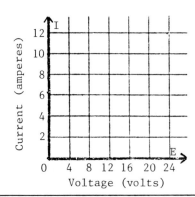

2. For a ball projected vertically upward, the
   graph shows the relation between the ball's
   height and the elapsed time.

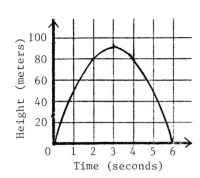

a) After 2 seconds, the height is
   _____ meters.

b) The greatest height is
   _____ meters.

c) The height is 50 meters for
   _____ second and _____ seconds.

d) The ball returns to the ground
   after _____ seconds.

3. The relation between applied force and stretch of a
   spring is given by the formula $\boxed{s = kF}$. For
   spring #1, k = 100; for spring #2, k = 50; for
   spring #3, k = 20. The resulting formulas
   $\boxed{s = 100F}$, $\boxed{s = 50F}$, and $\boxed{s = 20F}$ have
   been graphed at the right.

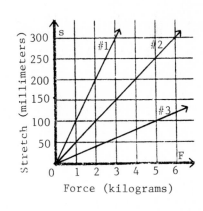

a) What stretch is produced in each spring by
   a force of $2\frac{1}{2}$ kilograms?

   #1: _____ mm   #2: _____ mm   #3: _____ mm

b) What force must be applied to stretch each
   spring 100 millimeters?

   #1: _____ kg   #2: _____ kg   #3: _____ kg

ANSWERS: 1.

E	I
0	--
1	12
2	6
4	3
6	2
8	$\frac{3}{2}$
16	$\frac{3}{4}$
24	$\frac{1}{2}$

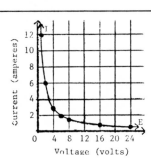

2. a) 80 meters

   b) 90 meters

   c) 1 second
      and 5 seconds

   d) 6 seconds

3. a) #1: 250 mm
      #2: 125 mm
      #3: 50 mm

   b) #1: 1 kg
      #2: 2 kg
      #3: 5 kg

SUPPLEMENTARY PROBLEMS - CHAPTER 6

<u>Assignment</u> <u>19</u>

Complete the solution-table for each equation.

1. | y = 12x | x | y |
   |---|---|---|
   | | 5 | |
   | | 1 | |
   | | 0 | |
   | | -2 | |
   | | -4 | |

2. | x - y = 0 | x | y |
   |---|---|---|
   | | 8 | |
   | | 3 | |
   | | 0 | |
   | | -1 | |
   | | -5 | |

3. | 2x + y = 4 | x | y |
   |---|---|---|
   | | 6 | |
   | | 2 | |
   | | 0 | |
   | | -3 | |
   | | -7 | |

4. | 4x - y = 10 | x | y |
   |---|---|---|
   | | 4 | |
   | | 2 | |
   | | 0 | |
   | | -2 | |
   | | -4 | |

5. | 5x + 2y = 20 | x | y |
   |---|---|---|
   | | 8 | |
   | | 4 | |
   | | 0 | |
   | | -2 | |
   | | -6 | |

6. | 5y - 3x = 15 | x | y |
   |---|---|---|
   | | 10 | |
   | | 5 | |
   | | 0 | |
   | | -5 | |
   | | -10 | |

7. Write the coordinates of each point.

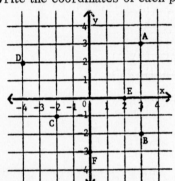

A ____
B ____
C ____
D ____
E ____
F ____

8. Write the coordinates of each point.

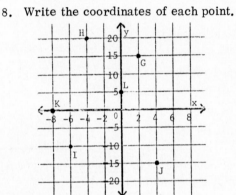

G ____
H ____
I ____
J ____
K ____
L ____

9. State the number of the quadrant in which each point lies.

    a) (-8, 1)   b) (20, -5)   c) (2, 13)   d) (-1, -3)   e) (-32, 50)

10. Which of the following points lie on the x-axis?

    a) (0, -7)   b) (-7, 0)   c) (0, 0)   d) (30, 0)   e) (0, 1)

11. Which of the following points lie on the y-axis?

    a) (0, 4)   b) (1, -1)   c) (-3, 0)   d) (0, 0)   e) (0, -50)

12. What name is given to the point (0, 0)?

13. The abscissa of (9, -4) is _____,   14. The ordinate of (-1, -10) is _____.

15. Write the coordinates of the point whose ordinate is -9 and whose abscissa is 17.

Assignment 20

Complete each solution-table and graph each equation.

1. | y = x |

x	y
-5	
-2	
0	
1	
4	

2. | y = x² - 4 |

x	y	x	y
$\frac{1}{2}$		$-\frac{1}{2}$	
1		-1	
2		-2	
3		-3	
0			

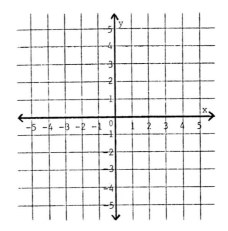

Complete each solution-table and graph each equation.

3. | 2x + y = 4 |

x	y
-3	
-1	
0	
1	
2	
5	

4. | y = 3x - x² |

x	y	x	y
-2		2	
-1		3	
0		4	
1		5	
$1\frac{1}{2}$			

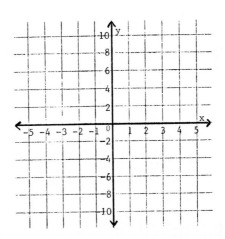

Complete each solution-table and graph each equation.

5. | 5y = 2x - 50 |

x	y
-25	
-15	
0	
10	
25	

6. | xy = 240 |

x	y	x	y
5		-5	
6		-6	
10		-10	
15		-15	
20		-20	
24		-24	
0		0	

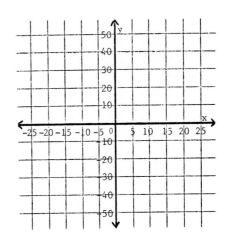

Assignment 21

Complete the solution-table and graph each formula.

1. $\boxed{w = 10p^2}$

p	w
0	
1	
2	
3	
4	
5	

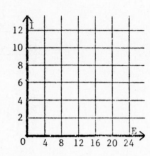

2. $\boxed{E + 2I = 20}$

E	I
0	
4	
8	
12	
16	
20	

3. The graph below shows the relation between the length and width of a rectangle whose area is constant.

a) If the length is 40 centimeters, the width is approximately _____ centimeters.

b) If the length is 15 centimeters, the width is approximately _____ centimeters.

c) If the width is 10 centimeters, the length is approximately _____ centimeters.

d) If the width is 5 centimeters, the length is approximately _____ centimeters.

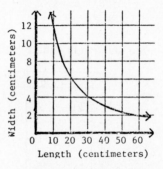

4. The graph below shows the relation between the diameter and the area of a circle.

a) If the diameter is 10 millimeters, the area is approximately _____ square millimeters.

b) If the diameter is 25 millimeters, the area is approximately _____ square millimeters.

c) If the area is 100 square millimeters, the diameter is approximately _____ millimeters.

d) If the area is 250 square millimeters, the diameter is approximately _____ millimeters.

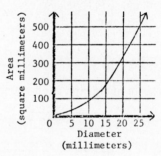

5. A car moving at constant "v" meters per second will travel "d" meters in "t" seconds according to the formula $\boxed{d = vt}$. For car #1, v = 30; for car #2, v = 20; for car #3, v = 10. The resulting formulas $\boxed{d = 30t}$, $\boxed{d = 20t}$, and $\boxed{d = 10t}$ are graphed below.

a) How far does each car travel in 6 seconds?

#1: _____ m   #2: _____ m   #3: _____ m

b) How long does it take each car to travel 120 meters?

#1: _____ sec   #2: _____ sec   #3: _____ sec

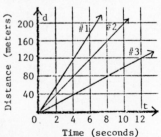

Chapter 7 ALGEBRAIC FRACTIONS

An algebraic fraction is a fraction that contains one or more variables in its numerator or denominator or both. In this chapter, we will discuss the four basic operations with algebraic fractions. We will also discuss equivalent forms for some algebraic fractions. A knowledge of the content of this chapter is especially useful for formula rearrangement and formula derivation, topics that are discussed in the following chapters.

7-1 MULTIPLICATIONS INVOLVING ALGEBRAIC FRACTIONS

In this section, we will discuss the procedure for multiplications involving algebraic fractions.

1. When a multiplication contains a numerical factor and a literal factor, we <u>always</u> write the numerical factor first. For example:

     Instead of "x5", we write "5x".

   When a multiplication contains more than one literal factor, <u>we usually write the literal factors in alphabetical order</u>. A numerical factor is <u>always</u> written first. For example:

     Instead of "5ta", we write "5at".

     Instead of "VP^{2}R", we write "P^{2}RV".

   Perform each multiplication. Write the factors of each product in the usual order.

   a) $(2m)(4b^2) = $ _____

   b) $(cS)(dT_1) = $ _____

2. As we saw earlier, we multiply fractions containing a single variable (or letter) by multiplying their numerators and denominators. For example:

   $$\left(\frac{2}{3}\right)\left(\frac{x}{7}\right) = \frac{2x}{21} \qquad \left(\frac{1}{5y}\right)\left(\frac{9}{2}\right) = \frac{9}{10y}$$

   The same procedure is used to multiply fractions containing more than one variable. For example:

   $$\left(\frac{x}{2}\right)\left(\frac{y}{3}\right) = \frac{xy}{6} \qquad \left(\frac{aF_1}{d}\right)\left(\frac{c^2}{bM}\right) = \frac{ac^2F_1}{bdM}$$

   Continued on following page.

---

a) $8b^2m$

b) $cdST_1$

2.   Continued

Complete these. <u>Write the literal factors in alphabetical order.</u>

a) $\left(\dfrac{3v}{t^2}\right)\left(\dfrac{k}{cy}\right) = $ _____

b) $\left(\dfrac{4V_1}{T}\right)\left(\dfrac{P_1}{7R}\right) = $ _____

---

3.   When one factor in a multiplication is a binomial, we usually write the binomial last. For example:

Instead of $(p + q)d,$ we write $d(p + q).$

Instead of $5(S + T)Q,$ we write $5Q(S + T).$

In the multiplication below, one numerator is a binomial.

$$\left(\frac{y + 9}{a}\right)\left(\frac{x}{t^2}\right) = \frac{x(y + 9)}{at^2}$$

Following the example, complete these:

a) $\left(\dfrac{1}{c}\right)\left(\dfrac{F}{R + T}\right) = $ _____

b) $\left[\dfrac{a(x + y)}{b}\right]\left(\dfrac{t}{v}\right) = $ _____

a) $\dfrac{3kv}{ct^2y}$

b) $\dfrac{4P_1V_1}{7RT}$

---

4.   When any expression is multiplied by "1", the product is <u>identical</u> to the original expression. For example:

$(1)(xy) = xy$ $\qquad\qquad (a + b)(1) = a + b$

Following the examples, complete these:

a) $(ct^2)(1) = $ _____

b) $(1)[b(c - d)] = $ _____

a) $\dfrac{F}{c(R + T)}$

b) $\dfrac{at(x + y)}{bv}$

---

5.   Any non-fractional expression containing a single variable can be converted to a fraction whose denominator is "1". For example:

$$x^2 = \frac{x^2}{1} \qquad\qquad 3a = \frac{3a}{1}$$

Using the facts above, we converted each multiplication below to a multiplication of two fractions.

$$\frac{b}{5}(x^2) = \frac{b}{5}\left(\frac{x^2}{1}\right) = \frac{bx^2}{5} \qquad\qquad 3a\left(\frac{c}{d}\right) = \left(\frac{3a}{1}\right)\left(\frac{c}{d}\right) = \frac{3ac}{d}$$

However, we ordinarily use the shorter method below for such multiplications. That is, we multiply the numerator and the non-fractional expression.

$$\frac{b}{5}(x^2) = \frac{b(x^2)}{5} = \frac{bx^2}{5} \qquad\qquad 3a\left(\frac{c}{d}\right) = \frac{3a(c)}{d} = $$ _____

a) ct^2

b) $b(c - d)$

$\dfrac{3ac}{d}$

6.   Any non-fractional expression containing more than one variable can also be converted to a fraction whose denominator is "1".   For example:

$$x^2y = \frac{x^2y}{1} \qquad\qquad M + T = \frac{M + T}{1}$$

Using the fact above, we converted each multiplication below to a multiplication of two fractions.

$$ab\left(\frac{5}{t}\right) = \left(\frac{ab}{1}\right)\left(\frac{5}{t}\right) = \frac{5ab}{t} \qquad\qquad \frac{1}{7}(c^2d) = \frac{1}{7}\left(\frac{c^2d}{1}\right) = \frac{c^2d}{7}$$

However, we ordinarily use the shorter method below for the same multiplications.

$$ab\left(\frac{5}{t}\right) = \frac{(ab)(5)}{t} = \frac{5ab}{t} \qquad\qquad \frac{1}{7}(c^2d) = \frac{(1)(c^2d)}{7} = \underline{\hspace{2cm}}$$

---

7.   Complete: a) $x^2\left(\dfrac{3}{b}\right) =$ \underline{\hspace{2cm}}   c) $\dfrac{m}{t}(s^2v) =$ \underline{\hspace{2cm}}

   b) $pq\left(\dfrac{1}{r}\right) =$ \underline{\hspace{2cm}}   d) $\dfrac{1}{2}(bh) =$ \underline{\hspace{2cm}}

(answer box: $\dfrac{c^2d}{7}$)

---

8.   We performed the multiplication below in two different ways.

   1) Converting the non-fractional expression to a fraction.

$$\left(\frac{1}{a+b}\right)(x+y) = \left(\frac{1}{a+b}\right)\left(\frac{x+y}{1}\right) = \frac{x+y}{a+b}$$

   2) Multiplying the numerator and the non-fractional expression.

$$\left(\frac{1}{a+b}\right)(x+y) = \frac{(1)(x+y)}{a+b} = \frac{x+y}{a+b}$$

Following the example, complete these:

   a) $\dfrac{a}{mt}(p-q) =$ \underline{\hspace{2cm}}   b) $\dfrac{1}{AK(t_2 - t_1)}(LW) =$ \underline{\hspace{2cm}}

(answer box:
a) $\dfrac{3x^2}{b}$ c) $\dfrac{ms^2v}{t}$

b) $\dfrac{pq}{r}$ d) $\dfrac{bh}{2}$
)

---

9.   When any algebraic fraction is multiplied by "1", the product is <u>identical</u> to the original fraction.   That is:

$$1\left(\frac{c}{d}\right) = \left(\frac{1}{1}\right)\left(\frac{c}{d}\right) = \frac{c}{d} \qquad\qquad \frac{ST}{P}(1) = \left(\frac{ST}{P}\right)\left(\frac{1}{1}\right) = \frac{ST}{P}$$

Using the fact above, complete these:

   a) $1\left(\dfrac{d^2}{v}\right) =$ \underline{\hspace{1.5cm}}   b) $1\left(\dfrac{m+t}{R}\right) =$ \underline{\hspace{1.5cm}}   c) $\dfrac{pq}{ad}(1) =$ \underline{\hspace{1.5cm}}

(answer box:
a) $\dfrac{a(p-q)}{mt}$

b) $\dfrac{LW}{AK(t_2 - t_1)}$
)

(answer box bottom:
a) $\dfrac{d^2}{v}$ b) $\dfrac{m+t}{R}$ c) $\dfrac{pq}{ad}$
)

7-2   REDUCING ALGEBRAIC FRACTIONS TO LOWEST TERMS

In this section, we will show how an algebraic fraction can be reduced to lowest terms whenever we can factor out a fraction that equals "1".

---

10.   When any expression is divided by itself, the quotient is "1".   For example:

$$\frac{x^2}{x^2} = 1 \qquad\qquad \frac{3ab}{3ab} = 1 \qquad\qquad \frac{p^2 - q}{p^2 - q} = 1$$

Using the fact above, complete these:

a) $\dfrac{S(T + V)}{S(T + V)} = $ _____

b) $\dfrac{a - b^2 + c}{a - b^2 + c} = $ _____

---

11.   To reduce an algebraic fraction to lowest terms, we factor out a fraction that equals "1".   For example:

$$\frac{3x}{7x} = \left(\frac{3}{7}\right)\left(\frac{x}{x}\right) = \left(\frac{3}{7}\right)(1) = \frac{3}{7}$$

$$\frac{bd}{cd} = \left(\frac{b}{c}\right)\left(\frac{d}{d}\right) = \left(\frac{b}{c}\right)(\quad) = \text{_____}$$

a) 1   b) 1

---

12.   Following the example, complete the other reduction to lowest terms.

$$\frac{p^2 q}{p^2 t} = \left(\frac{p^2}{p^2}\right)\left(\frac{q}{t}\right) = (1)\left(\frac{q}{t}\right) = \frac{q}{t}$$

$$\frac{aSV}{aST} = \left(\frac{aS}{aS}\right)\left(\frac{V}{T}\right) = (\quad)\left(\frac{V}{T}\right) = \text{_____}$$

$\left(\dfrac{b}{c}\right)(1) = \dfrac{b}{c}$

---

13.   Following the example, complete the other reduction to lowest terms.

$$\frac{3(x + 5)}{7(x + 5)} = \left(\frac{3}{7}\right)\left(\frac{x + 5}{x + 5}\right) = \left(\frac{3}{7}\right)(1) = \frac{3}{7}$$

$$\frac{2a(b - c)}{3d(b - c)} = \left(\frac{2a}{3d}\right)\left(\frac{b - c}{b - c}\right) = \left(\frac{2a}{3d}\right)(\quad) = \text{_____}$$

$(1)\left(\dfrac{V}{T}\right) = \dfrac{V}{T}$

---

14.   If one of the factors is a numerical fraction, it should also be reduced to lowest terms.   For example:

$$\frac{8cd}{12cd} = \left(\frac{8}{12}\right)\left(\frac{cd}{cd}\right) = \left(\frac{2}{3}\right)(1) = \frac{2}{3}$$

$$\frac{9PV}{3PV} = \left(\frac{9}{3}\right)\left(\frac{PV}{PV}\right) = (\quad)(\quad) = \text{_____}$$

$\left(\dfrac{2a}{3d}\right)(1) = \dfrac{2a}{3d}$

$(3)(1) = 3$

15. Following the example, complete the other reduction.

$$\frac{7ab}{abd} = \left(\frac{7}{d}\right)\left(\frac{ab}{ab}\right) = \left(\frac{7}{d}\right)(1) = \frac{7}{d}$$

$$\frac{PQR}{FPR} = \left(\frac{Q}{F}\right)\left(\frac{PR}{PR}\right) = \left(\frac{Q}{F}\right)(\quad) = \underline{\quad\quad}$$

16. A fraction can be reduced to lowest terms only if we can factor out a fraction that equals "1". For example, none of the fractions below can be reduced.

$$\frac{H}{T} \qquad\qquad \frac{3mp}{7fq} \qquad\qquad \frac{a(x+y)}{b(x-y)}$$

Reduce to lowest terms if possible:

a) $\dfrac{cp^2}{p^2 t} = \underline{\quad}$

c) $\dfrac{m(p+q)}{v(p+q)} = \underline{\quad}$

b) $\dfrac{PV}{RT} = \underline{\quad}$

d) $\dfrac{20xy}{10xy} = \underline{\quad}$

$\left(\dfrac{Q}{F}\right)(1) = \dfrac{Q}{F}$

17. To reduce the fractions below to lowest terms, we began by substituting "1x" for "x" and "1m" for "m".

$$\frac{x}{7x} = \frac{1x}{7x} = \left(\frac{1}{7}\right)\left(\frac{x}{x}\right) = \left(\frac{1}{7}\right)(1) = \frac{1}{7}$$

$$\frac{m}{cm} = \frac{1m}{cm} = \left(\frac{1}{c}\right)\left(\frac{m}{m}\right) = \left(\frac{1}{c}\right)(1) = \frac{1}{c}$$

Following the examples, reduce these to lowest terms.

a) $\dfrac{t}{9t} = \underline{\quad}$ b) $\dfrac{S}{ST} = \underline{\quad}$ c) $\dfrac{a}{3ab} = \underline{\quad}$

a) $\dfrac{c}{t}$

b) not possible

c) $\dfrac{m}{v}$

d) 2

18. To reduce the fractions below to lowest terms, we began by substituting "1y" for "y" and "1a" for "a".

$$\frac{4y}{y} = \frac{4y}{1y} = \left(\frac{4}{1}\right)\left(\frac{y}{y}\right) = (4)(1) = 4$$

$$\frac{abt}{a} = \frac{abt}{1a} = \left(\frac{a}{a}\right)\left(\frac{bt}{1}\right) = (1)(bt) = bt$$

Following the examples, reduce these to lowest terms.

a) $\dfrac{7d}{d} = \underline{\quad}$ b) $\dfrac{bs}{s} = \underline{\quad}$ c) $\dfrac{8RQ}{R} = \underline{\quad}$

a) $\dfrac{1}{9}$

b) $\dfrac{1}{T}$

c) $\dfrac{1}{3b}$

a) 7

b) b

c) 8Q

19. In the reduction below, we began by inserting "1" as a factor in the numerator.

$$\frac{5}{15pq} = \frac{(5)(1)}{15pq} = \left(\frac{5}{15}\right)\left(\frac{1}{pq}\right) = \left(\frac{1}{3}\right)\left(\frac{1}{pq}\right) = \frac{1}{3pq}$$

Following the example, reduce these to lowest terms.

a) $\dfrac{7}{14F_1T_1} = $ _____

b) $\dfrac{9}{9cm} = $ _____

---

20. In the reduction below, we also began by inserting "1" as a factor in the numerator.

$$\frac{5c^2}{10ac^2} = \frac{5c^2(1)}{10ac^2} = \left(\frac{5}{10}\right)\left(\frac{c^2}{c^2}\right)\left(\frac{1}{a}\right) = \left(\frac{1}{2}\right)(1)\left(\frac{1}{a}\right) = \frac{1}{2a}$$

Following the example, reduce these to lowest terms.

a) $\dfrac{2m}{8mt} = $ _____

b) $\dfrac{6pq^2}{12hpq^2T} = $ _____

a) $\dfrac{1}{2F_1T_1}$

b) $\dfrac{1}{cm}$

---

a) $\dfrac{1}{4t}$ b) $\dfrac{1}{2hT}$

7-3   CANCELLING TO REDUCE ALGEBRAIC FRACTIONS TO LOWEST TERMS

Instead of reducing algebraic fractions to lowest terms by factoring out a fraction that equals "1", we can use a shorter equivalent method called "cancelling". We will discuss "cancelling" in this section.

---

21. The fraction below contains the same factor "t" in both its numerator and denominator. We reduced it to lowest terms by cancelling the "t's". Use cancelling to reduce the other fractions to lowest terms.

$\dfrac{a\cancel{t}}{b\cancel{t}} = \dfrac{a}{b}$ a) $\dfrac{3x}{5x} = $ _____   b) $\dfrac{7m^2}{cm^2} = $ _____

---

22. To reduce the fraction below to lowest terms, we cancelled both the "c's" and the "d's". Use cancelling to reduce the other fractions to lowest terms.

$\dfrac{5\cancel{c}\cancel{d}}{8\cancel{c}\cancel{d}} = \dfrac{5}{8}$ a) $\dfrac{kPV}{9PV} = $ _____   b) $\dfrac{cp^2q}{hp^2q} = $ _____

a) $\dfrac{3\cancel{x}}{5\cancel{x}} = \dfrac{3}{5}$

b) $\dfrac{7\cancel{m}^2}{c\cancel{m}^2} = \dfrac{7}{c}$

---

23. Following the example, reduce the other fractions to lowest terms.

$\dfrac{3\cancel{c}\cancel{t}}{\cancel{c}s\cancel{t}} = \dfrac{3}{s}$ a) $\dfrac{HSV}{AHV} = $ _____   b) $\dfrac{ab^2t}{b^2tv} = $ _____

a) $\dfrac{k\cancel{P}\cancel{V}}{9\cancel{P}\cancel{V}} = \dfrac{k}{9}$

b) $\dfrac{c\cancel{p}^2\cancel{q}}{h\cancel{p}^2\cancel{q}} = \dfrac{c}{h}$

24. The fraction below contains the same binomial factor (y – 4) in both its numerator and denominator. We reduced it to lowest terms by cancelling both binomials. Use cancelling to reduce the other fractions to lowest terms.

$$\frac{5a(y-4)}{3b(y-4)} = \frac{5a}{3b}$$

a) $\frac{4(c+d)}{7(c+d)} =$ _____   b) $\frac{ct(p-q)}{kt(p-q)} =$ _____

a) $\frac{AS Y}{A Y Y} = \frac{S}{A}$

b) $\frac{ab^2 Y}{b^2 Y v} = \frac{a}{v}$

25. To reduce the fraction below to lowest terms, we cancelled the "x's" and divided both the 6 and 8 by 2. Reduce the other fractions to lowest terms.

$$\frac{\overset{3}{6x}}{\underset{4}{8x}} = \frac{3}{4}$$

a) $\frac{12cd}{8ac} =$ _____   b) $\frac{7pq}{14pq} =$ _____

a) $\frac{4(c+d)}{7(c+d)} = \frac{4}{7}$

b) $\frac{c(p-q)}{k(p-q)} = \frac{c}{k}$

26. Following the example, reduce the other fractions to lowest terms.

$$\frac{\overset{4}{8m}}{\underset{1}{2m}} = 4$$

a) $\frac{12xy}{6xy} =$ _____   b) $\frac{9t}{3} =$ _____

a) $\frac{\overset{3}{12cd}}{\underset{2}{8ac}} = \frac{3d}{2a}$

b) $\frac{\overset{1}{7pq}}{\underset{2}{14pq}} = \frac{1}{2}$

27. Before cancelling the "b's" below, we inserted "1" as a factor in the numerator.

$$\frac{b}{ab} = \frac{1b}{ab} = \frac{1}{a}$$

Using the same method, reduce these to lowest terms.

a) $\frac{x}{dx} =$ _____   b) $\frac{T^2}{9T^2} =$ _____   c) $\frac{s}{5sv} =$ _____

a) $\frac{\overset{2}{12xy}}{\underset{1}{6xy}} = 2$

b) $\frac{\overset{3}{9t}}{\underset{1}{3}} = 3t$

28. Before cancelling the "x's" below, we inserted "1" as a factor in the denominator.

$$\frac{8x}{x} = \frac{8x}{1x} = 8$$

Using the same method, reduce these to lowest terms.

a) $\frac{ab}{b} =$ _____   b) $\frac{7ct^2}{ct^2} =$ _____   c) $\frac{5QR}{Q} =$ _____

a) $\frac{1x}{dx} = \frac{1}{d}$

b) $\frac{1T^2}{9T^2} = \frac{1}{9}$

c) $\frac{1s}{5sv} = \frac{1}{5v}$

29. Following the example, reduce the other fractions to lowest terms.

$$\frac{\overset{1}{4}}{\underset{2}{8ab}} = \frac{1}{2ab}$$

a) $\frac{3}{9c^2d} =$ _____   b) $\frac{7}{7xy} =$ _____

a) $\frac{ab}{1b} = a$

b) $\frac{7ct^2}{1ct^2} = 7$

c) $\frac{5QR}{1Q} = 5R$

30. Following the example, reduce the other fractions to lowest terms.

$$\frac{\overset{1}{\cancel{2}x^2}}{\underset{2}{\cancel{4}bx^{\cancel{2}}}} = \frac{1}{2b}$$

a) $\dfrac{5a}{15at} = $ _____

b) $\dfrac{4pq^2}{16pq^2R} = $ _____

a) $\dfrac{\overset{1}{\cancel{3}}}{\cancel{9}c^2d} = \dfrac{1}{3c^2d}$

b) $\dfrac{\overset{1}{\cancel{7}}}{\underset{1}{\cancel{7}}xy} = \dfrac{1}{xy}$

---

31. A fraction can be reduced to lowest terms only if we can cancel common factors. Reduce these to lowest terms if possible.

a) $\dfrac{ax^2}{x^2y} = $ _____

c) $\dfrac{4a(x-y)}{8b(x-y)} = $ _____

b) $\dfrac{5PQ}{7HR} = $ _____

d) $\dfrac{bc^2v}{c^2v} = $ _____

a) $\dfrac{\overset{1}{\cancel{5}a}}{\underset{3}{\cancel{15}at}} = \dfrac{1}{3t}$

b) $\dfrac{\overset{1}{\cancel{4}pq^2}}{\underset{4}{\cancel{16}pq^2R}} = \dfrac{1}{4R}$

a) $\dfrac{a}{y}$ b) not possible   c) $\dfrac{a}{2b}$ d) b

7-4   CANCELLING IN MULTIPLICATIONS

When multiplying fractions, the product should always be reduced to lowest terms. To avoid products that are not in lowest terms, we can cancel before multiplying. We will discuss the method in this section.

32. When multiplying algebraic fractions, the product should always be reduced to lowest terms. For example:

$$\left(\frac{cd}{m}\right)\left(\frac{m}{dt}\right) = \frac{cd m}{d m t} = \frac{c}{t}$$

To avoid having to reduce the product above to lowest terms, we can cancel before multiplying. That is:

$$\left(\frac{c\cancel{d}}{\cancel{m}}\right)\left(\frac{\cancel{m}}{\cancel{d}t}\right) = \frac{c}{t}$$

Do these. Cancel before multiplying.

a) $\left(\dfrac{P}{S}\right)\left(\dfrac{S}{Q}\right) = $ _____

b) $\left(\dfrac{4a}{b}\right)\left(\dfrac{bd}{2a}\right) = $ _____

33. We did the same multiplication below in two different ways: 1) by multiplying and then cancelling, and 2) by cancelling before multiplying. The second way is ordinarily used because it is shorter.

$$\frac{1}{bd^2}(ad^2) = \frac{ad^2}{bd^2} = \frac{a}{b} \qquad \frac{1}{bd^{\cancel{2}}}(ad^{\cancel{2}}) = \frac{a}{b}$$

Do these. Cancel before multiplying.

a) $y\left(\dfrac{vx}{cy}\right) = $ _____

b) $\left(\dfrac{Hq}{p}\right)(ap) = $ _____

a) $\left(\dfrac{P}{\cancel{S}}\right)\left(\dfrac{\cancel{S}}{Q}\right) = \dfrac{P}{Q}$

b) $\left(\dfrac{\overset{2}{\cancel{4}a}}{\cancel{b}}\right)\left(\dfrac{\cancel{b}d}{\underset{1}{\cancel{2}a}}\right) = 2d$

34.   Following the example, complete the other multiplication.

$$\left(\frac{t^2-1}{\cancel{R}}\right)\left(\frac{\cancel{R}}{F}\right) = \frac{t^2-1}{F} \qquad \left(\frac{a}{b}\right)\left(\frac{b}{c+d}\right) = \underline{\quad\quad}$$

a) $\cancel{y}\left(\frac{xv}{c\cancel{y}}\right) = \frac{xv}{c}$

b) $\left(\frac{Hq}{\cancel{p}}\right)(a\cancel{p}) = aHq$

---

35.   Before multiplying below, we cancelled each $(x + y)$.

$$\left(\frac{1}{\cancel{x+y}}\right)\left(\frac{\cancel{x+y}}{c}\right) = \frac{1}{c}$$

Following the example, complete these:

a) $\left(\frac{p-7}{d}\right)\left(\frac{H}{p-7}\right) = \underline{\quad\quad}$ b) $\left(\frac{1}{c+d}\right)[t(c+d)] = \underline{\quad\quad}$

$\left(\frac{a}{\cancel{b}}\right)\left(\frac{\cancel{b}}{c+d}\right) = \frac{a}{c+d}$

---

36.   When cancelling letters, we get "1" as a factor. Though we do not ordinarily show the "1", we did so below so that it is clear that the numerator of the product is "1".

$$\left(\frac{\overset{1}{\cancel{y}}}{k\cancel{m}}\right)\left(\frac{\overset{1}{\cancel{m}}}{\cancel{y}}\right) = \frac{1}{k}$$
$$\underset{1}{}\quad\underset{1}{}$$

Following the example, complete these.

a) $\left(\frac{V}{T}\right)\left(\frac{T}{V}\right) = \underline{\quad\quad}$ b) $\left(\frac{a-b}{cd}\right)\left(\frac{c}{a-b}\right) = \underline{\quad\quad}$

a) $\left(\frac{\cancel{p-7}}{d}\right)\left(\frac{H}{\cancel{p-7}}\right) = \frac{H}{d}$

b) $\left(\frac{1}{\cancel{c+d}}\right)[t\cancel{(c+d)}] = t$

a) $\left(\dfrac{\overset{1}{\cancel{V}}}{\underset{1}{\cancel{T}}}\right)\left(\dfrac{\overset{1}{\cancel{T}}}{\underset{1}{\cancel{V}}}\right) = 1$ b) $\left(\dfrac{\overset{1}{\cancel{a-b}}}{\underset{1}{\cancel{c}d}}\right)\left(\dfrac{\overset{1}{\cancel{c}}}{\underset{1}{\cancel{a-b}}}\right) = \dfrac{1}{d}$

7-5   RECIPROCALS

In this section, we will discuss the reciprocals of algebraic fractions and non-fractional expressions. The concept of reciprocals is then used to perform divisions involving algebraic fractions in the next section.

37.   Two quantities are reciprocals if their product is +1. For example:

Since $x\left(\frac{1}{x}\right) = \frac{x}{x} = 1$:   the reciprocal of x is $\frac{1}{x}$.

the reciprocal of $\frac{1}{x}$ is x.

Since $ab\left(\frac{1}{ab}\right) = \frac{ab}{ab} = 1$:   the reciprocal of ab is $\frac{1}{ab}$.

the reciprocal of $\frac{1}{ab} = ab$.

Write the reciprocal of each quantity.

a) t $\underline{\quad\quad}$ b) $\frac{1}{y}$ $\underline{\quad\quad}$ c) cd $\underline{\quad\quad}$ d) $\frac{1}{PV}$ $\underline{\quad\quad}$

38. Two fractions are reciprocals if their product is +1. For example:

Since $\left(\dfrac{c}{d}\right)\left(\dfrac{d}{c}\right) = \dfrac{cd}{cd} = 1$:   the reciprocal of $\dfrac{c}{d}$ is $\dfrac{d}{c}$.

the reciprocal of $\dfrac{d}{c}$ is $\dfrac{c}{d}$.

Write the reciprocal of each fraction.

a) $\dfrac{v}{t}$ _____   b) $\dfrac{ab}{h}$ _____   c) $\dfrac{k}{mp}$ _____   d) $\dfrac{abf}{ct}$ _____

a) $\dfrac{1}{t}$ c) $\dfrac{1}{cd}$

b) y d) PV

---

39. Write the reciprocal of each quantity.

a) $3y$ _____   b) $\dfrac{1}{4x}$ _____   c) $\dfrac{6m}{p}$ _____   d) $\dfrac{5t}{2ab}$ _____

a) $\dfrac{t}{v}$ c) $\dfrac{mp}{k}$

b) $\dfrac{h}{ab}$ d) $\dfrac{ct}{abf}$

---

40. Since $(p + q)\left(\dfrac{1}{p + q}\right) = \dfrac{p + q}{p + q} = 1$:

a) the reciprocal of $(p + q)$ is _____ .

b) the reciprocal of $\dfrac{1}{p + q}$ is _____ .

a) $\dfrac{1}{3y}$ c) $\dfrac{p}{6m}$

b) $4x$ d) $\dfrac{2ab}{5t}$

---

41. Since $m(a - c)\left[\dfrac{1}{m(a - c)}\right] = \dfrac{m(a - c)}{m(a - c)} = 1$:

a) the reciprocal of $m(a - c)$ is _____ .

b) the reciprocal of $\dfrac{1}{m(a - c)}$ is _____ .

a) $\dfrac{1}{p + q}$

b) $p + q$

---

42. Since $\left(\dfrac{x - 1}{y}\right)\left(\dfrac{y}{x - 1}\right) = \dfrac{y(x - 1)}{y(x - 1)} = 1$:

a) the reciprocal of $\dfrac{x - 1}{y}$ is _____ .

b) the reciprocal of $\dfrac{y}{x - 1}$ is _____ .

a) $\dfrac{1}{m(a - c)}$

b) $m(a - c)$

---

43. Write the reciprocal of each quantity.

a) $x - y$ _____   c) $\dfrac{m}{p + q}$ _____

b) $\dfrac{1}{a(b + c)}$ _____   d) $\dfrac{h - 1}{t - 7}$ _____

a) $\dfrac{y}{x - 1}$

b) $\dfrac{x - 1}{y}$

---

a) $\dfrac{1}{x - y}$

b) $a(b + c)$

c) $\dfrac{p + q}{m}$

d) $\dfrac{t - 7}{h - 1}$

7-6   DIVISIONS INVOLVING ALGEBRAIC FRACTIONS

To perform a division involving one or two algebraic fractions, we multiply the numerator by the reciprocal of the denominator. We will discuss divisions of that type in this section.

---

44.   Divisions involving algebraic fractions are written in complex-fraction form. The "major" fraction line separates the numerator and denominator. For example:

$$\frac{\frac{t}{4}}{\frac{t}{3}} \quad \text{means: divide } \frac{t}{4} \text{ by } \frac{t}{3} \qquad \frac{m}{\frac{a}{b}} \quad \text{means: divide } m \text{ by } \frac{a}{b}$$

$$\frac{\frac{P}{Q}}{S} \quad \text{means: divide } \underline{\quad\quad} \text{ by } \underline{\quad\quad}$$

---

45.   To perform divisions involving algebraic fractions, we multiply the numerator by the reciprocal of the denominator.   That is:

$$\frac{\frac{x}{4}}{\frac{2}{3}} = \left(\frac{x}{4}\right)\left(\text{the reciprocal of } \frac{2}{3}\right) = \left(\frac{x}{4}\right)\left(\frac{3}{2}\right) = \frac{3x}{8}$$

a) $\dfrac{h}{\frac{p}{q}} = (h)\left(\text{the reciprocal of } \frac{p}{q}\right) = (h)\left(\quad\right) = \underline{\quad\quad}$

b) $\dfrac{\frac{F}{S}}{V} = \left(\frac{F}{S}\right)\left(\text{the reciprocal of } V\right) = \left(\frac{F}{S}\right)\left(\quad\right) = \underline{\quad\quad}$

> $\dfrac{P}{Q}$ by S

---

46.   Complete each division.

a) $\dfrac{\frac{m}{2}}{\frac{v}{3}} = \left(\quad\right)\left(\quad\right) = \underline{\quad\quad}$ b) $\dfrac{\frac{pt}{d}}{\frac{a}{bc}} = \left(\quad\right)\left(\quad\right) = \underline{\quad\quad}$

> a) $(h)\left(\dfrac{q}{p}\right) = \dfrac{hq}{p}$
>
> b) $\left(\dfrac{F}{S}\right)\left(\dfrac{1}{V}\right) = \dfrac{F}{SV}$

---

47.   Complete each division.

a) $\dfrac{p}{\frac{1}{8}} = \left(\quad\right)\left(\quad\right) = \underline{\quad\quad}$ b) $\dfrac{\frac{5c}{d}}{4a} = \left(\quad\right)\left(\quad\right) = \underline{\quad\quad}$

> a) $\left(\dfrac{m}{2}\right)\left(\dfrac{3}{v}\right) = \dfrac{3m}{2v}$
>
> b) $\left(\dfrac{pt}{d}\right)\left(\dfrac{bc}{a}\right) = \dfrac{bcpt}{ad}$

> a) $(p)(8) = 8p$
>
> b) $\left(\dfrac{5c}{d}\right)\left(\dfrac{1}{4a}\right) = \dfrac{5c}{4ad}$

48.   When dividing algebraic fractions, the quotient should always be reduced to lowest terms. We cancel before multiplying to do so. For example:

$$\frac{\dfrac{hs}{4}}{\dfrac{ah}{2}} = \left(\frac{\cancel{h}s}{\cancel{4}_2}\right)\left(\frac{\cancel{2}^1}{a\cancel{h}}\right) = \frac{s}{2a}$$

Complete. Cancel to reduce to lowest terms.

a) $\dfrac{\dfrac{9t}{5}}{\dfrac{3t}{10}} = $

b) $\dfrac{\dfrac{ab}{c}}{\dfrac{bd}{m}} = $

---

49.   Complete. Cancel to reduce to lowest terms.

a) $\dfrac{\dfrac{15t}{5}}{7} = $

b) $\dfrac{\dfrac{P}{P}}{T} = $

a) $\left(\dfrac{\cancel{9}^3 t}{\cancel{5}_1}\right)\left(\dfrac{\cancel{10}^2}{\cancel{3}_1 t}\right) = 6$

b) $\left(\dfrac{a\cancel{b}}{c}\right)\left(\dfrac{m}{\cancel{b}d}\right) = \dfrac{am}{cd}$

---

50.   Complete. Cancel to reduce to lowest terms.

a) $\dfrac{\dfrac{2v}{3}}{6v} = $

b) $\dfrac{\dfrac{7m}{d}}{7d} = $

a) $(\cancel{15}^3 t)\left(\dfrac{7}{\cancel{3}_1}\right) = 21t$

b) $\cancel{P}\left(\dfrac{T}{\cancel{P}}\right) = T$

---

51.   Complete. Cancel when possible.

a) $\dfrac{\dfrac{b}{c}}{\dfrac{a}{2t}} = $

b) $\dfrac{\dfrac{1}{PV}}{\dfrac{1}{P}} = $

a) $\left(\dfrac{\cancel{2}^1 y}{3}\right)\left(\dfrac{1}{\cancel{6}y}\right) = \dfrac{1}{9}$

b) $\left(\dfrac{\cancel{7}^1 m}{d}\right)\left(\dfrac{1}{\cancel{7}d}\right)_1 = \dfrac{m}{d^2}$

---

52.   Complete. Cancel when possible.

a) $\dfrac{t}{\dfrac{3}{4}} = $

b) $\dfrac{\dfrac{2r}{w}}{r} = $

a) $\dfrac{2bt}{ac}$

b) $\dfrac{1}{V}$

---

53.   The division below involves a binomial.

$$\frac{\dfrac{a-b}{d}}{\dfrac{c}{t}} = \left(\frac{a-b}{d}\right)\left(\frac{t}{c}\right) = \frac{t(a-b)}{cd}$$

Following the example, complete each division.

a) $\dfrac{\dfrac{V}{S-A}}{H} = $

b) $\dfrac{x+y}{\dfrac{a}{b}} = $

a) $\dfrac{4t}{3}$

b) $\dfrac{2}{w}$

54. In the division below, we were able to cancel binomials.

$$\frac{\dfrac{c+d}{m}}{\dfrac{c+d}{t}} = \left(\frac{\cancel{c+d}}{m}\right)\left(\frac{t}{\cancel{c+d}}\right) = \frac{t}{m}$$

Following the example, complete each division.

a) $\dfrac{p-q}{\dfrac{p-q}{v}} =$

b) $\dfrac{\dfrac{t^2-1}{h}}{t^2-1} =$

a) $\dfrac{HV}{S-A}$

b) $\dfrac{b(x+y)}{a}$

---

55. The numerator of the complex fraction below is "1".

$$\frac{1}{\dfrac{a}{b}} = (1)\left(\text{the reciprocal of } \frac{a}{b}\right) = (1)\left(\frac{b}{a}\right) = \frac{b}{a}$$

Complete: a) $\dfrac{1}{\dfrac{m}{pq}} =$

b) $\dfrac{1}{\dfrac{x+y}{c}} =$

a) v

b) $\dfrac{1}{h}$

---

56. Three special properties of division are given below.

1) When a fraction is divided by itself, the quotient is "1".

$$\frac{\dfrac{m}{t}}{\dfrac{m}{t}} = \left(\frac{m}{t}\right)\left(\frac{t}{m}\right) = 1$$

2) When "0" is divided by a fraction, the quotient is "0".

$$\frac{0}{\dfrac{x}{3}} = (0)\left(\frac{3}{x}\right) = 0$$

3) Division of a fraction by "0" is an impossible operation because "0" has no reciprocal.

$$\frac{\dfrac{7}{P}}{0} = \frac{7}{P} \text{ (the reciprocal of 0)}$$

Using the properties above, complete these:

a) $\dfrac{\dfrac{ab}{c}}{\dfrac{ab}{c}} =$

b) $\dfrac{0}{\dfrac{R}{S}} =$

c) $\dfrac{\dfrac{x}{7y}}{0} =$

d) $\dfrac{\dfrac{p+q}{5}}{\dfrac{p+q}{5}} =$

a) $\dfrac{pq}{m}$

b) $\dfrac{c}{x+y}$

a) 1   b) 0   c) impossible   d) 1

SELF-TEST 22 (pages 273-286)

Reduce each fraction to lowest terms.

1. $\dfrac{2ab}{6bd} = $ _____

2. $\dfrac{R^2}{3R^2} = $ _____

3. $\dfrac{dp^2v}{p^2vt} = $ _____

4. $\dfrac{8w(x-y)}{4(x-y)} = $ _____

Multiply. Report each product in lowest terms.

5. $\left(\dfrac{r}{2t}\right)(2at) = $ _____

6. $\left(\dfrac{3k}{bh^2}\right)\left(\dfrac{h^2}{2k}\right) = $ _____

7. $\left(\dfrac{n}{c+s}\right)\left(\dfrac{c+s}{dn}\right) = $ _____

8. Write the reciprocal of $\dfrac{P+1}{H}$.

Divide. Report each quotient in lowest terms.

9. $\dfrac{\frac{2p}{w}}{\frac{r}{2w}} = $ _____

10. $\dfrac{hs}{\frac{s}{b}} = $ _____

11. $\dfrac{\frac{r^2+1}{at}}{r^2+1} = $ _____

ANSWERS: 1. $\dfrac{a}{3d}$ 3. $\dfrac{d}{t}$ 5. ar 7. $\dfrac{1}{d}$ 9. $\dfrac{4p}{r}$ 11. $\dfrac{1}{at}$

2. $\dfrac{1}{3}$ 4. $2w$ 6. $\dfrac{3}{2b}$ 8. $\dfrac{H}{P+1}$ 10. bh

---

7-7   ADDING ALGEBRAIC FRACTIONS WITH LIKE DENOMINATORS

In this section, we will discuss the procedure for adding algebraic fractions with like (or identical) denominators.

57. To add algebraic fractions with like denominators, we add their numerators and keep the same denominator. For example:

$$\frac{2x}{5} + \frac{3y}{5} = \frac{2x+3y}{5} \qquad \frac{m}{ab} + \frac{n}{ab} = \frac{m+n}{ab}$$

Following the examples, complete these.

a) $\dfrac{p}{3} + \dfrac{5}{3} = $ _____

b) $\dfrac{5S}{PV} + \dfrac{7R}{PV} = $ _____

58. When adding algebraic fractions, we combine like terms in the numerator of the sum if possible. For example:

$$\frac{6}{t} + \frac{3}{t} = \frac{6+3}{t} = \frac{9}{t} \qquad \frac{4m}{a} + \frac{3m}{a} = \frac{4m+3m}{a} = \frac{7m}{a}$$

Following the examples, complete these:

a) $\dfrac{3x}{7} + \dfrac{2x}{7} = $ _____

b) $\dfrac{b}{cd} + \dfrac{b}{cd} = $ _____

a) $\dfrac{p+5}{3}$

b) $\dfrac{5S+7R}{PV}$

59. Complete. Combine like terms in the sum if possible.

a) $\dfrac{4}{3x} + \dfrac{1}{3x} =$ _____

b) $\dfrac{a}{c} + \dfrac{b}{c} =$ _____

c) $\dfrac{2y}{5} + \dfrac{y}{5} =$ _____

d) $\dfrac{v}{st} + \dfrac{3m}{st} =$ _____

Answers:
a) $\dfrac{5x}{7}$

b) $\dfrac{2b}{cd}$

60. After combining like terms in the numerator of the sum, we can sometimes reduce the sum to lowest terms. For example:

$$\dfrac{4}{3x} + \dfrac{5}{3x} = \dfrac{4+5}{3x} = \dfrac{\overset{3}{\cancel{9}}}{\underset{1}{\cancel{3}x}} = \dfrac{3}{x}$$

Complete. Reduce to lowest terms if possible.

a) $\dfrac{3y}{10} + \dfrac{3y}{10} =$ _____

b) $\dfrac{x}{9} + \dfrac{4x}{9} =$ _____

c) $\dfrac{7}{2t} + \dfrac{3}{2t} =$ _____

d) $\dfrac{m}{6b} + \dfrac{5m}{6b} =$ _____

Answers:
a) $\dfrac{5}{3x}$ c) $\dfrac{3y}{5}$

b) $\dfrac{a+b}{c}$ d) $\dfrac{v+3m}{st}$

61. To add the fractions below, we added their numerators and then combined like terms.

$$\dfrac{x+5}{7} + \dfrac{3}{7} = \dfrac{(x+5)+3}{7} = \dfrac{x+8}{7}$$

$$\dfrac{m}{t} + \dfrac{m+s}{t} = \dfrac{m+(m+s)}{t} = \dfrac{2m+s}{t}$$

Following the examples, complete these.

a) $\dfrac{7a+b}{9} + \dfrac{b}{9} =$ _____

b) $\dfrac{x+y}{cd} + \dfrac{x-y}{cd} =$ _____

Answers:
a) $\dfrac{3y}{5}$ c) $\dfrac{5}{t}$

b) $\dfrac{5x}{9}$ d) $\dfrac{m}{b}$

62. We use the same method to add fractions with like denominators when the denominators are more complicated. For example:

$$\dfrac{v}{p-q} + \dfrac{t}{p-q} = \dfrac{v+t}{p-q} \qquad \dfrac{1}{3(a+b)} + \dfrac{h}{3(a+b)} = \dfrac{1+h}{3(a+b)}$$

Following the examples, complete these:

a) $\dfrac{x}{c+d} + \dfrac{5}{c+d} =$ _____

b) $\dfrac{S}{2(x-y)} + \dfrac{T}{2(x-y)} =$ _____

Answers:
a) $\dfrac{7a+2b}{9}$

b) $\dfrac{2x}{cd}$

a) $\dfrac{x+5}{c+d}$

b) $\dfrac{S+T}{2(x-y)}$

63.   Following the example, complete the other addition.

$$\frac{x+y}{m+t} + \frac{x}{m+t} = \frac{(x+y)+x}{m+t} = \frac{2x+y}{m+t}$$

$$\frac{2a+3b}{v-7} + \frac{2a-3b}{v-7} = \underline{\hspace{3cm}}$$

64.   To add more than two fractions with like denominators, we also add their numerators. For example:

$$\frac{a}{x} + \frac{b}{x} + \frac{c}{x} = \frac{a+b+c}{x} \qquad \frac{s}{7} + \frac{t}{7} + \frac{3}{7} = \underline{\hspace{2cm}}$$

$\dfrac{4a}{v-7}$

$\dfrac{s+t+3}{7}$

7-8   EQUIVALENT FORMS OF ALGEBRAIC FRACTIONS

To add algebraic fractions with unlike denominators, we must convert one or both fractions to an equivalent form. We will discuss the method in this section.

65.   Below we converted $\frac{x}{2}$ to an equivalent fraction whose denominator is 8. To do so, we multiplied by $\frac{4}{4}$, a fraction that equals "1". We used $\frac{4}{4}$ since $\frac{8}{2} = 4$.

$$\frac{x}{2} = \frac{x}{2}\left(\frac{4}{4}\right) = \frac{4x}{8}$$

To convert an algebraic fraction to an equivalent form, we always multiply by a fraction that equals "1". To find that fraction, we divide the new denominator by the original denominator. That is:

For $\dfrac{2y}{3} = \dfrac{(\quad)}{6}$, we multiply by $\dfrac{2}{2}$ since $\dfrac{6}{3} = 2$.

For $\dfrac{m}{5} = \dfrac{(\quad)}{15}$, we multiply by $\dfrac{3}{3}$ since $\dfrac{15}{5} = 3$.

Multiplying by a fraction that equals "1", complete each conversion below.

a) $\dfrac{3y}{4}\left(\dfrac{\quad}{\quad}\right) = \dfrac{(\quad)}{12}$ b) $\dfrac{ab}{2}\left(\dfrac{\quad}{\quad}\right) = \dfrac{(\quad)}{10}$

a) $\dfrac{3y}{4}\left(\dfrac{3}{3}\right) = \dfrac{9y}{12}$

b) $\dfrac{ab}{2}\left(\dfrac{5}{5}\right) = \dfrac{5ab}{10}$

66. Below we converted $\frac{3}{t}$ to an equivalent fraction whose denominator is $5t$. To do so, we multiplied by $\frac{5}{5}$ since $\frac{5t}{t} = 5$.

$$\frac{3}{t} = \frac{3}{t}\left(\frac{5}{5}\right) = \frac{15}{5t}$$

To find the fraction that equals "1" to multiply by, we divide the new denominator by the original denominator. For example:

For $\frac{7}{x} = \frac{(\ \ )}{2x}$, we multiply by $\frac{2}{2}$ since $\frac{2x}{x} = 2$.

For $\frac{a}{9} = \frac{(\ \ )}{9b}$, we multiply by $\frac{b}{b}$ since $\frac{9b}{9} = b$.

Multiplying by a fraction that equals "1", complete each conversion below.

a) $\frac{5}{R}\left(\frac{\ \ }{\ \ }\right) = \frac{(\ \ )}{3R}$ 　　　　b) $\frac{m}{8}\left(\frac{\ \ }{\ \ }\right) = \frac{(\ \ )}{8d}$

---

67. Using the correct fraction that equals "1", complete each conversion below.

a) $\frac{x}{2}\left(\frac{\ \ }{\ \ }\right) = \frac{(\ \ )}{10}$ 　　　　c) $\frac{1}{y}\left(\frac{\ \ }{\ \ }\right) = \frac{(\ \ )}{2y}$

b) $\frac{4d}{3}\left(\frac{\ \ }{\ \ }\right) = \frac{(\ \ )}{12}$ 　　　　d) $\frac{v}{7}\left(\frac{\ \ }{\ \ }\right) = \frac{(\ \ )}{7m}$

a) $\frac{5}{R}\left(\frac{3}{3}\right) = \frac{15}{3R}$

b) $\frac{m}{8}\left(\frac{d}{d}\right) = \frac{dm}{8d}$

---

68. To find the fraction that equals "1" to multiply by, we always divide the new denominator by the original denominator. That is:

For $\frac{1}{2x} = \frac{(\ \ )}{10x}$, we multiply by $\frac{5}{5}$ since $\frac{10x}{2x} = 5$.

For $\frac{a}{5} = \frac{(\ \ )}{20t}$, we multiply by $\frac{4t}{4t}$ since $\frac{20t}{5} = 4t$.

Multiplying by a fraction that equals "1", complete each conversion below.

a) $\frac{3a}{4x}\left(\frac{\ \ }{\ \ }\right) = \frac{(\ \ )}{8x}$ 　　　　b) $\frac{1}{3}\left(\frac{\ \ }{\ \ }\right) = \frac{(\ \ )}{9y}$

a) $\frac{x}{2}\left(\frac{5}{5}\right) = \frac{5x}{10}$

b) $\frac{4d}{3}\left(\frac{4}{4}\right) = \frac{16d}{12}$

c) $\frac{1}{y}\left(\frac{2}{2}\right) = \frac{2}{2y}$

d) $\frac{v}{7}\left(\frac{m}{m}\right) = \frac{mv}{7m}$

---

69. Two more examples of finding the correct fraction that equals "1" to multiply by are given below.

For $\frac{a}{b} = \frac{(\ \ )}{bd}$, we multiply by $\frac{d}{d}$ since $\frac{bd}{b} = d$.

For $\frac{m}{x} = \frac{(\ \ )}{3xy}$, we multiply by $\frac{3y}{3y}$ since $\frac{3xy}{x} = 3y$.

Multiplying by a fraction that equals "1", complete each conversion.

a) $\frac{1}{m}\left(\frac{\ \ }{\ \ }\right) = \frac{(\ \ )}{mt}$ 　　　　b) $\frac{a}{q}\left(\frac{\ \ }{\ \ }\right) = \frac{(\ \ )}{pqr}$

a) $\frac{3a}{4x}\left(\frac{2}{2}\right) = \frac{6a}{8x}$

b) $\frac{1}{3}\left(\frac{3y}{3y}\right) = \frac{3y}{9y}$

70. Using the correct fraction that equals "1", complete each conversion below.

a) $\dfrac{1}{4y}\left(\dfrac{\quad}{\quad}\right) = \dfrac{(\quad)}{8y}$

b) $\dfrac{2b}{3}\left(\dfrac{\quad}{\quad}\right) = \dfrac{(\quad)}{15c}$

c) $\dfrac{5}{t}\left(\dfrac{\quad}{\quad}\right) = \dfrac{(\quad)}{at}$

d) $\dfrac{V}{R}\left(\dfrac{\quad}{\quad}\right) = \dfrac{(\quad)}{RST}$

a) $\dfrac{1}{m}\left(\dfrac{t}{t}\right) = \dfrac{t}{mt}$

b) $\dfrac{a}{q}\left(\dfrac{pr}{pr}\right) = \dfrac{apr}{pqr}$

---

71. You should be able to make conversions without writing the fraction that equals "1". Do so below.

a) $\dfrac{y}{5} = \dfrac{(\quad)}{25}$

b) $\dfrac{2x}{7} = \dfrac{(\quad)}{14}$

c) $\dfrac{9}{a} = \dfrac{(\quad)}{2a}$

d) $\dfrac{H}{4} = \dfrac{(\quad)}{12T}$

a) $\dfrac{1}{4y}\left(\dfrac{2}{2}\right) = \dfrac{2}{8y}$

b) $\dfrac{2b}{3}\left(\dfrac{5c}{5c}\right) = \dfrac{10bc}{15c}$

c) $\dfrac{5}{t}\left(\dfrac{a}{a}\right) = \dfrac{5a}{at}$

d) $\dfrac{V}{R}\left(\dfrac{ST}{ST}\right) = \dfrac{STV}{RST}$

---

72. Do these without writing the fraction that equals "1".

a) $\dfrac{3}{2x} = \dfrac{(\quad)}{6x}$

b) $\dfrac{t}{4} = \dfrac{(\quad)}{20v}$

c) $\dfrac{1}{k} = \dfrac{(\quad)}{kt}$

d) $\dfrac{a}{b} = \dfrac{(\quad)}{4bc}$

a) $\dfrac{5y}{25}$ c) $\dfrac{18}{2a}$

b) $\dfrac{4x}{14}$ d) $\dfrac{3HT}{12T}$

a) $\dfrac{9}{6x}$ b) $\dfrac{5tv}{20v}$ c) $\dfrac{t}{kt}$ d) $\dfrac{4ac}{4bc}$

7-9 FINDING LOWEST COMMON DENOMINATORS

To add algebraic fractions with unlike denominators, we use the lowest common denominator. We will discuss a method for finding lowest common denominators in this section.

---

73. We have already discussed a method for finding the lowest common denominator (or LCD) when both denominators are numbers.

If the larger denominator is a multiple of the smaller, the larger denominator is the LCD.

For $\dfrac{x}{8} + \dfrac{y}{2}$, the LCD is 8.

If the larger denominator is not a multiple of the smaller, we check multiples of the larger until we find the smallest one that is also a multiple of the smaller.

For $\dfrac{t}{6} + \dfrac{1}{8}$, the LCD is 24.   For $\dfrac{m}{5} + \dfrac{d}{7}$, the LCD is 35.

Find the lowest common denominator for each addition.

a) $\dfrac{1}{5} + \dfrac{y}{20}$ _____   b) $\dfrac{x}{6} + \dfrac{3}{4}$ _____   c) $\dfrac{3x}{4} + \dfrac{2y}{7}$ _____

74. When one denominator is a number and the other a letter, the LCD is the product of the number and the letter. That is

      For $\dfrac{3}{p} + \dfrac{1}{4}$, the LCD is 4p.   For $\dfrac{h}{7} + \dfrac{1}{t}$, the LCD is _____.

    a) 20
    b) 12
    c) 28

---

75. When one denominator is a letter and the other is the same letter with a numerical coefficient, the LCD equals the denominator with the numerical coefficient. That is:

      For $\dfrac{7}{5x} + \dfrac{1}{x}$, the LCD is 5x.   For $\dfrac{a}{y} + \dfrac{b}{9y}$, the LCD is _____.

    7t

---

76. When one denominator is a number and the other is a letter with a numerical coefficient, the LCD is the product of the letter and the smallest common multiple of the two numbers. That is:

      For $\dfrac{7}{3x} + \dfrac{1}{3}$, the LCD is 3x.

      For $\dfrac{7}{12} + \dfrac{5}{3a}$, the LCD is 12a.

      For $\dfrac{3}{2y} + \dfrac{4}{7}$, the LCD is 14y.

Find the lowest common denominator for each addition.

    a) $\dfrac{5}{9} + \dfrac{1}{9x}$ _____   b) $\dfrac{x}{4} + \dfrac{y}{8d}$ _____   c) $\dfrac{1}{5t} + \dfrac{5}{6}$ _____

    9y

---

77. When both denominators contain the same letter with a numerical coefficient, the LCD is the product of that letter and the smallest common multiple of the coefficients. That is:

      For $\dfrac{3}{2d} + \dfrac{11}{16d}$, the LCD is 16d.

      For $\dfrac{a}{4b} + \dfrac{c}{3b}$, the LCD is 12b.

Find the lowest common denominator for each addition.

    a) $\dfrac{5}{9x} + \dfrac{1}{3x}$ _____   b) $\dfrac{x}{2a} + \dfrac{y}{5a}$ _____   c) $\dfrac{t}{10m} + \dfrac{1}{4m}$ _____

    a) 9x
    b) 8d
    c) 30t

---

78. When the denominators are two different letters, the LCD is the product of the letters. That is:

      For $\dfrac{5}{x} + \dfrac{3}{y}$, the LCD is xy.   For $\dfrac{a}{d} + \dfrac{c}{b}$, the LCD is _____.

    a) 9x
    b) 10a
    c) 20m

---

    bd

79.   When the denominators contain different letters with numerical coefficients, the LCD is the product of the letters and the smallest common multiple of the coefficients. That is:

For $\dfrac{5}{3a} + \dfrac{1}{6b}$, the LCD is 6ab.

For $\dfrac{m}{12k} + \dfrac{9}{8d}$, the LCD is 24dk.

Find the lowest common denominator for each addition.

a) $\dfrac{a}{10x} + \dfrac{b}{2y}$ _____   b) $\dfrac{1}{3P} + \dfrac{1}{7R}$ _____   c) $\dfrac{y}{15m} + \dfrac{9}{10t}$ _____

80.   When the denominators contain various letters, the LCD is the product of each letter <u>used</u> <u>once</u>. That is:

For $\dfrac{m}{ab} + \dfrac{t}{c}$, the LCD is abc.

For $\dfrac{h}{pq} + \dfrac{k}{pqr}$, the LCD is pqr.

Find the lowest common denominator for each addition.

a) $\dfrac{1}{xy} + \dfrac{1}{t}$ _____   b) $\dfrac{a}{cd} + \dfrac{b}{d}$ _____   c) $\dfrac{T}{AS} + \dfrac{R}{ASV}$ _____

a) 10xy

b) 21PR

c) 30mt

81.   When the denominators contain various letters and numerical coefficients, the LCD is the product of each letter <u>used</u> <u>once</u> and the smallest common multiple of the coefficients. That is:

For $\dfrac{5}{2a} + \dfrac{7}{ms}$, the LCD is 2ams.

For $\dfrac{H}{4PV} + \dfrac{T}{8P}$, the LCD is 8PV.

Find the lowest common denominator for each addition.

a) $\dfrac{9}{2b} + \dfrac{8}{3cd}$ _____   b) $\dfrac{x}{m} + \dfrac{y}{3mp}$ _____   c) $\dfrac{1}{12RT} + \dfrac{1}{3R}$ _____

a) txy

b) cd

c) ASV

82.   The following general principle is used to find lowest common denominators.

1) Use the smallest common multiple of the numerical factors.

2) Use each literal factor <u>once</u>.

Using the general principle above, find the lowest common denominator for each addition.

a) $\dfrac{3x}{4} + \dfrac{5x}{16}$ _____

b) $\dfrac{a}{8y} + \dfrac{b}{3}$ _____

c) $\dfrac{1}{8} + \dfrac{5}{t}$ _____

d) $\dfrac{v}{a} + \dfrac{d}{c}$ _____

a) 6bcd

b) 3mp

c) 12RT

83. For some additions, the LCD equals one of the denominators. For example:

$$\text{For } \frac{c}{4y} + \frac{d}{y}, \text{ the LCD is } 4y.$$

$$\text{For } \frac{1}{P} + \frac{1}{PV}, \text{ the LCD is PV.}$$

For other additions, the LCD does not equal one of the denominators. For example:

$$\text{For } \frac{x}{2d} + \frac{y}{3d}, \text{ the LCD is } 6d.$$

$$\text{For } \frac{3}{st} + \frac{5}{v}, \text{ the LCD is stv.}$$

Find the LCD for each addition.

a) $\dfrac{m}{7} + \dfrac{t}{6}$ _____   c) $\dfrac{1}{bcd} + \dfrac{1}{c}$ _____

b) $\dfrac{R}{V} + \dfrac{S}{8V}$ _____   d) $\dfrac{d}{xy} + \dfrac{f}{a}$ _____

a) 16   c) 8t

b) 24y   d) ac

---

84. Find the LCD for each addition.

a) $\dfrac{1}{4m} + \dfrac{1}{6m}$ _____   c) $\dfrac{a}{ms} + \dfrac{1}{2}$ _____

b) $\dfrac{3}{5} + \dfrac{x}{3y}$ _____   d) $\dfrac{v}{8PT} + \dfrac{m}{2T}$ _____

a) 42   c) bcd

b) 8V   d) axy

---

a) 12m   b) 15y   c) 2ms   d) 8PT

7-10   ADDING ALGEBRAIC FRACTIONS WITH UNLIKE DENOMINATORS

In this section, we will discuss the procedure for adding algebraic fractions with unlike denominators.

85. To add fractions by adding their numerators, they must have like or common denominators. Therefore, to add algebraic fractions with unlike denominators, we must get an equivalent addition with like denominators first. When one of the denominators is the LCD, we can get like denominators by substituting for the other fraction. Two examples are discussed below.

The LCD is 6. We substituted $\dfrac{3}{6}$ for $\dfrac{1}{2}$.

$$\frac{x}{6} + \frac{1}{2} = \frac{x}{6} + \frac{3}{6} = \frac{x+3}{6}$$

The LCD is 2y. We substituted $\dfrac{10}{2y}$ for $\dfrac{5}{y}$.

$$\frac{5}{y} + \frac{3}{2y} = \frac{10}{2y} + \frac{3}{2y} = \frac{13}{2y}$$

Continued on following page.

85.   Continued

Following the examples, complete these additions.

a) $\dfrac{7}{10} + \dfrac{t}{5} = \dfrac{7}{10} + \underline{\hphantom{xxxx}} = \underline{\hphantom{xxxxxxxx}}$

b) $\dfrac{3}{2} + \dfrac{5}{8d} = \underline{\hphantom{xxxx}} + \dfrac{5}{8d} = \underline{\hphantom{xxxxxxxx}}$

---

86.   Two more examples of the same type of addition are shown below.

The LCD is 12x. We substituted $\dfrac{3m}{12x}$ for $\dfrac{m}{4x}$.

$$\dfrac{1}{12x} + \dfrac{m}{4x} = \dfrac{1}{12x} + \dfrac{3m}{12x} = \dfrac{1 + 3m}{12x}$$

The LCD is bc. We substituted $\dfrac{ac}{bc}$ for $\dfrac{a}{b}$.

$$\dfrac{a}{b} + \dfrac{d}{bc} = \dfrac{ac}{bc} + \dfrac{d}{bc} = \dfrac{ac + d}{bc}$$

Following the examples, complete these additions.

a) $\dfrac{5}{3x} + \dfrac{1}{9x} = \underline{\hphantom{xxxx}} + \dfrac{1}{9x} = \underline{\hphantom{xxxx}}$

b) $\dfrac{R}{PV} + \dfrac{S}{P} = \dfrac{R}{PV} + \underline{\hphantom{xxxx}} = \underline{\hphantom{xxxx}}$

Answers (86):

a) $\dfrac{7}{10} + \dfrac{2t}{10} = \dfrac{7 + 2t}{10}$

b) $\dfrac{12d}{8d} + \dfrac{5}{8d} = \dfrac{12d + 5}{8d}$

---

87.   Using the same method, complete each addition.

a) $\dfrac{x}{8} + \dfrac{3x}{4} = \dfrac{x}{8} + \underline{\hphantom{xxxx}} = \underline{\hphantom{xxxx}}$

b) $\dfrac{1}{5} + \dfrac{7}{5y} = \underline{\hphantom{xxxx}} + \dfrac{7}{5y} = \underline{\hphantom{xxxx}}$

c) $\dfrac{a}{mt} + \dfrac{v}{m} = \dfrac{a}{mt} + \underline{\hphantom{xxxx}} = \underline{\hphantom{xxxx}}$

d) $\dfrac{1}{2pq} + \dfrac{5}{4pq} = \underline{\hphantom{xxxx}} + \dfrac{5}{4pq} = \underline{\hphantom{xxxx}}$

Answers (87):

a) $\dfrac{15}{9x} + \dfrac{1}{9x} = \dfrac{16}{9x}$

b) $\dfrac{R}{PV} + \dfrac{SV}{PV} = \dfrac{R + SV}{PV}$

---

88.   In the addition below, we had to reduce the sum to lowest terms.

$$\dfrac{1}{a} + \dfrac{12}{2a} = \dfrac{2}{2a} + \dfrac{12}{2a} = \dfrac{\overset{7}{\cancel{14}}}{\underset{1}{\cancel{2a}}} = \dfrac{7}{a}$$

Complete these. Reduce each sum to lowest terms.

a) $\dfrac{7}{12x} + \dfrac{3}{4x} = \underline{\hphantom{xxxxxxxxxxxxxxxx}}$

b) $\dfrac{7a}{8y} + \dfrac{9a}{24y} = \underline{\hphantom{xxxxxxxxxxxxxxxx}}$

Answers (88):

a) $\dfrac{x}{8} + \dfrac{6x}{8} = \dfrac{7x}{8}$

b) $\dfrac{y}{5y} + \dfrac{7}{5y} = \dfrac{y + 7}{5y}$

c) $\dfrac{a}{mt} + \dfrac{tv}{mt} = \dfrac{a + tv}{mt}$

d) $\dfrac{2}{4pq} + \dfrac{5}{4pq} = \dfrac{7}{4pq}$

89.   When neither denominator is the LCD, we have to substitute for both fractions to get an equivalent addition with like denominators.   Two examples are discussed below.

The LCD is 12.   We substituted $\frac{3t}{12}$ for $\frac{t}{4}$ and $\frac{4t}{12}$ for $\frac{t}{3}$.

$$\frac{t}{4} + \frac{t}{3} = \frac{3t}{12} + \frac{4t}{12} = \frac{7t}{12}$$

The LCD is 5x.   We substituted $\frac{3x}{5x}$ for $\frac{3}{5}$ and $\frac{5}{5x}$ for $\frac{1}{x}$.

$$\frac{3}{5} + \frac{1}{x} = \frac{3x}{5x} + \frac{5}{5x} = \frac{3x + 5}{5x}$$

Following the examples, complete each addition below.

a) $\frac{a}{6} + \frac{a}{4} = $ _____ + _____ = _____

b) $\frac{1}{2x} + \frac{5}{3} = $ _____ + _____ = _____

a) $\frac{4}{3x}$, from $\frac{16}{12x}$

b) $\frac{5a}{4y}$, from $\frac{30a}{24y}$

---

90.   Two more examples of the same type of addition are discussed below.

The LCD is cd.   We substituted $\frac{ad}{cd}$ for $\frac{a}{c}$ and $\frac{bc}{cd}$ for $\frac{b}{d}$.

$$\frac{a}{c} + \frac{b}{d} = \frac{ad}{cd} + \frac{bc}{cd} = \frac{ad + bc}{cd}$$

The LCD is 2xy.   We substituted $\frac{y}{2xy}$ for $\frac{1}{2x}$ and $\frac{2}{2xy}$ for $\frac{1}{xy}$.

$$\frac{1}{2x} + \frac{1}{xy} = \frac{y}{2xy} + \frac{2}{2xy} = \frac{y + 2}{2xy}$$

Following the examples, complete each addition below.

a) $\frac{2}{p} + \frac{3}{q} = $ _____ + _____ = _____

b) $\frac{t}{my} + \frac{1}{v} = $ _____ + _____ = _____

a) $\frac{2a}{12} + \frac{3a}{12} = \frac{5a}{12}$

b) $\frac{3}{6x} + \frac{10x}{6x} = \frac{3 + 10x}{6x}$

---

91.   Using the same method, complete each addition.

a) $\frac{m}{4} + \frac{t}{5} = $ _____ + _____ = _____

b) $\frac{1}{2x} + \frac{1}{3x} = $ _____ + _____ = _____

c) $\frac{R}{S} + \frac{3}{T} = $ _____ + _____ = _____

d) $\frac{3b}{c} + \frac{2a}{d} = $ _____ + _____ = _____

a) $\frac{2q}{pq} + \frac{3p}{pq} = \frac{2q + 3p}{pq}$

b) $\frac{tv}{mvy} + \frac{my}{mvy} = \frac{tv + my}{mvy}$

92. When adding algebraic fractions, <u>check first to see if one of the</u> <u>denominators is the LCD</u>.

     If one of the denominators is the LCD, you only have to substitute for <u>one</u> fraction.

     If neither denominator is the LCD, you have to substitute for <u>both</u> fractions.

Complete: a) $\dfrac{1}{4t} + \dfrac{1}{t} =$ _____

                b) $\dfrac{a}{3} + \dfrac{b}{2} =$ _____

                c) $\dfrac{c}{xy} + \dfrac{d}{x} =$ _____

                d) $\dfrac{4}{v} + \dfrac{3}{t} =$ _____

a) $\dfrac{5m}{20} + \dfrac{4t}{20} = \dfrac{5m + 4t}{20}$

b) $\dfrac{3}{6x} + \dfrac{2}{6x} = \dfrac{5}{6x}$

c) $\dfrac{RT}{ST} + \dfrac{3S}{ST} = \dfrac{RT + 3S}{ST}$

d) $\dfrac{3bd}{cd} + \dfrac{2ac}{cd} = \dfrac{3bd + 2ac}{cd}$

a) $\dfrac{5}{4t}$ b) $\dfrac{2a + 3b}{6}$ c) $\dfrac{c + dy}{xy}$ d) $\dfrac{4t + 3v}{tv}$

7-11   ADDING AN ALGEBRAIC FRACTION AND A NON-FRACTION

In this section, we will discuss the procedure for adding an algebraic fraction and a non-fraction.

93. To convert a non-fraction to a fraction with a specific denominator, we multiply by a fraction that equals "1". The terms of the fraction that equals "1" are determined by the denominator of the desired fraction. For example:

     To convert 2 to a fraction whose denominator is 4, we multiply by $\dfrac{4}{4}$.

$$2\left(\dfrac{4}{4}\right) = \dfrac{8}{4}$$

     To convert 5 to a fraction whose denominator is "x", we multiply by $\dfrac{x}{x}$.

$$5\left(\dfrac{x}{x}\right) = \dfrac{5x}{x}$$

Following the examples, complete these.

    a) $7\left(\dfrac{\quad}{\quad}\right) = \dfrac{(\quad)}{2}$ b) $3\left(\dfrac{\quad}{\quad}\right) = \dfrac{(\quad)}{2y}$

a) $7\left(\dfrac{2}{2}\right) = \dfrac{14}{2}$

b) $3\left(\dfrac{2y}{2y}\right) = \dfrac{6y}{2y}$

94.   Two more examples of the same type of conversion are discussed below.

To convert "y" to a fraction whose denominator is 7, we multiply by $\frac{7}{7}$.

$$y\left(\frac{7}{7}\right) = \frac{7y}{7}$$

To convert "2t" to a fraction whose denominator is b, we multiply by $\frac{b}{b}$.

$$2t\left(\frac{b}{b}\right) = \frac{2bt}{b}$$

Following the examples, complete these

a) $3x\left(\dfrac{\quad}{\quad}\right) = \dfrac{(\quad)}{4}$ b) $p\left(\dfrac{\quad}{\quad}\right) = \dfrac{(\quad)}{qr}$

---

95.   The same type of conversion can be made <u>by</u> <u>multiplying</u> the <u>non-fraction</u> <u>and the desired denominator</u>.   That is:

To convert 3 to a fraction whose denominator is "y":

$$3 = \frac{(3y)}{y} \qquad [\text{since } (3)(y) = 3y]$$

$$2m = \frac{(8m)}{4} \qquad [\text{since } (2m)(4) = 8m]$$

Using the shorter method above, complete these:

a) $1 = \dfrac{(\quad)}{9}$ b) $x = \dfrac{(\quad)}{5}$ c) $2t = \dfrac{(\quad)}{7}$

Answers:

a) $3x\left(\dfrac{4}{4}\right) = \dfrac{12x}{4}$

b) $p\left(\dfrac{qr}{qr}\right) = \dfrac{pqr}{qr}$

---

96.   Using the shorter method, complete these:

a) $7 = \dfrac{(\quad)}{b}$ b) $5 = \dfrac{(\quad)}{3d}$ c) $4y = \dfrac{(\quad)}{cd}$

Answers:

a) $\dfrac{9}{9}$ c) $\dfrac{14t}{7}$

b) $\dfrac{5x}{5}$

---

97.   To add an algebraic fraction and a non-fraction, we convert the non-fraction to a fraction with the same denominator.   Two examples are shown below.

$$\frac{x}{3} + 2 = \frac{x}{3} + \frac{6}{3} = \frac{x+6}{3} \qquad 2m + \frac{3}{5} = \frac{10m}{5} + \frac{3}{5} = \frac{10m+3}{5}$$

Following the examples, complete these.

a) $1 + \dfrac{2y}{7} = \underline{\qquad} + \dfrac{2y}{7} = \underline{\qquad}$

b) $\dfrac{1}{2} + b = \dfrac{1}{2} + \underline{\qquad} = \underline{\qquad}$

Answers:

a) $\dfrac{7b}{b}$

b) $\dfrac{15d}{3d}$

c) $\dfrac{4cdy}{cd}$

Answers (97):

a) $\dfrac{7}{7} + \dfrac{2y}{7} = \dfrac{7+2y}{7}$

b) $\dfrac{1}{2} + \dfrac{2b}{2} = \dfrac{1+2b}{2}$

98.   Two more examples of the same type of addition are shown below.

$$\frac{8}{8} + \frac{1}{x} = \frac{8x}{x} + \frac{1}{x} = \frac{8x + 1}{x} \qquad \frac{b}{cd} + R = \frac{b}{cd} + \frac{cdR}{cd} = \frac{b + cdR}{cd}$$

Following the examples, complete these.

a) $\dfrac{3}{t} + 1 = \dfrac{3}{t} + \underline{\hspace{1.5cm}} = \underline{\hspace{1.5cm}}$

b) $h + \dfrac{P}{V} = \underline{\hspace{1.5cm}} + \dfrac{P}{V} = \underline{\hspace{1.5cm}}$

---

99.   Use the same method for these.

a) $\dfrac{x}{8} + 7 = \underline{\hspace{2cm}}$ c) $\dfrac{1}{4} + d = \underline{\hspace{2cm}}$

b) $4y + \dfrac{2}{3} = \underline{\hspace{2cm}}$ d) $1 + \dfrac{R}{5} = \underline{\hspace{2cm}}$

a) $\dfrac{3}{t} + \dfrac{t}{t} = \dfrac{3 + t}{t}$

b) $\dfrac{hV}{V} + \dfrac{P}{V} = \dfrac{hV + P}{V}$

---

100.   Use the same method for these.

a) $4 + \dfrac{5}{x} = \underline{\hspace{2cm}}$ c) $a + \dfrac{b}{c} = \underline{\hspace{2cm}}$

b) $\dfrac{m}{v} + 1 = \underline{\hspace{2cm}}$ d) $\dfrac{H}{PV} + 2T = \underline{\hspace{2cm}}$

a) $\dfrac{x + 56}{8}$

b) $\dfrac{12y + 2}{3}$

c) $\dfrac{1 + 4d}{4}$

d) $\dfrac{5 + R}{5}$

a) $\dfrac{4x + 5}{x}$ b) $\dfrac{m + v}{v}$ c) $\dfrac{ac + b}{c}$ d) $\dfrac{H + 2PTV}{PV}$

SELF-TEST 23 (pages 286-299)

Add.   Report each sum in lowest terms.

1. $\dfrac{b}{2a} + \dfrac{5b}{2a} = \underline{\hspace{3cm}}$ 2. $\dfrac{a + t}{p + r} + \dfrac{a - t}{p + r} = \underline{\hspace{3cm}}$

3. $\dfrac{y}{2} + \dfrac{5y}{6} = \underline{\hspace{3cm}}$ 4. $\dfrac{h}{6} + \dfrac{v}{4} = \underline{\hspace{3cm}}$

5. $\dfrac{d}{3w} + \dfrac{2}{w} = \underline{\hspace{3cm}}$ 6. $\dfrac{c}{r} + \dfrac{s}{t} = \underline{\hspace{3cm}}$

Continued on following page.

SELF-TEST 23 (pages 286-299) - continued

7. $\dfrac{1}{B} + \dfrac{1}{A} =$ _____

8. $\dfrac{a}{px} + \dfrac{1}{2x} =$ _____

9. $1 + \dfrac{1}{F} =$ _____

10. $3t + \dfrac{d}{p} =$ _____

ANSWERS:

1. $\dfrac{3b}{a}$

2. $\dfrac{2a}{p+r}$

3. $\dfrac{4y}{3}$

4. $\dfrac{2h+3v}{12}$

5. $\dfrac{d+6}{3w}$

6. $\dfrac{ct+rs}{rt}$

7. $\dfrac{A+B}{AB}$

8. $\dfrac{2a+p}{2px}$

9. $\dfrac{F+1}{F}$

10. $\dfrac{3pt+d}{p}$

7-12   SUBTRACTIONS INVOLVING ALGEBRAIC FRACTIONS

In this section, we will discuss the procedure for subtractions involving algebraic fractions.

101.   To subtract algebraic fractions with like denominators, we subtract the numerators and keep the same denominator. For example:

$$\frac{2x}{5} - \frac{7}{5} = \frac{2x-7}{5} \qquad \frac{M}{P} - \frac{T}{P} = \frac{M-T}{P}$$

Following the examples, complete these:

a) $\dfrac{b}{y} - \dfrac{1}{y} =$ _____

b) $\dfrac{3p}{ab} - \dfrac{2q}{ab} =$ _____

102.   When subtracting algebraic fractions, we combine like terms in the numerator if possible. For example:

$$\frac{4x}{7} - \frac{2x}{7} = \frac{4x-2x}{7} = \frac{2x}{7} \qquad \frac{m}{bc} - \frac{m}{bc} = \frac{m-m}{bc} = \frac{0}{bc} = 0$$

Following the examples, complete these.

a) $\dfrac{9}{a} - \dfrac{5}{a} =$ _____

b) $\dfrac{8d}{V} - \dfrac{7d}{V} =$ _____

a) $\dfrac{b-1}{y}$

b) $\dfrac{3p-2q}{ab}$

103.   In the subtraction below, we reduced the difference to lowest terms.

$$\frac{7}{5x} - \frac{2}{5x} = \frac{7-2}{5x} = \frac{\overset{1}{\cancel{5}}}{\underset{1}{\cancel{5}}x} = \frac{1}{x}$$

Following the example, complete these.

a) $\dfrac{9y}{10} - \dfrac{y}{10} =$ _____

b) $\dfrac{11P}{4m} - \dfrac{3P}{4m} =$ _____

a) $\dfrac{4}{a}$

b) $\dfrac{d}{V}$

104. In the subtraction below, the LCD is 6. Therefore, we had to substitute for only <u>one</u> fraction to get like denominators.

$$\frac{x}{2} - \frac{5}{6} = \frac{3x}{6} - \frac{5}{6} = \frac{3x - 5}{6}$$

Complete these. Reduce to lowest terms if possible.

a) $\dfrac{y}{cd} - \dfrac{t}{c} =$ _____

b) $\dfrac{1}{3x} - \dfrac{1}{12x} =$ _____

a) $\dfrac{4y}{5}$, from $\dfrac{8y}{10}$

b) $\dfrac{2P}{m}$, from $\dfrac{8P}{4m}$

---

105. In the subtraction below, the LCD is 15. Therefore, we had to substitute for <u>both</u> fractions to get like denominators.

$$\frac{x}{3} - \frac{y}{5} = \frac{5x}{15} - \frac{3y}{15} = \frac{5x - 3y}{15}$$

Complete: a) $\dfrac{a}{b} - \dfrac{c}{pq} =$ _____

b) $\dfrac{m}{3x} - \dfrac{t}{4x} =$ _____

a) $\dfrac{y - dt}{cd}$

b) $\dfrac{1}{4x}$, from $\dfrac{3}{12x}$

---

106. We had to convert the non-fraction below to a fraction before subtracting.

$$1 - \frac{p}{q} = \frac{q}{q} - \frac{p}{q} = \frac{q - p}{q}$$

Complete: a) $\dfrac{h}{t} - 2x =$ _____

b) $a - \dfrac{b}{cd} =$ _____

a) $\dfrac{apq - bc}{bpq}$

b) $\dfrac{4m - 3t}{12x}$

---

107. Complete: a) $\dfrac{1}{x} - \dfrac{1}{y} =$ _____

b) $\dfrac{7}{t} - \dfrac{9}{2t} =$ _____

c) $\dfrac{R}{PQ} - \dfrac{S}{Q} =$ _____

d) $\dfrac{a}{bc} - 1 =$ _____

a) $\dfrac{h - 2tx}{t}$

b) $\dfrac{acd - b}{cd}$

---

108. The same method is used to subtract fractions with binomial denominators. For example:

$$\frac{2x}{c + d} - \frac{y}{c + d} = \frac{2x - y}{c + d} \qquad \frac{H}{P - Q} - \frac{T}{P - Q} =$$ _____

a) $\dfrac{y - x}{xy}$

b) $\dfrac{5}{2t}$

c) $\dfrac{R - PS}{PQ}$

d) $\dfrac{a - bc}{bc}$

109. The same method is used to subtract fractions with binomial numerators. For example:

$$\frac{t+5}{v} - \frac{1}{v} = \frac{(t+5)-1}{v} = \frac{t+4}{v}$$

Complete: a) $\dfrac{a+4b}{c} - \dfrac{b}{c} =$ _____

b) $\dfrac{k+1}{P-p} - \dfrac{k}{P-p} =$ _____

$\dfrac{H-T}{P-Q}$

110. The expression below contains an addition and a subtraction.

$$\frac{x}{7} + \frac{y}{7} - \frac{5}{7} = \frac{x+y-5}{7}$$

Complete: a) $\dfrac{a}{d} - \dfrac{b}{d} - \dfrac{c}{d} =$ _____

b) $\dfrac{h}{2x} + \dfrac{3}{2x} - \dfrac{t}{2x} =$ _____

a) $\dfrac{a+3b}{c}$

b) $\dfrac{1}{P-p}$

a) $\dfrac{a-b-c}{d}$ b) $\dfrac{h+3-t}{2x}$

7-13 EQUIVALENT FORMS FOR SUMS OF FRACTIONS

Any fraction whose numerator is a two-term addition is a sum of two fractions. In this section, we will show how fractions of that type can be written in an equivalent form by breaking up the sum into the two original fractions.

111. Any fraction whose numerator is a two-term addition is a sum of two fractions. It can be broken up into the two original fractions. For example:

$$\frac{y+5}{7} = \frac{y}{7} + \frac{5}{7} \qquad\qquad \frac{a+b}{c} = \frac{a}{c} + \frac{b}{c}$$

Break up each sum into the original fractions.

a) $\dfrac{5m+3t}{8} =$ _____ + _____   b) $\dfrac{cd+pq}{mt} =$ _____ + _____

112. When a sum of two fractions is broken up, sometimes one or both of the original fractions can be reduced to lowest terms.

In the example below, one fraction can be reduced.

$$\frac{y+2}{6} = \frac{y}{6} + \frac{2}{6} = \frac{y}{6} + \frac{1}{3}$$

In the example below, both fractions can be reduced.

$$\frac{3a+4b}{12} = \frac{3a}{12} + \frac{4b}{12} = \frac{a}{4} + \frac{b}{3}$$

a) $\dfrac{5m}{8} + \dfrac{3t}{8}$

b) $\dfrac{cd}{mt} + \dfrac{pq}{mt}$

Continued on following page.

112.   Continued

Break up each sum into the two original fractions and then reduce to lowest terms where possible.

a) $\dfrac{t + 5}{10}$ = _____ + _____ = _____ + _____

b) $\dfrac{4m + 2t}{8}$ = _____ + _____ = _____ + _____

---

113.   Break up each sum into the two original fractions and then reduce to lowest terms where possible.

a) $\dfrac{m + tv}{ct}$ = _____ + _____ = _____ + _____

b) $\dfrac{ay + cx}{bxy}$ = _____ + _____ = _____ + _____

a) $\dfrac{t}{10} + \dfrac{5}{10} = \dfrac{t}{10} + \dfrac{1}{2}$

b) $\dfrac{4m}{8} + \dfrac{2t}{8} = \dfrac{m}{2} + \dfrac{t}{4}$

---

114.   In the example below, one of the fractions reduces to a non-fraction.

$$\dfrac{y + 7}{7} = \dfrac{y}{7} + \dfrac{7}{7} = \dfrac{y}{7} + 1$$

Following the example, complete these:

a) $\dfrac{a + 12}{2}$ = _____ + _____ = _____ + _____

b) $\dfrac{2x + 8y}{4}$ = _____ + _____ = _____ + _____

a) $\dfrac{m}{ct} + \dfrac{tv}{ct} = \dfrac{m}{ct} + \dfrac{v}{c}$

b) $\dfrac{ay}{bxy} + \dfrac{cx}{bxy} = \dfrac{a}{bx} + \dfrac{c}{by}$

---

115.   In the example below, one of the fractions also reduces to a non-fraction.

$$\dfrac{P + QR}{Q} = \dfrac{P}{Q} + \dfrac{QR}{Q} = \dfrac{P}{Q} + R$$

Following the example, complete these.

a) $\dfrac{bd + m}{d}$ = _____ + _____ = _____ + _____

b) $\dfrac{LT + AKT}{AKT}$ = _____ + _____ = _____ + _____

a) $\dfrac{a}{2} + \dfrac{12}{2} = \dfrac{a}{2} + 6$

b) $\dfrac{2x}{4} + \dfrac{8y}{4} = \dfrac{x}{2} + 2y$

---

116.   Write each of these in an equivalent form with both parts reduced to lowest terms.

a) $\dfrac{4d + 7}{7}$ = _____

b) $\dfrac{cH + S}{c}$ = _____

a) $\dfrac{bd}{d} + \dfrac{m}{d} = b + \dfrac{m}{d}$

b) $\dfrac{LT}{AKT} + \dfrac{AKT}{AKT} = \dfrac{L}{AK} + 1$

---

117.   Which of the following is equivalent to $\dfrac{5y + 9}{9}$?   _____

a) $5y + 1$ b) $\dfrac{5y}{9} + 1$ c) $5y$

a) $\dfrac{4d}{7} + 1$

b) $H + \dfrac{S}{c}$

118. Which of the following is equivalent to $\dfrac{a + bc}{b}$? _____

    a) $a + c$ b) $a + \dfrac{c}{b}$ c) $\dfrac{a}{b} + c$

(b)

119. Which of the following is equivalent to $\dfrac{PQR + T}{QR}$? _____

    a) $P + \dfrac{T}{QR}$ b) $P + T$ c) $\dfrac{P}{QR} + T$

(c)

120. Which of the following is equivalent to $\dfrac{2x}{3} + 3$? _____

    a) $2x$ b) $\dfrac{2x + 9}{3}$ c) $\dfrac{5x}{3}$

(a)

121. Which of the following is equivalent to $S + \dfrac{T}{V}$? _____

    a) $\dfrac{SV + T}{V}$ b) $\dfrac{S + T}{V}$ c) $\dfrac{S + TV}{V}$

(b)

122. Which of the following is equivalent to $\dfrac{33,000H}{\pi dR} + F$? _____

    a) $\dfrac{33,000H + F}{\pi dR}$ b) $\dfrac{33,000\pi dRH + F}{\pi dR}$ c) $\dfrac{33,000H + \pi dRF}{\pi dR}$

(a)

123. **Though the fractions** below have binomial denominators, each is the sum of two fractions. Therefore, each can be broken up into the original fractions. That is:

$$\frac{p + q}{m + t} = \frac{p}{m + t} + \frac{q}{m + t}$$

$$\frac{K + T}{S - V} = \underline{\quad\quad} + \underline{\quad\quad}$$

(c)

124. The numerator of the fraction below is a three-term addition. Therefore, it is the sum of three fractions. That is:

$$\frac{p + q + r}{t} = \frac{p}{t} + \frac{q}{t} + \frac{r}{t}$$

Break up each of these into the original fractions with each reduced to lowest terms.

    a) $\dfrac{a + b + 4}{4} = $ _____

    b) $\dfrac{cV + dS + c}{c} = $ _____

$\dfrac{K}{S - V} + \dfrac{T}{S - V}$

a) $\dfrac{a}{4} + \dfrac{b}{4} + 1$ b) $V + \dfrac{dS}{c} + 1$

7-14   EQUIVALENT FORMS FOR DIFFERENCES OF FRACTIONS

Any fraction whose numerator is a subtraction is a difference of two fractions. In this section, we will show how fractions of that type can be written in an equivalent form by breaking up the difference into the two original fractions.

---

125.   Any fraction whose numerator is a subtraction is the difference of two fractions. It can be broken up into the original fractions. For example:

$$\frac{2t - 3p}{m} = \frac{2t}{m} - \frac{3p}{m} \qquad\qquad \frac{R - S}{P} = \underline{\quad\quad} - \underline{\quad\quad}$$

---

126.   Break up each fraction below into the two original fractions and then reduce to lowest terms.

a) $\dfrac{m - 14}{7} = \underline{\quad\quad} - \underline{\quad\quad} = \underline{\quad\quad} - \underline{\quad\quad}$

b) $\dfrac{P - Q}{P} = \underline{\quad\quad} - \underline{\quad\quad} = \underline{\quad\quad} - \underline{\quad\quad}$

$\dfrac{R}{P} - \dfrac{S}{P}$

---

127.   Write each fraction in an equivalent form with both parts reduced to lowest terms.

a) $\dfrac{5 - y}{5} = \underline{\quad\quad\quad}$ b) $\dfrac{a - bx}{b} = \underline{\quad\quad\quad}$

a) $\dfrac{m}{7} - \dfrac{14}{7} = \dfrac{m}{7} - 2$

b) $\dfrac{P}{P} - \dfrac{Q}{P} = 1 - \dfrac{Q}{P}$

---

128.   Write each fraction in an equivalent form with both parts reduced to lowest terms.

a) $\dfrac{mt - v}{t} = \underline{\quad\quad\quad}$ b) $\dfrac{KV - BPQ}{BP} = \underline{\quad\quad\quad}$

a) $1 - \dfrac{y}{5}$

b) $\dfrac{a}{b} - x$

---

129.   Which of the following is equivalent to $\dfrac{abc - ky}{ac}$? $\underline{\quad\quad}$

a) $abc - \dfrac{ky}{bc}$ b) $b - \dfrac{ky}{ac}$ c) $b - ky$

a) $m - \dfrac{v}{t}$

b) $\dfrac{KV}{BP} - Q$

---

130.   Which of the following is equivalent to $\dfrac{M - STV}{TV}$? $\underline{\quad\quad}$

a) $\dfrac{M}{TV} - S$ b) $M - S$ c) $\dfrac{M}{TV} - STV$

(b)

---

131.   Which of the following is equivalent to $x - \dfrac{ay}{ct}$? $\underline{\quad\quad}$

a) $\dfrac{x - ay}{ct}$ b) $\dfrac{ctx - acty}{ct}$ c) $\dfrac{ctx - ay}{ct}$

(a)

(c)

132. Though the fractions below have binomial denominators, each is the difference of two fractions. Therefore, we can break them up into the original fractions. That is:

$$\frac{m - 7}{t + 5} = \frac{m}{t + 5} - \frac{7}{t + 5}$$

$$\frac{P - Q}{R - S} = \underline{} - \underline{}$$

133. Following the example, break up the other fractions into three original fractions.

$$\frac{m + y - 9}{5} = \frac{m}{5} + \frac{y}{5} - \frac{9}{5}$$

a) $\dfrac{a - b + c}{d} = \underline{}$ b) $\dfrac{D - S - T}{M} = \underline{}$

> $\dfrac{P}{R - S} - \dfrac{Q}{R - S}$

> a) $\dfrac{a}{d} - \dfrac{b}{d} + \dfrac{c}{d}$ b) $\dfrac{D}{M} - \dfrac{S}{M} - \dfrac{T}{M}$

7-15   CONTRASTING TWO TYPES OF COMPLICATED FRACTIONS

When a fraction contains a binomial in its numerator, it is the sum or difference of two fractions. However, when a fraction contains a binomial only in its denominator, it is not a sum or difference of two fractions. We will contrast those two types of fractions in this section.

134. When a fraction contains a binomial addition in its numerator, it is the sum of two fractions and can be broken up into the original fractions. For example:

$$\frac{x + 3}{7} = \frac{x}{7} + \frac{3}{7}$$

However, when a fraction contains a binomial addition only in its denominator, it is not the sum of two fractions. Therefore, it cannot be broken up into two fractions. That is:

$$\frac{10}{y + 10} \neq \frac{10}{y} + \frac{10}{10}$$

To show that the expressions above are not equal, we can substitute 2 for "y" in each and evaluate.

a) $\dfrac{10}{y + 10} = \dfrac{10}{2 + 10} = \underline{}$ b) $\dfrac{10}{y} + \dfrac{10}{10} = \dfrac{10}{2} + \dfrac{10}{10} = \underline{}$

> a) $\dfrac{10}{12} = \dfrac{5}{6}$
>
> b) $5 + 1 = 6$

135.   When a fraction contains a binomial subtraction in its numerator, it is the difference of two fractions and can be broken up into the original fractions. For example:

$$\frac{w - 5}{w} = \frac{w}{w} - \frac{5}{w}$$

However, when a fraction contains a binomial subtraction <u>only in its denominator</u>, it is not the difference of two fractions. Therefore, it cannot be broken up into two fractions. That is:

$$\frac{x}{x - 1} \neq \frac{x}{x} - \frac{x}{1}$$

To show that the expressions above are not equal, we can substitute 5 for "x" in each and evaluate.

a) $\dfrac{x}{x - 1} = \dfrac{5}{5 - 1} =$ _____   b) $\dfrac{x}{x} - \dfrac{x}{1} = \dfrac{5}{5} - \dfrac{5}{1} =$ _____

---

136.   Which of the following is either a sum or difference of two fractions? _____

a) $\dfrac{M + 5}{M}$ b) $\dfrac{t}{t + 7}$ c) $\dfrac{ab}{a + b}$ d) $\dfrac{a - b}{ab}$

a) $\dfrac{5}{4}$

b) $1 - 5 = -4$

---

137.   Fractions like those below cannot be written in an equivalent form by breaking them up into two fractions.

$$\frac{P}{P + 6} \qquad\qquad \frac{d}{d - a}$$

If possible, write each of these in an equivalent form by breaking it up into two fractions.

a) $\dfrac{k}{k + 1} =$ _____   c) $\dfrac{R + T}{T} =$ _____

b) $\dfrac{k - 1}{k} =$ _____   d) $\dfrac{T}{R - T} =$ _____

Only (a) and (d)

---

138.   A fraction with a binomial <u>in</u> <u>both</u> <u>its</u> <u>numerator</u> <u>and</u> <u>denominator</u> is the sum or difference of two fractions. Therefore, it can be broken up into the original fractions. That is:

$$\frac{a + b}{c + d} = \frac{a}{c + d} + \frac{b}{c + d}$$

$$\frac{I - CV}{M - T} =$$ _____

a) Not possible

b) $\dfrac{k}{k} - \dfrac{1}{k} = 1 - \dfrac{1}{k}$

c) $\dfrac{R}{T} + \dfrac{T}{T} = \dfrac{R}{T} + 1$

d) Not possible

$$\frac{I}{M - T} - \frac{CV}{M - T}$$

7-16   MULTIPLYING AND FACTORING BY THE DISTRIBUTIVE PRINCIPLE

In this section, we will review the procedures for multiplying and factoring by the distributive principle and extend them to expressions containing more than one variable.   Both procedures will be used in the discussion of the equivalent forms of complicated fractions in the next section.

139.   Two examples of multiplying by the distributive principle are shown below.

$$3(x + 5) = 3x + 15 \qquad 7(y - 1) = 7y - 7$$

The same procedure is used when the expression contains more than one variable.   That is:

$$2(x + y) = 2x + 2y \qquad a(b - c) = ab - ac$$

Following the examples, complete these:

a) $9(y + 7)$ = _____   c) $2(R + 3P)$ = _____

b) $4(m - t)$ = _____   d) $M(T - V)$ = _____

---

140.   When a binomial expression contains the same factor in each term, we can factor by the distributive principle.   For example:

$$2x + 6 = 2(x) + 2(3) = 2(x + 3)$$

$$10y - 5 = 5(2y) - 5(1) = 5(2y - 1)$$

The same procedure is used when the expression contains more than one variable.   That is:

$$7a + 7b = 7(a) + 7(b) = 7(a + b)$$

$$6d - 3h = 3(2d) - 3(h) = 3(2d - h)$$

Following the examples, factor each expression.

a) $9p - 9q$ = _____   c) $4d - 20$ = _____

b) $8x + 4$ = _____   d) $2h + 8t$ = _____

Answers to frame 139:
a) $9y + 63$
b) $4m - 4t$
c) $2R + 6P$
d) $MT - MV$

---

141.   In each factoring below, the common factor is a letter.

$$ab + ac = a(b + c)$$

$$cP - dP = P(c - d)$$

Factor out the common factor in these.

a) $mt + mv$ = _____   c) $pq + kq$ = _____

b) $H_1T_1 - H_1S_2$ = _____   d) $MR - NR$ = _____

Answers to frame 140:
a) $9(p - q)$
b) $4(2x + 1)$
c) $4(d - 5)$
d) $2(h + 4t)$

142. Two more examples of factoring out a common factor are shown below.

$$bs + st = s(b + t) \qquad CM - AC = C(M - A)$$

Following the examples, factor these.

a) $mt + km =$ _____   b) $C_1D_1 - D_1T_1 =$ _____

---

143. In the factorings below, we substituted "1S" for "S' and "1b" for "b" so that the "1" was not forgotten.

$$ST + S = ST + 1S = S(T + 1)$$

$$b - bc = 1b - bc = b(1 - c)$$

Following the examples, factor these:

a) $MT + T =$ _____   c) $cd - c =$ _____

b) $p + pP =$ _____   d) $R - QR =$ _____

---

144. To factor by the distributive principle, each term must have a common factor. That is:

$bd + cd$ can be factored because "d" is a common factor.

$HM + TV$ cannot be factored because there is no common factor.

Factor by the distributive principle if possible.

a) $aH - bH =$ _____   c) $KT - SV =$ _____

b) $cp + dq =$ _____   d) $c_1t + c_1v =$ _____

---

145. Factor by the distributive principle if possible.

a) $ab - c =$ _____   c) $VT - V =$ _____

b) $mM + m =$ _____   d) $RS - Q =$ _____

Answer column:

142.
a) $m(t + v)$
b) $H_1(T_1 - S_2)$
c) $q(p + k)$
d) $R(M - N)$

143.
a) $m(t + k)$
b) $D_1(C_1 - T_1)$

144.
a) $T(M + 1)$
b) $p(1 + P)$
c) $c(d - 1)$
d) $R(1 - Q)$

145.
a) $H(a - b)$
b) Not possible
c) Not possible
d) $c_1(t + v)$

a) Not possible
b) $m(M + 1)$
c) $V(T - 1)$
d) Not possible

SELF-TEST 24 (pages 299-309)

Do these subtractions. Report each difference in lowest terms.

1. $\dfrac{8}{3P} - \dfrac{2}{3P} =$ _____

2. $\dfrac{3t}{d} - \dfrac{r}{a} =$ _____

3. $\dfrac{h+1}{c-w} - \dfrac{h}{c-w} =$ _____

4. $P - \dfrac{1}{F} =$ _____

Break up each fraction into an addition or a subtraction of two fractions in lowest terms.

5. $\dfrac{5x+2y}{10x} =$ _____

6. $\dfrac{3A+B}{AB} =$ _____

7. $\dfrac{hs-2pr}{ps} =$ _____

8. $\dfrac{V-TW}{W} =$ _____

9. Which of the following can be broken up into two fractions. _____

  a) $\dfrac{b}{d+2}$ b) $\dfrac{G+1}{G-1}$ c) $\dfrac{t-3a}{r}$ d) $\dfrac{1}{x-y}$

Factor each expression by the distributive principle if possible.

10. $4BR + 8R =$ _____

11. $2hp - dt =$ _____

12. $s - ks =$ _____

ANSWERS:

1. $\dfrac{2}{P}$

2. $\dfrac{3at-dr}{ad}$

3. $\dfrac{1}{c-w}$

4. $\dfrac{FP-1}{F}$

5. $\dfrac{1}{2} + \dfrac{y}{5x}$

6. $\dfrac{3}{B} + \dfrac{1}{A}$

7. $\dfrac{h}{p} - \dfrac{2r}{s}$

8. $\dfrac{V}{W} - T$

9. b, c

10. $4R(B+2)$

11. Not possible

12. $s(1-k)$

7-17 REDUCING COMPLICATED FRACTIONS TO LOWEST TERMS

In this section, we will show how complicated fractions can be reduced to lowest terms by factoring out a fraction that equals "1". By "complicated" fractions, we mean fractions with a binomial numerator or denominator or both.

---

146.   When a fraction has a binomial numerator, we can generate an equivalent fraction by multiplying it by a fraction that equals "1". For example, we multiplied the fraction below by $\dfrac{a}{a}$. Notice that we multiplied by the distributive principle in the numerator.

$$\frac{p+q}{r} = \left(\frac{a}{a}\right)\left(\frac{p+q}{r}\right) = \frac{a(p+q)}{ar} = \frac{ap+aq}{ar}$$

Continued on following page.

146.   Continued

Following the example, complete these:

a) $\dfrac{a + b}{c} = \left(\dfrac{3}{3}\right)\left(\dfrac{a + b}{c}\right) = \dfrac{3(a + b)}{3c} = $ _____

b) $\dfrac{x - 2y}{5} = \left(\dfrac{2}{2}\right)\left(\dfrac{x - 2y}{5}\right) = \dfrac{2(x - 2y)}{10} = $ _____

---

147.   When a fraction has a binomial denominator, we can generate an equivalent fraction by multiplying it by a fraction that equals "1".   For example, we multiplied the fraction below by $\dfrac{4}{4}$. Notice that we multiplied by the distributive principle in the denominator.

$$\dfrac{c}{d - t} = \left(\dfrac{4}{4}\right)\left(\dfrac{c}{d - t}\right) = \dfrac{4c}{4(d - t)} = \dfrac{4c}{4d - 4t}$$

Following the example, complete these:

a) $\dfrac{M}{P + Q} = \left(\dfrac{b}{b}\right)\left(\dfrac{M}{P + Q}\right) = \dfrac{bM}{b(P + Q)} = $ _____

b) $\dfrac{1}{3a - 2b} = \left(\dfrac{3}{3}\right)\left(\dfrac{1}{3a - 2b}\right) = \dfrac{3}{3(3a - 2b)} = $ _____

a) $\dfrac{3a + 3b}{3c}$

b) $\dfrac{2x - 4y}{10}$

---

148.   When both terms of a fraction are binomials, we can also generate an equivalent fraction by multiplying it by a fraction that equals "1".   For example, we multiplied the fraction below by $\dfrac{2}{2}$. Notice that we multiplied by the distributive principle in both terms.

$$\dfrac{x + 3}{y - 2} = \left(\dfrac{2}{2}\right)\left(\dfrac{x + 3}{y - 2}\right) = \dfrac{2(x + 3)}{2(y - 2)} = \dfrac{2x + 6}{2y - 4}$$

Following the example, complete these:

a) $\dfrac{a + b}{c - d} = \left(\dfrac{s}{s}\right)\left(\dfrac{a + b}{c - d}\right) = \dfrac{s(a + b)}{s(c - d)} = $ _____

b) $\dfrac{t - 1}{t + 3} = \left(\dfrac{4}{4}\right)\left(\dfrac{t - 1}{t + 3}\right) = \dfrac{4(t - 1)}{4(t + 3)} = $ _____

a) $\dfrac{bM}{bP + bQ}$

b) $\dfrac{3}{9a - 6b}$

---

149.   To generate an equivalent fraction below, we multiplied by $\dfrac{b}{b}$. Notice that <u>each</u> <u>term</u> in the numerator and denominator of the new fraction has "b" as a common factor.

$$\dfrac{m}{s - v} = \left(\dfrac{b}{b}\right)\left(\dfrac{m}{s - v}\right) = \dfrac{bm}{bs - bv}$$

Identify the common factor in each term of the numerator and denominator of these:

a) $\dfrac{mp + mq}{mt}$ _____   b) $\dfrac{ax}{ay - 3a}$ _____   c) $\dfrac{at - bt}{ct + dt}$ _____

a) $\dfrac{as + bs}{cs - ds}$

b) $\dfrac{4t - 4}{4t + 12}$

150. To generate an equivalent fraction below, we multiplied by $\frac{3}{3}$. Notice that each <u>term</u> in the numerator and denominator of the new fraction has "3" as a common factor.

$$\frac{x + 2}{y} = \left(\frac{3}{3}\right)\left(\frac{x + 2}{y}\right) = \frac{3x + 6}{3y}$$

Identify the common factor in each term of the numerator and denominator of these:

a) $\dfrac{2p}{2q - 2t}$ _____   b) $\dfrac{5R - 10}{15Q}$ _____   c) $\dfrac{3y + 6}{3x - 9}$ _____

a) m

b) a

c) t

---

151. When a complicated fraction has a common literal factor in each term of its numerator and denominator, we can reduce it to lowest terms by factoring out a fraction that equals "1". For example, we factored out a "B" from each term below. Notice that we factored by the distributive principle in the numerator.

$$\frac{BD + BF}{BT} = \frac{B(D + F)}{BT} = \left(\frac{B}{B}\right)\left(\frac{D + F}{T}\right) = (1)\left(\frac{D + F}{T}\right) = \frac{D + F}{T}$$

Following the example, complete these:

a) $\dfrac{ac}{ap - aq} = \dfrac{ac}{a(p - q)} = \left(\dfrac{a}{a}\right)\left(\dfrac{c}{p - q}\right) =$ _____

b) $\dfrac{BT - KT}{MT + ST} = \dfrac{T(B - K)}{T(M + S)} = \left(\dfrac{T}{T}\right)\left(\dfrac{B - K}{M + S}\right) =$ _____

a) 2

b) 5

c) 3

---

152. Reduce to lowest terms by factoring out the common literal factor.

a) $\dfrac{ax + bx}{cx} =$ _____

b) $\dfrac{cP - cQ}{cR + cV} =$ _____

a) $\dfrac{c}{p - q}$

b) $\dfrac{B - K}{M + S}$

---

153. When a complicated fraction has a common numerical factor in each term of its numerator and denominator, we can also reduce it to lowest terms by factoring out a fraction that equals "1". For example, we factored out a "$\frac{4}{4}$" below. Notice that we factored by the distributive principle in the denominator.

$$\frac{4x}{4y - 8} = \frac{4x}{4(y - 2)} = \left(\frac{4}{4}\right)\left(\frac{x}{y - 2}\right) = (1)\left(\frac{x}{y - 2}\right) = \frac{x}{y - 2}$$

Following the example, complete these:

a) $\dfrac{7m + 7}{7t} = \dfrac{7(m + 1)}{7t} = \left(\dfrac{7}{7}\right)\left(\dfrac{m + 1}{t}\right) =$ _____

b) $\dfrac{2a - 6b}{4c + 2} = \dfrac{2(a - 3b)}{2(2c + 1)} = \left(\dfrac{2}{2}\right)\left(\dfrac{a - 3b}{2c + 1}\right) =$ _____

a) $\dfrac{a + b}{c}$

b) $\dfrac{P - Q}{R + V}$

154. Reduce to lowest terms by factoring out the common numerical factor.

a) $\dfrac{5P}{10Q - 15}$ = _____

b) $\dfrac{3x - 6}{3y + 9}$ = _____

a) $\dfrac{m + 1}{t}$

b) $\dfrac{a - 3b}{2c + 1}$

---

155. When the fraction below is reduced to lowest terms, we get a non-fraction.

$$\frac{a + ab}{a} = \frac{a(1 + b)}{a} = \left(\frac{a}{a}\right)(1 + b) = 1 + b$$

Following the example, reduce these to lowest terms.

a) $\dfrac{4 + 4x}{4}$ = _____

b) $\dfrac{cd - d}{d}$ = _____

a) $\dfrac{P}{2Q - 3}$

b) $\dfrac{x - 2}{y + 3}$

---

156. To reduce the fraction below, we substituted "1P" for "P" in the numerator. Notice that the numerator of the reduced fraction is "1".

$$\frac{P}{P - PQ} = \frac{1P}{P(1 - Q)} = \left(\frac{P}{P}\right)\left(\frac{1}{1 - Q}\right) = \frac{1}{1 - Q}$$

Following the example, reduce these:

a) $\dfrac{y}{y + ay}$ = _____

b) $\dfrac{m}{dm - m}$ = _____

a) $1 + x$

b) $c - 1$

---

157. A complicated fraction can be reduced to lowest terms <u>only</u> <u>if</u> <u>it</u> <u>contains</u> <u>a</u> <u>common</u> <u>factor</u> <u>in</u> <u>each</u> <u>term</u> <u>of</u> <u>its</u> <u>numerator</u> <u>and</u> <u>denominator</u>. Therefore, neither fraction below can be reduced.

$$\frac{a}{b - c} \qquad \frac{x + 7}{y - 3}$$

Reduce to lowest terms if possible.

a) $\dfrac{x - t}{b}$ = _____

c) $\dfrac{2m + 4}{2m - 8}$ = _____

b) $\dfrac{CP}{AP + BP}$ = _____

d) $\dfrac{5b}{3a - 7}$ = _____

a) $\dfrac{1}{1 + a}$

b) $\dfrac{1}{d - 1}$

---

a) Not possible

b) $\dfrac{C}{A + B}$

c) $\dfrac{m + 2}{m - 4}$

d) Not possible

158. Though we can factor out "t" in the numerator below, we cannot reduce to lowest terms because "t" is not a factor in the denominator.

$$\frac{ct + dt}{am} = \frac{t(c + d)}{am}$$

If possible, reduce to lowest terms.

a) $\dfrac{3B}{6S - 3} = \dfrac{3B}{3(2S - 1)} = $ _____

b) $\dfrac{5x + 5y}{3p - 3q} = \dfrac{5(x + y)}{3(p - q)} = $ _____

a) $\dfrac{B}{2S - 1}$ b) Not possible

7-18   CANCELLING TO REDUCE COMPLICATED FRACTIONS TO LOWEST TERMS

Instead of reducing complicated fractions to lowest terms by factoring out a fraction that equals "1", we can use a shorter "cancelling" method. We will discuss the cancelling method in this section.

159. Each term in the numerator and denominator of the fraction below contains "a" as a common factor. Instead of reducing it to lowest terms by factoring out $\frac{a}{a}$, we can simply cancel the "a's". We get:

$$\frac{\cancel{a}m + \cancel{a}t}{\cancel{a}p} = \frac{m + t}{p}$$

Following the example, cancel to reduce these to lowest terms.

a) $\dfrac{CD}{CX - CY} = $ _____

b) $\dfrac{bm - bv}{bs - bt} = $ _____

a) $\dfrac{\cancel{C}D}{\cancel{C}X - \cancel{C}Y} = \dfrac{D}{X - Y}$

b) $\dfrac{\cancel{b}m - \cancel{b}v}{\cancel{b}s - \cancel{b}t} = \dfrac{m - v}{s - t}$

160. Each term in the numerator and denominator of the fraction below contains "2" as a common factor. Instead of reducing it to lowest terms by factoring out $\frac{2}{2}$, we can simply cancel the 2's. We get:

$$\frac{\overset{2}{\cancel{4}}x}{\underset{1}{\cancel{2}}y - \underset{3}{\cancel{6}}} = \frac{2x}{y - 3}$$

Following the example, cancel to reduce these to lowest terms.

a) $\dfrac{8a + 8b}{8c} = $ _____

b) $\dfrac{3t - 9}{3s + 6} = $ _____

a) $\dfrac{\overset{1}{\cancel{8}}a + \overset{1}{\cancel{8}}b}{\underset{1}{\cancel{8}}c} = \dfrac{a + b}{c}$

b) $\dfrac{\overset{1}{\cancel{3}}t - \overset{3}{\cancel{9}}}{\underset{1}{\cancel{3}}s + \underset{2}{\cancel{6}}} = \dfrac{t - 3}{s + 2}$

161. Cancel to reduce these to lowest terms.

a) $\dfrac{cS - cT}{cV} = $ _____

b) $\dfrac{5x}{5y + 10} = $ _____

c) $\dfrac{FM + KM}{GM - HM} = $ _____

d) $\dfrac{2d - 4}{4p + 6} = $ _____

162.   After cancelling below, we got a non-fraction.

$$\frac{\overset{1}{\cancel{m}} - \overset{1}{\cancel{m}}t}{\underset{1}{\cancel{m}}} = \frac{1 - t}{1} = 1 - t$$

Cancel to reduce these to lowest terms.

a) $\dfrac{4 + 4x}{4} =$ _____   b) $\dfrac{ay - y}{y} =$ _____

a) $\dfrac{\cancel{d}S - \cancel{d}T}{\cancel{d}V} = \dfrac{S - T}{V}$

b) $\dfrac{\overset{1}{\cancel{5}}x}{\underset{1}{\cancel{5}}y + \underset{2}{\cancel{10}}} = \dfrac{x}{y + 2}$

c) $\dfrac{F\cancel{M} + K\cancel{M}}{G\cancel{M} - H\cancel{M}} = \dfrac{F + K}{G - H}$

d) $\dfrac{\overset{1}{\cancel{2}}d - \overset{2}{\cancel{4}}}{\underset{2}{\cancel{4}}p + \underset{3}{\cancel{6}}} = \dfrac{d - 2}{2p + 3}$

---

163.   After cancelling below, we got "1" as the numerator of the reduced fraction.

$$\frac{\overset{1}{\cancel{x}}}{B\underset{1}{\cancel{x}} + \underset{1}{\cancel{x}}} = \frac{1}{B + 1}$$

Following the example, reduce these:

a) $\dfrac{3}{3t + 3} =$ _____   b) $\dfrac{d}{d - bd} =$ _____

a) $1 + x$

b) $a - 1$

---

164.   A complicated fraction can be reduced to lowest terms by cancelling only if each term in its numerator and denominator contains a common factor. Therefore, neither fraction below can be reduced.

$$\frac{P - Q}{R} \qquad \frac{3x + 7}{2y - 5}$$

Reduce to lowest terms if possible.

a) $\dfrac{d}{dS + dF} =$ _____   c) $\dfrac{5t + 10}{5t - 10} =$ _____

b) $\dfrac{2a - 3}{5y} =$ _____   d) $\dfrac{M - ST}{V} =$ _____

a) $\dfrac{1}{t + 1}$

b) $\dfrac{1}{1 - b}$

---

165.   One of the most common errors with fractions is to cancel the 3's or "m's" in the fractions below. However, cancelling is not possible because each term of the fraction does not contain a "3" or "m" as a factor. Therefore, neither fraction can be reduced.

$$\frac{y + 3}{3} \qquad \text{(The "y" term does not have a "3" factor.)}$$

$$\frac{m - t}{m} \qquad \text{(The "t" term does not have an "m" factor.)}$$

Reduce to lowest terms if possible.

a) $\dfrac{4x + 4}{4} =$ _____   c) $\dfrac{a - b}{a} =$ _____

b) $\dfrac{x + 4}{4} =$ _____   d) $\dfrac{a - ab}{a} =$ _____

a) $\dfrac{1}{S + F}$

b) Not possible

c) $\dfrac{t + 2}{t - 2}$

d) Not possible

166. The same type of common error is to cancel the 2's or "x's" in the fractions below. However, cancelling is again not possible because each term of the fraction does not contain a "2" or "x" as a factor. Therefore, <u>neither</u> <u>fraction</u> <u>can</u> <u>be</u> <u>reduced</u>.

$$\frac{2}{y + 2}$$ (The "y" term does not have a "2" factor.)

$$\frac{x}{x - 7}$$ (The "7" term does not have an "x" factor.)

Reduce to lowest terms if possible.

a) $\dfrac{8}{8t + 8}$ = _____   c) $\dfrac{d}{a - d}$ = _____

b) $\dfrac{8}{8 - t}$ = _____   d) $\dfrac{d}{d + ad}$ = _____

a) $x + 1$

b) Not possible

c) Not possible

d) $1 - b$

---

167. If a complicated fraction is the sum or difference of two fractions and is reducible, it can be written in an equivalent form in two ways:

1) By reducing to lowest terms.

$$\frac{ad + bd}{cd} = \frac{a\cancel{d} + b\cancel{d}}{c\cancel{d}} = \frac{a + b}{c}$$

2) By breaking it up into the two original fractions and reducing each part to lowest terms.

$$\frac{ad + bd}{cd} = \frac{ad}{cd} + \frac{bd}{cd} = \frac{a}{c} + \frac{b}{c}$$

Using the two methods above, write the fraction below in two equivalent forms.

$$\frac{3x - 6y}{9} =$$ _____ or _____

a) $\dfrac{1}{t + 1}$

b) Not possible

c) Not possible

d) $\dfrac{1}{1 + a}$

---

168. When the same two methods are used to write the fraction below in an equivalent form, we get the same non-fractional expression.

$$\frac{5y - 5}{5} = \frac{\cancel{5}^1 y - \cancel{5}^1}{\cancel{5}_1} = y - 1$$

$$\frac{5y - 5}{5} = \frac{5y}{5} - \frac{5}{5} = y - 1$$

Using either method above, write each fraction below in an equivalent form.

a) $\dfrac{3 + 3x}{3}$ = _____   b) $\dfrac{PQ - P}{P}$ = _____

$\dfrac{x - 2y}{3}$ or $\dfrac{x}{3} - \dfrac{2y}{3}$

---

a) $1 + x$

b) $Q - 1$

169. Though the following fraction cannot be reduced, it can be written in an equivalent form. That is:

$$\frac{C + DF}{C} = \frac{C}{C} + \frac{DF}{C} = 1 + \frac{DF}{C}$$

Write each of these in an equivalent form.

a) $\frac{x + 3y}{3}$ = _____

b) $\frac{m - 2t}{m}$ = _____

170. If a complicated fraction is not the sum or difference of two fractions, it can only be written in a simpler equivalent form if it can be reduced. That is:

For $\frac{AP}{AC - AF}$, a simpler equivalent form is $\frac{P}{C - F}$.

For $\frac{H}{1 + CH}$, there is no simpler equivalent form, since there are no common factors and the binomial is in the denominator.

Write each of these in a simpler equivalent form if possible.

a) $\frac{3t}{6m + 9}$ = _____

c) $\frac{mM}{M - m}$ = _____

b) $\frac{2x}{1 - 2x}$ = _____

d) $\frac{ab}{ac + a}$ = _____

171. Which of the following are equivalent to $\frac{H + CT}{T}$? _____

a) $H + C$ b) $\frac{H + C}{T}$ c) $\frac{H}{T} + C$ d) $\frac{H}{T} + \frac{C}{T}$

172. Which of the following are equivalent to $\frac{AKT + KT}{AT}$? _____

a) $K + \frac{K}{A}$ b) $K + \frac{KT}{A}$ c) $\frac{AK + K}{A}$ d) $\frac{K + T}{A}$

173. Which of the following are equivalent to $\frac{KV}{KS - KV}$? _____

a) $\frac{1}{KS}$ b) $\frac{V}{S - V}$ c) $\frac{1}{S - V}$ d) $\frac{V}{S} - 1$

Answer column (right side):

a) $\frac{x}{3} + y$

b) $1 - \frac{2t}{m}$

a) $\frac{t}{2m + 3}$

b) Not possible

c) Not possible

d) $\frac{b}{c + 1}$

Only (c)

Both (a) and (c)

Only (b)

174. Which of the following are equivalent to $\dfrac{ab}{ac + bc}$? _____

    a) $\dfrac{b}{c + bc}$ b) $\dfrac{a}{ac + c}$ c) $\dfrac{1}{2c}$ d) $\dfrac{b}{c} + \dfrac{a}{c}$

175. Though the fraction below cannot be reduced, it can be written as an addition of two fractions. That is:

$$\frac{C_1 + C_T}{C_1 - C_2} = \frac{C_1}{C_1 - C_2} + \frac{C_T}{C_1 - C_2}$$

Which of the following are equivalent to $\dfrac{R_1 - R_2}{R_1 + R_t}$? _____

    a) $\dfrac{1 - R_2}{1 - R_t}$ b) $\dfrac{R_1}{R_1 + R_t} - \dfrac{R_2}{R_1 + R_t}$ c) $\dfrac{1}{R_t} - \dfrac{R_2}{R_1 + R_t}$

176. After factoring below, we were able to reduce to lowest terms by cancelling the binomial factors.

$$\frac{ac + a}{bc + b} = \frac{a(c + 1)}{b(c + 1)} = \frac{a}{b}$$

Following the example, reduce this one to lowest terms.

$$\frac{7x + 7}{3x + 3} = \underline{\hspace{4cm}}$$

None of them

Only (b)

$\dfrac{7}{3}$

7-19   SIMPLIFYING FRACTIONS WHOSE NUMERATOR OR DENOMINATOR CONTAINS A FRACTION

Any fraction whose numerator or denominator contains a fraction can be simplified to a fraction that does not contain a fraction in its numerator or denominator. We will discuss the method in this section.

177. In the complex fraction below, the numerator is a fraction. To simplify the complex fraction, we performed the division.

$$\frac{\frac{a + b}{a}}{d} = \left(\frac{a + b}{a}\right)\left(\frac{1}{d}\right) = \frac{a + b}{ad}$$

Following the example, do this division.

$$\frac{\frac{m - t}{t}}{v} = \underline{\hspace{4cm}}$$

$\dfrac{m - t}{tv}$

178. Each complex fraction below has a fraction in either its numerator or denominator or both.

$$\frac{P + V}{\dfrac{Q}{R}} \qquad\qquad \frac{\dfrac{c}{d}}{\dfrac{k + r}{t}} \qquad\qquad \frac{\dfrac{b}{m}}{1 - h}$$

To simplify fractions like those above, we perform the division. To do so, we multiply the numerator by the reciprocal of the denominator. Remember that reciprocals are two quantities whose product is +1. Therefore:

The reciprocal of $\dfrac{Q}{R}$ is $\dfrac{R}{Q}$.

The reciprocal of $\dfrac{k + r}{t}$ is $\dfrac{t}{k + r}$.

The reciprocal of $1 - h$ is $\dfrac{1}{1 - h}$.

Write the reciprocal of each expression.

a) $\dfrac{a - b}{c}$ _____   b) $V - T$ _____   c) $\dfrac{1}{M + 3p}$ _____

---

179. Convert each division to a multiplication by multiplying the numerator by the reciprocal of the denominator.

a) $\dfrac{\dfrac{k}{x}}{\dfrac{c + d}{a}} = \Big(\quad\Big)\Big(\quad\Big)$ c) $\dfrac{\dfrac{q}{r}}{\dfrac{1}{x - y}} = \Big(\quad\Big)\Big(\quad\Big)$

b) $\dfrac{\dfrac{t - v}{b}}{p - q} = \Big(\quad\Big)\Big(\quad\Big)$

Answers:

a) $\dfrac{c}{a - b}$

b) $\dfrac{1}{V - T}$

c) $M + 3p$

---

180. When performing multiplications like those in the last frame, we can write the product in either of two forms. For example:

$$\left(\frac{k}{x}\right)\left(\frac{a}{c + d}\right) = \frac{ak}{x(c + d)} \quad\text{or}\quad \frac{ak}{cx + dx}$$

$$\left(\frac{q}{r}\right)(x - y) = \frac{q(x - y)}{r} \quad\text{or}\quad \frac{qx - qy}{r}$$

In the products above, $x(c + d)$ and $q(x - y)$ are instances of the distributive principle. When writing products like those above, we ordinarily leave instances of the distributive principle in factored form without multiplying. Therefore, we write:

$$\left(\frac{k}{x}\right)\left(\frac{a}{c + d}\right) = \frac{ak}{x(c + d)} \qquad \left(\frac{q}{r}\right)(x - y) = \underline{\qquad}$$

Answers:

a) $\left(\dfrac{k}{x}\right)\left(\dfrac{a}{c + d}\right)$

b) $\left(\dfrac{t - v}{b}\right)\left(\dfrac{1}{p - q}\right)$

c) $\left(\dfrac{q}{r}\right)(x - y)$

$\dfrac{q(x - y)}{r}$

181.   Perform these multiplications. Write each product in the preferred form.

a) $\left(\dfrac{p-q}{t}\right)\left(\dfrac{1}{a-b}\right) = $ _____

c) $\left(\dfrac{V}{1-T}\right)\left(\dfrac{1-R}{P}\right) = $ _____

b) $\left(\dfrac{c}{d}\right)(m-v) = $ _____

---

182.   Complete each division. Write each answer in the preferred form.

a) $\dfrac{x+y}{\dfrac{v}{t}} = \Big(\quad\Big)\Big(\quad\Big) = $ _____

b) $\dfrac{\dfrac{p+q}{a}}{\dfrac{p-q}{b}} = \Big(\quad\Big)\Big(\quad\Big) = $ _____

c) $\dfrac{\dfrac{m-b}{m}}{1+r} = \Big(\quad\Big)\Big(\quad\Big) = $ _____

a) $\dfrac{p-q}{t(a-b)}$

b) $\dfrac{c(m-v)}{d}$

c) $\dfrac{V(1-R)}{P(1-T)}$

---

183.   In the answer below, we cancelled to reduce to lowest terms.

$$\dfrac{\dfrac{a+b}{c}}{\dfrac{a-b}{c}} = \left(\dfrac{a+b}{c}\right)\left(\dfrac{c}{a-b}\right) = \dfrac{\cancel{c}(a+b)}{\cancel{c}(a-b)} = \dfrac{a+b}{a-b}$$

Following the example, complete this division.

$$\dfrac{\dfrac{t-1}{v}}{\dfrac{t-1}{h}} = $$

a) $(x+y)\left(\dfrac{t}{v}\right) = \dfrac{t(x+y)}{v}$

b) $\left(\dfrac{p+q}{a}\right)\left(\dfrac{b}{p-q}\right) = $

$\dfrac{b(p+q)}{a(p-q)}$

c) $\left(\dfrac{m-b}{m}\right)\left(\dfrac{1}{1+r}\right) = $

$\dfrac{m-b}{m(1+r)}$

---

184.   In the complex fraction below, the numerator contains a fraction. To
simplify the complex fraction, we performed the addition in the
numerator and then divided.

$$\dfrac{1+\dfrac{a}{b}}{t} = \dfrac{\dfrac{b+a}{b}}{t} + \left(\dfrac{b+a}{b}\right)\left(\dfrac{1}{t}\right) = \dfrac{b+a}{bt}$$

Following the example, simplify this fraction.

$$\dfrac{1-\dfrac{c}{d}}{m} = $$

$\dfrac{h}{v}$, from $\dfrac{h\cancel{(t-1)}}{v\cancel{(t-1)}}$

$\dfrac{d-c}{dm}$

185. To simplify the complex fraction below, we performed the subtraction in the denominator and then divided.

$$\frac{P}{1 - \dfrac{Q}{R}} = \frac{P}{\dfrac{R - Q}{R}} = P\left(\frac{R}{R - Q}\right) = \frac{PR}{R - Q}$$

Following the example, simplify this fraction.

$$\frac{h}{m + \dfrac{h}{t}} =$$

---

186. To simplify the fraction below, we performed the additions in the numerator and denominator and then divided.

$$\frac{1 + \dfrac{1}{c}}{1 + \dfrac{1}{d}} = \frac{\dfrac{c + 1}{c}}{\dfrac{d + 1}{d}} = \left(\frac{c + 1}{c}\right)\left(\frac{d}{d + 1}\right) = \frac{d(c + 1)}{c(d + 1)}$$

Following the example, simplify this fraction.

$$\frac{h - \dfrac{p}{q}}{h - \dfrac{v}{t}} =$$

$\dfrac{ht}{mt + h}$

---

187. In the simplification below, we were able to cancel to reduce the answer to lowest terms.

$$\frac{x - \dfrac{2x}{b}}{3x} = \frac{\dfrac{bx - 2x}{b}}{3x} = \left(\frac{bx - 2x}{b}\right)\left(\frac{1}{3x}\right) = \frac{b\cancel{x} - 2\cancel{x}}{3b\cancel{x}} = \frac{b - 2}{3b}$$

Following the example, simplify this fraction.

$$\frac{y + \dfrac{5y}{c}}{7y} =$$

$\dfrac{t(hq - p)}{q(ht - v)}$

---

188. Simplify this one. Cancel to reduce to lowest terms.

$$\frac{p}{p - \dfrac{p}{q}} =$$

$\dfrac{c + 5}{7c}$

$\dfrac{q}{q - 1}$

SELF-TEST 25 (pp. 309-321)

Reduce each fraction to lowest terms.

1. $\dfrac{4t}{2r - 6w}$ = _____

2. $\dfrac{GP - PV}{KP}$ = _____

3. $\dfrac{8R - 12}{4}$ = _____

4. $\dfrac{cm - mx}{dm + my}$ = _____

5. $\dfrac{x}{x - xy}$ = _____

6. $\dfrac{AH + 2H}{HR - H}$ = _____

7. Which of the following are equivalent to $\dfrac{hk + kw}{bk}$? _____

   a) $\dfrac{h}{b} + kw$ b) $\dfrac{h}{b} + \dfrac{w}{b}$ c) $\dfrac{h + w}{b}$ d) $hk + \dfrac{w}{b}$

Simplify each expression. Write each answer as a single fraction in lowest terms that does not contain a fraction in its numerator or denominator.

8. $\dfrac{\frac{p - t}{d}}{2a}$ = _____

9. $\dfrac{E}{R + \frac{F}{H}}$ = _____

10. $\dfrac{\frac{p}{v} - 1}{\frac{2r}{v}}$ = _____

11. $\dfrac{1 + \frac{1}{N}}{1 - \frac{1}{N}}$ = _____

ANSWERS: 1. $\dfrac{2t}{r - 3w}$ 4. $\dfrac{c - x}{d + y}$ 7. b, c   10. $\dfrac{p - v}{2r}$

2. $\dfrac{G - V}{K}$ 5. $\dfrac{1}{1 - y}$ 8. $\dfrac{p - t}{2ad}$ 11. $\dfrac{N + 1}{N - 1}$

3. $2R - 3$ 6. $\dfrac{A + 2}{R - 1}$ 9. $\dfrac{EH}{HR + F}$

SUPPLEMENTARY PROBLEMS - CHAPTER 7

Assignment 22

Reduce each fraction to lowest terms.

1. $\dfrac{x^2 y}{x^2 y}$ 2. $\dfrac{ab}{2bd}$ 3. $\dfrac{2G(R-1)}{6P(R-1)}$ 4. $\dfrac{6tw}{8tw}$

5. $\dfrac{3r}{3cr}$ 6. $\dfrac{4PT}{P}$ 7. $\dfrac{d^2 t}{bd^2}$ 8. $\dfrac{2ks}{10krs}$

Do these multiplications. Report each product in lowest terms.

9. $\left(\dfrac{1}{x}\right)\left(\dfrac{h}{y}\right)$ 10. $\left(\dfrac{a}{2}\right)\left(\dfrac{4}{ab}\right)$ 11. $\dfrac{1}{k}(p^2 t)$ 12. $\left(\dfrac{p}{r^2}\right)\left(\dfrac{ar^2}{p}\right)$

13. $\left(\dfrac{L+R}{G}\right)\left(\dfrac{E}{L+R}\right)$ 14. $\left(\dfrac{s}{m}\right)\left(\dfrac{m}{st}\right)$ 15. $xy\left(\dfrac{2t}{y}\right)$ 16. $\left(\dfrac{v}{k-1}\right)\left(\dfrac{1}{v}\right)$

Write the reciprocal of: 17. $\dfrac{1}{h+rw}$ 18. $\dfrac{x-y}{v}$

Do these divisions. Report each quotient in lowest terms.

19. $\dfrac{\frac{a}{d}}{\frac{a}{b}}$ 20. $\dfrac{\frac{3y}{4x}}{\frac{9y}{12}}$ 21. $\dfrac{2R}{\frac{R}{P}}$ 22. $\dfrac{1}{\frac{t}{ks}}$

23. $\dfrac{\frac{4m}{r}}{8p}$ 24. $\dfrac{\frac{F}{T}}{1-A}$ 25. $\dfrac{1}{\frac{1}{c+k}}$ 26. $\dfrac{2t^2}{\frac{mt^2}{2d}}$

27. $\dfrac{\frac{x}{y+1}}{2x}$ 28. $\dfrac{\frac{h+2}{3}}{h+2}$ 29. $\dfrac{\frac{1}{h}}{\frac{1}{2h}}$ 30. $\dfrac{\frac{r}{dv}}{\frac{r}{dv}}$

Assignment 23

Do these additions. Report each sum in lowest terms.

1. $\dfrac{3w}{8}+\dfrac{w}{8}$ 2. $\dfrac{v}{r}+\dfrac{w}{r}$ 3. $\dfrac{ab}{2}+\dfrac{5ab}{2}$ 4. $\dfrac{3t}{w}+\dfrac{2}{w}$

5. $\dfrac{F}{E-2}+\dfrac{H}{E-2}$ 6. $\dfrac{h}{x+y}+\dfrac{h+1}{x+y}$ 7. $\dfrac{c}{h}+\dfrac{d}{h}+\dfrac{p}{h}$ 8. $\dfrac{1}{F}+\dfrac{1}{F}+\dfrac{1}{F}$

9. $\dfrac{3r}{8}+\dfrac{r}{2}$ 10. $\dfrac{2}{G}+\dfrac{1}{3G}$ 11. $\dfrac{d}{x}+\dfrac{r}{5x}$ 12. $\dfrac{5a}{6t}+\dfrac{a}{2t}$

13. $\dfrac{m}{3}+\dfrac{1}{2}$ 14. $\dfrac{d}{8}+\dfrac{r}{6}$ 15. $\dfrac{k}{r}+\dfrac{p}{v}$ 16. $\dfrac{1}{Q}+\dfrac{1}{P}$

17. $\dfrac{h}{rw}+\dfrac{s}{r}$ 18. $\dfrac{5}{6}+\dfrac{1}{2m}$ 19. $\dfrac{w}{a}+\dfrac{t}{2b}$ 20. $\dfrac{1}{pv}+\dfrac{1}{cp}$

Do these additions of a fraction and a non-fraction.

21. $\dfrac{E}{2} + 1$ 22. $1 + \dfrac{w}{r}$ 23. $\dfrac{2}{3} + P$ 24. $5 + \dfrac{4}{B}$

25. $4 + \dfrac{1}{t}$ 26. $m + \dfrac{w}{k}$ 27. $\dfrac{r}{av} + 2d$ 28. $\dfrac{1}{x} + 1$

Assignment 24

Do these subtractions.   Report each difference in lowest terms.

1. $\dfrac{gh}{m} - \dfrac{d}{m}$ 2. $\dfrac{9}{4p} - \dfrac{7}{4p}$ 3. $\dfrac{a}{rt} - \dfrac{a}{rt}$ 4. $\dfrac{8R}{3} - \dfrac{2R}{3}$

5. $\dfrac{P+1}{T} - \dfrac{P}{T}$ 6. $\dfrac{d+2}{a+b} - \dfrac{1}{a+b}$ 7. $\dfrac{P}{2} - \dfrac{P}{3}$ 8. $\dfrac{r}{t} - \dfrac{1}{4}$

9. $\dfrac{2}{x} - \dfrac{1}{2x}$ 10. $\dfrac{1}{d} - \dfrac{1}{a}$ 11. $\dfrac{r}{s} - \dfrac{t}{w}$ 12. $\dfrac{2v}{pw} - \dfrac{b}{w}$

Do these subtractions involving a fraction and a non-fraction.

13. $a - \dfrac{1}{r}$ 14. $\dfrac{h}{3} - 2$ 15. $1 - \dfrac{B}{A}$ 16. $\dfrac{m}{t} - w$

Break up each fraction into a sum or difference of two fractions in lowest terms.

17. $\dfrac{F+H}{FH}$ 18. $\dfrac{4r-3}{6r}$ 19. $\dfrac{AB-C}{AC}$ 20. $\dfrac{ct-br}{rt}$

21. $\dfrac{r+s+t}{d}$ 22. $\dfrac{x+1}{y+1}$ 23. $\dfrac{a+8}{4}$ 24. $\dfrac{1-r}{r}$

State whether each of the following is "True or "False".

25. $\dfrac{R-W}{W} = \dfrac{R}{W} - 1$ 26. $\dfrac{H}{H+K} = 1 + \dfrac{H}{K}$ 27. $\dfrac{6}{4x-8} = \dfrac{3}{2x} - \dfrac{3}{4}$

Factor each expression by the distributive principle if possible.

28. $18at + 12t$ 29. $F - FP$ 30. $cd + dh$ 31. $2rs - 3tw$

Assignment 25

Reduce each fraction to lowest terms.

1. $\dfrac{6x+12y}{9w}$ 2. $\dfrac{rt}{pr-2r}$ 3. $\dfrac{dm}{mw+m}$

4. $\dfrac{15s}{5s-10v}$ 5. $\dfrac{bp+ab}{4b}$ 6. $\dfrac{V-PV}{RV}$

7. $\dfrac{xy+x}{x-xy}$ 8. $\dfrac{2k-4}{4r-2}$ 9. $\dfrac{6t+3w}{3p+9v}$

10. $\dfrac{F}{AF+F}$ 11. $\dfrac{2}{4y-6}$ 12. $\dfrac{bd-d}{d}$

State whether each of the following is "True" or "False".

13. $\dfrac{P}{P - KP} = \dfrac{1}{1 - K}$

14. $\dfrac{cr + ct}{ct} = \dfrac{r}{t} + 1$

15. $\dfrac{2x}{2x + 2} = 1 + x$

Simplify each expression. Write each answer as a single fraction in lowest terms that does not contain a fraction in its numerator or denominator.

16. $\dfrac{\dfrac{km}{m}}{r - a}$

17. $\dfrac{d}{\dfrac{t + w}{b}}$

18. $\dfrac{\dfrac{x - y}{p}}{2v}$

19. $\dfrac{\dfrac{F}{D + H}}{F}$

20. $\dfrac{1 + \dfrac{1}{x}}{2y}$

21. $\dfrac{at}{\dfrac{a}{r} - a}$

22. $\dfrac{P}{P + \dfrac{2P}{R}}$

23. $\dfrac{1 - \dfrac{1}{w}}{b}$

24. $\dfrac{x + y}{\dfrac{t}{3}}$

25. $\dfrac{\dfrac{d}{c - k}}{\dfrac{d}{p - t}}$

26. $\dfrac{1 + \dfrac{1}{R}}{1 - \dfrac{1}{R}}$

27. $\dfrac{\dfrac{a}{w} - 1}{\dfrac{b}{t} + 1}$

Chapter 8 FORMULA REARRANGEMENT

Formulas are frequently rearranged to solve for a variable (or letter). Three of the purposes of formula rearrangement are:

1) to make formula evaluations easier

2) to emphasize a different meaning or relationship

3) to eliminate a variable or variables from a system of formulas and thereby derive new formulas.

We will discuss formula rearrangement in this chapter. We will show that the same algebraic principles used to solve equations are used to rearrange formulas.

8-1   FORMULA REARRANGEMENT AND FORMULA EVALUATIONS

In this section, we will show how formula rearrangement can be used to avoid equation-solving in formula evaluations.

1.   In the formula below, N is <u>not solved-for</u>. Therefore to find the value of N when $D = 10$ and $B = 70$, we have to solve an equation. We get:

$$D = \frac{B}{N}$$

$$10 = \frac{70}{N}$$

$$N(10) = \cancel{N}\left(\frac{70}{\cancel{N}}\right)$$

$$10N = 70$$

$$N = \frac{70}{10} = 7$$

However, we can also use algebraic principles to rearrange the above formula to solve for N. That is:

Rearranging $D = \frac{B}{N}$, we get $N = \frac{B}{D}$

Using the rearranged formula in which N is <u>solved-for</u>, we can perform the same evaluation without solving an equation. We get:

$$N = \frac{B}{D} = \frac{70}{10} = 7$$

Did we get the same value for N with both methods? _____

2.   In the formula below, "v_1" is <u>not-solved-for</u>. Therefore to find the value of v_1 when $a = 10$, $v_2 = 60$, and $t = 2$, we must solve an equation. We get:

$$a = \frac{v_2 - v_1}{t}$$

$$10 = \frac{60 - v_1}{2}$$

$$2(10) = \cancel{2}\left(\frac{60 - v_1}{\cancel{2}}\right)$$

$$20 = 60 - v_1$$

$$-40 = -v_1$$

$$v_1 = \frac{-40}{-1} = 40$$

Using algebraic principles, we can also rearrange the above formula to solve for "v_1". That is:

Rearranging $a = \dfrac{v_2 - v_1}{t}$, we get $v_1 = v_2 - at$

Using the rearranged formula in which "v_1" is <u>solved-for</u>, we can perform the same evaluation without solving an equation. We get:

$$v_1 = v_2 - at = 60 - (10)(2) = 60 - 20 = 40$$

Did we get the same value for "v_1" with both methods? _____

Yes. It is 7.

> One of the major purposes of formula rearrangement is to avoid the need for equation-solving in formula evaluations. Rearrangements are especially useful when they avoid types of equation-solving that are not simple.

Yes. It is 40.

8-2   IDENTIFYING TERMS IN FORMULAS

Each side of a formula contains one or more terms. In this section, we will discuss the types of terms in formulas and identify terms in formulas.

3.   In a formula, terms are separated by addition or subtraction symbols. <u>Any</u> <u>single</u> <u>letter</u> <u>or</u> <u>squared</u> <u>letter</u> (<u>with</u> <u>or</u> <u>without</u> <u>a</u> <u>subscript</u>) is a term. For example, we have drawn boxes around each term in the formulas below.

$$\boxed{F_T} = \boxed{F_1} + \boxed{F_2} + \boxed{F_3} \qquad\qquad \boxed{a^2} = \boxed{c^2} - \boxed{b^2}$$

Draw a box around each term in these formulas.

a) $I_C = I_E - I_B$ b) $A = s^2$ c) $R = C + M$

4.   <u>Any</u> <u>multiplication</u> <u>of</u> <u>two</u> <u>or</u> <u>more</u> <u>factors</u> <u>is</u> <u>one</u> <u>term</u>. For example, we have drawn boxes around each term in the formulas below.

$$\boxed{E} \;=\; \boxed{\dfrac{1}{2}mv^2} \qquad\qquad \boxed{I_A} \;=\; \boxed{I_C} \;+\; \boxed{md^2}$$

Draw a box around each term in these formulas.

   a) $V_1 T_2 \;=\; T_1 V_2$ b) $E \;=\; IR_1 \;+\; IR_2$

a) $\boxed{I_C} = \boxed{I_E} - \boxed{I_B}$

b) $\boxed{A} = \boxed{s^2}$

c) $\boxed{R} = \boxed{C} + \boxed{M}$

---

5.   <u>A</u> <u>multiplication</u> <u>is</u> <u>one</u> <u>term</u> <u>even</u> <u>if</u> <u>one</u> <u>of</u> <u>the</u> <u>factors</u> <u>is</u> <u>a</u> <u>binomial</u>. For example, we have drawn boxes around each term in the formulas below.

$$\boxed{W_s} \;=\; \boxed{P(V_1 - V_2)} \qquad\qquad \boxed{A} \;=\; \boxed{\dfrac{1}{2}h(b_1 + b_2)}$$

Draw a box around each term in these formulas.

   a) $C \;=\; \dfrac{5}{9}(F - 32)$ b) $H \;=\; ms(t_2 - t_1)$

a) $\boxed{V_1 T_2} \;=\; \boxed{T_1 V_2}$

b) $\boxed{E} = \boxed{IR_1} + \boxed{IR_2}$

---

6.   <u>Any</u> <u>fraction</u> (<u>simple</u> <u>or</u> <u>complex</u>) <u>is</u> <u>one</u> <u>term</u>. For example, we have drawn boxes around each term in the formulas below.

$$\boxed{P} \;=\; \boxed{\dfrac{E^2}{R}} \qquad\qquad \boxed{H} \;=\; \boxed{\dfrac{\pi dR(F_1 - F_2)}{33,000}}$$

Draw a box around each term in these formulas.

   a) $\dfrac{P_1}{P_2} \;=\; \dfrac{V_2}{V_1}$ b) $m \;=\; \dfrac{y_2 - y_1}{x_2 - x_1}$

a) $\boxed{C} = \boxed{\dfrac{5}{9}(F - 32)}$

b) $\boxed{H} \;=\; \boxed{ms(t_2 - t_1)}$

---

7.   Draw a box around each term.

   a) $y \;=\; mx \;+\; b$ c) $M \;=\; \dfrac{WLX}{2} \;-\; \dfrac{WX^2}{2}$

   b) $A \;=\; \dfrac{1}{2}bh$ d) $D \;=\; P(Q + R)$

a) $\boxed{\dfrac{P_1}{P_2}} \;=\; \boxed{\dfrac{V_2}{V_1}}$

b) $\boxed{m} \;=\; \boxed{\dfrac{y_2 - y_1}{x_2 - x_1}}$

---

8.   Draw a box around each term.

   a) $C_t C_2 \;=\; C_1 C_2 \;-\; C_1 C_t$ c) $v_o^2 \;=\; v_f^2 \;-\; 2gs$

   b) $v_{av} \;=\; \dfrac{v_o + v_f}{2}$ d) $P \;=\; p_1 \;+\; w(h - h_1)$

a) $\boxed{y} = \boxed{mx} + \boxed{b}$

b) $\boxed{A} = \boxed{\dfrac{1}{2}bh}$

c) $\boxed{M} = \boxed{\dfrac{WLX}{2}} - \boxed{\dfrac{WX^2}{2}}$

d) $\boxed{D} = \boxed{P(Q + R)}$

---

a) $\boxed{C_t C_2} = \boxed{C_1 C_2} - \boxed{C_1 C_t}$ b) $\boxed{v_{av}} = \boxed{\dfrac{v_o + v_f}{2}}$ c) $\boxed{v_o^2} = \boxed{v_f^2} - \boxed{2gs}$ d) $\boxed{P} = \boxed{p_1} + \boxed{w(h - h_1)}$

8-3   COEFFICIENTS OF LETTERS IN NON-FRACTIONAL TERMS

In this section, we will define what is meant by the coefficient of a letter in a non-fractional term.

9.   In any two-factor term containing a number and a letter, the <u>number</u> is the coefficient of the letter. That is:
 	
    In 4s , the coefficient of "s" is 4 .
 	
    In 1.5t , the coefficient of "t" is _____ .	

10.   In any two-factor multiplication containing two letters, the coefficient of each factor is <u>the other</u> factor. That is:
 	
    In LW , the coefficient of W is L .
 	
    In $V_1 T_2$, the coefficient of V_1 is _____ .
	1.5

11.   In any multiplication with more than two factors, the coefficient of each factor is <u>the other</u> factors. That is:
 	
    In LWH , the coefficient of H is LW .
 	
    In $.24 I^2 Rt$, the coefficient of I^2 is _____ .
	T_2

12.   In any term, the coefficient of any one factor is <u>the other</u> factor <u>or</u> factors. That is:
 	
    In P(Q + R) : the coefficient of (Q + R) is P .
 	
    the coefficient of P is (Q + R) .
 	
    In $ms(t_2 - t_1)$: a) the coefficient of $(t_2 - t_1)$ is _____ .
 	
    b) the coefficient of "s" is _____ .
 	
    c) the coefficient of "m" is _____ .
	.24Rt

13.   a) In $.24 I^2 Rt$, the coefficient of R is _____ .
 	
    b) In W(R - r) , the coefficient of W is _____ .
 	
    c) In $AKT(t_2 - t_1)$, the coefficient of T is _____ .
	a) ms
	b) $m(t_2 - t_1)$
	c) $s(t_2 - t_1)$

	a) $.24 I^2 t$ b) (R - r) c) $AK(t_2 - t_1)$

8-4   THE MULTIPLICATION AXIOM AND FORMULA REARRANGEMENT

When a formula has only one non-fractional term on each side, we can use the multiplication axiom to solve for any letter that is a factor. We will discuss the method in this section. A shortcut for the multiplication axiom is shown.

---

14.   We used the multiplication axiom to solve for "h" below. That is, we multiplied both sides by $\frac{1}{.433}$, the <u>reciprocal</u> of <u>the</u> <u>coefficient</u> of "h". Solve for "t" in the other formula by multiplying both sides by $\frac{1}{10}$.

$$P = .433h \qquad\qquad v = 10t$$

$$\frac{1}{.433}(P) = \frac{1}{\boxed{.433}}(.433h)$$

$$\frac{P}{.433} = \quad 1h$$

$$h = \frac{P}{.433}$$

---

15.   To solve for R below, we multiplied both sides by $\frac{1}{I^2}$, the <u>reciprocal</u> of <u>the</u> <u>coefficient</u> of R . Solve for V_1 in the other formula by multiplying both sides by $\frac{1}{P_1}$.

$$P = I^2 R \qquad\qquad P_1 V_1 = P_2 V_2$$

$$\frac{1}{I^2}(P) = \frac{1}{\boxed{I^2}}(I^2 R)$$

$$\frac{P}{I^2} = \quad 1R$$

$$R = \frac{P}{I^2}$$

Answer (box 14):

$t = \dfrac{v}{10}$, from:

$$\frac{1}{10}(v) = \frac{1}{\boxed{10}}(10t)$$

$$\frac{v}{10} = \quad 1t$$

---

16.   To solve for "s" below, we multiplied both sides by $\frac{1}{2a}$, the <u>reciprocal</u> of the <u>coefficient</u> of "s". Solve for L in the other formula by multiplying both sides by $\frac{1}{WH}$.

$$v^2 = 2as \qquad\qquad V = LWH$$

$$\frac{1}{2a}(v^2) = \frac{1}{\boxed{2a}}(2as)$$

$$\frac{v^2}{2a} = \quad 1s$$

$$s = \frac{v^2}{2a}$$

Answer (box 15):

$V_1 = \dfrac{P_2 V_2}{P_1}$, from:

$$\frac{1}{\boxed{P_1}}(P_1 V_1) = \frac{1}{P_1}(P_2 V_2)$$

$$1V_1 = \frac{P_2 V_2}{P_1}$$

17.   To solve for P below, we multiplied both sides by $\frac{1}{Q+R}$, the <u>reciprocal</u> <u>of</u> <u>the</u> <u>coefficient</u> <u>of</u> P . Solve for "b" in the other formula by multiplying both sides by $\frac{1}{c-d}$.

$$D = P(Q + R) \qquad\qquad a = b(c - d)$$

$$\left(\frac{1}{Q+R}\right)(D) = P(Q+R)\left(\frac{1}{Q+R}\right)$$

$$\frac{D}{Q+R} = P \cdot 1$$

$$P = \frac{D}{Q+R}$$

$L = \dfrac{V}{WH}$, from:

$$\frac{1}{WH}(V) = LWH\left(\frac{1}{WH}\right)$$

$$\frac{V}{WH} = L \cdot 1$$

---

18.   Two solutions from the preceding frames are shown below.

If $P = .433h$,

$$h = \frac{P}{.433}$$

If $V = LWH$,

$$L = \frac{V}{WH}$$

As you can see from the above examples, there is a shortcut for the multiplication axiom. That is, to solve for each letter, <u>we</u> <u>can</u> <u>simply</u> <u>divide</u> <u>the</u> <u>other</u> <u>side</u> <u>of</u> <u>the</u> <u>formula</u> <u>by</u> <u>the</u> <u>coefficient</u> <u>of</u> <u>that</u> <u>letter</u>. For example:

In $E = IR$: The coefficient of I is R .

$$\text{Therefore,}\quad I = \frac{E}{R}$$

In $2Pt^2 = rw$: The coefficient of P is $2t^2$.

$$\text{Therefore,}\quad P = \underline{\qquad\qquad}$$

$b = \dfrac{a}{c-d}$, from:

$$\frac{1}{c-d}(a) = b(c-d)\left(\frac{1}{c-d}\right)$$

$$\frac{a}{c-d} = b \cdot 1$$

---

19.   Two more solutions from the preceding frames are shown below.

If $D = P(Q - r)$,

$$P = \frac{D}{Q+R}$$

If $a = b(c - d)$,

$$b = \frac{a}{c-d}$$

Again you can see that we can solve for each letter <u>by</u> <u>simply</u> <u>dividing</u> <u>the</u> <u>other</u> <u>side</u> <u>of</u> <u>the</u> <u>formula</u> <u>by</u> <u>the</u> <u>coefficient</u> <u>of</u> <u>the</u> <u>letter</u>.

In $2RF = W(R - r)$: The coefficient of W is $(R - r)$.

$$\text{Therefore,}\quad W = \frac{2RF}{R-r}$$

In $H = ms(t_2 - t_1)$: The coefficient of "s" is $m(t_2 - t_1)$.

$$\text{Therefore,}\quad s = \underline{\qquad\qquad}$$

$P = \dfrac{rw}{2t^2}$

20. When rearranging a formula, a capital letter should not be changed to a small letter and a small letter should not be changed to a capital letter. That is:

$$s = \frac{H}{m(t_2 - t_1)}$$

Don't change "A" to "a".

Don't change "e" to "E".

Using the shortcut of the multiplication axiom, solve for the indicated letters in this frame and the next.

a) Solve for F .

$$d^2 Fr = m_1 m_2$$

F = _____

b) Solve for R .

$$H = .24I^2 Rt$$

R = _____

---

21.   a) Solve for "b".

$$2ab = c(d - f)$$

b = _____

b) Solve for "m".

$$H = ms(t_2 - t_1)$$

m = _____

a) $F = \dfrac{m_1 m_2}{d^2 r}$

b) $R = \dfrac{H}{.24I^2 t}$

---

a) $b = \dfrac{c(d - f)}{2a}$ b) $m = \dfrac{H}{s(t_2 - t_1)}$

8-5   REARRANGING FORMULAS CONTAINING ONE FRACTION

In this section, we will discuss the method for rearranging formulas that contain one fraction. The formulas are limited to those with only one term on each side.

---

22.   To rearrange a formula containing one fraction, we begin by "clearing the fraction". To do so, we use the multiplication axiom, multiplying both sides by the denominator of the fraction. An example is shown. Clear the fraction in the other formula by multiplying both sides by 2.5 .

$$v = \frac{s}{t}$$

$$t(v) = \cancel{t}\left(\frac{s}{\cancel{t}}\right)$$

$$tv = s$$

$$H = \frac{D^2 N}{2.5}$$

$2.5H = D^2 N$, from:

$$2.5(H) = \cancel{2.5}\left(\frac{D^2 N}{\cancel{2.5}}\right)$$

23. Following the example, clear the fraction in the other formula by multiplying both sides by $33,000$.

$$H = \frac{AKT(t_2 - t_1)}{L} \qquad\qquad H = \frac{\pi dR(F_1 - F_2)}{33,000}$$

$$L(H) = \not{L}\left[\frac{AKT(t_2 - t_1)}{\not{L}}\right]$$

$$HL = AKT(t_2 - t_1)$$

$33,000H = \pi dR(F_1 - F_2)$

from:

$33,000(H) =$
$$\not{33,000}\left[\frac{\pi dR(F_1 - F_2)}{\not{33,000}}\right]$$

---

24. Following the example, clear the fraction in the other formula by multiplying both sides by "$2\pi fC$".

$$F = \frac{m_1 m_2}{rd^2} \qquad\qquad X_c = \frac{1}{2\pi fC}$$

$$rd^2(F) = \not{rd^2}\left(\frac{m_1 m_2}{\not{rd^2}}\right)$$

$$rd^2 F = m_1 m_2$$

$2\pi fC X_c = 1$, from:

$$2\pi fC(X_c) = \not{2\pi fC}\left(\frac{1}{\not{2\pi fC}}\right)$$

---

25. Following the example, clear the fraction in the other formula.

$$v_{av} = \frac{v_o + v_f}{2} \qquad\qquad a = \frac{v_2 - v_1}{t}$$

$$2(v_{av}) = \not{2}\left(\frac{v_o + v_f}{\not{2}}\right)$$

$$2v_{av} = v_o + v_f$$

$at = v_2 - v_1$, from:

$$t(a) = \not{t}\left(\frac{v_2 - v_1}{\not{t}}\right)$$

---

26. To solve for B below, we simply cleared the fraction. Complete the other rearrangement.

Solve for B .   Solve for "a".

$$D = \frac{B}{N} \qquad\qquad\qquad h = \frac{a}{s^2}$$

$$N(D) = \not{N}\left(\frac{B}{\not{N}}\right)$$

$$DN = B$$

$$B = DN$$

$a = hs^2$

27.   To solve for "s" below, we cleared the fraction and then used the shortcut of the multiplication axiom. Complete the other rearrangement.

Solve for "s".

$$P = \frac{Fs}{t}$$

$$t(P) = \cancel{t}\left(\frac{Fs}{\cancel{t}}\right)$$

$$Pt = Fs$$

$$s = \frac{Pt}{F}$$

Solve for "m_1".

$$F = \frac{m_1 m_2}{rd^2}$$

---

28.   Following the example, complete the other rearrangement.

Solve for R .

$$I = \frac{E}{R}$$

$$R(I) = \cancel{R}\left(\frac{E}{\cancel{R}}\right)$$

$$IR = E$$

$$R = \frac{E}{I}$$

Solve for P_1 .

$$V_1 = \frac{P_2 V_2}{P_1}$$

$$m_1 = \frac{Frd^2}{m_2}$$

---

29.   Following the example, complete the other rearrangement.

Solve for "t".

$$a = \frac{v_2 - v_1}{t}$$

$$t(a) = \cancel{t}\left(\frac{v_2 - v_1}{\cancel{t}}\right)$$

$$at = v_2 - v_1$$

$$t = \frac{v_2 - v_1}{a}$$

Solve for "r".

$$m = \frac{c - d}{2r}$$

$$P_1 = \frac{P_2 V_2}{V_1}$$

$$r = \frac{c - d}{2m}$$

30.   Following the example, complete the other rearrangement.

Solve for A .

$$H = \frac{AKT(t_2 - t_1)}{L}$$

$$L(H) = \cancel{L}\left[\frac{AKT(t_2 - t_1)}{\cancel{L}}\right]$$

$$HL = AKT(t_2 - t_1)$$

$$A = \frac{HL}{KT(t_2 - t_1)}$$

Solve for R .

$$H = \frac{\pi dR(F_1 - F_2)}{33,000}$$

---

31.   a) Solve for "t".

$$H = \frac{M[(V_1)^2 - (V_2)^2]}{1100gt}$$

b) Solve for C .

$$X_c = \frac{1}{2\pi fC}$$

$$R = \frac{33,000H}{\pi d(F_1 - F_2)}$$

---

32.   a) Solve for L .

$$H = \frac{AKT(t_2 - t_1)}{L}$$

b) Solve for F .

$$Ft = \frac{mv}{g}$$

a) $$t = \frac{M[(V_1)^2 - (V_2)^2]}{1100gH}$$

b) $$C = \frac{1}{2\pi fX_c}$$

---

a) $$L = \frac{AKT(t_2 - t_1)}{H}$$

b) $$F = \frac{mv}{gt}$$

33.   In the formula at the left below, "t x 10^8" means "t times 10^8". Therefore, we can rewrite the formula as we have done at the right below.

$$E = \frac{N\varphi}{t \times 10^8} \qquad or \qquad E = \frac{N\varphi}{t(10^8)}$$

a) Solve for φ :

b) Solve for t :

---

34.   Since $\frac{1}{2}bh = \frac{1}{2}(bh) = \frac{bh}{2}$, we can write the formula below in two equivalent forms.

$$A = \frac{1}{2}bh \qquad or \qquad A = \frac{bh}{2}$$

It is easier to clear the fraction in the form at the right above. Therefore, before clearing the fraction in formulas of that type, we should write them in that form. That is:

$$s = \frac{1}{2}at^2 \quad should\ be\ written \quad s = \frac{at^2}{2}$$

$$A = \frac{1}{2}h(b_1 + b_2) \quad should\ be\ written \quad A = \underline{\qquad\qquad}$$

a) $\varphi = \dfrac{Et(10^8)}{N}$

b) $t = \dfrac{N\varphi}{E(10^8)}$

---

35.   Following the example, complete the other rearrangement.

Solve for "h".

$$A = \frac{1}{2}bh$$

$$A = \frac{bh}{2}$$

$$2(A) = \cancel{2}\left(\frac{bh}{\cancel{2}}\right)$$

$$2A = bh$$

$$h = \frac{2A}{b}$$

Solve for "a".

$$s = \frac{1}{2}at^2$$

$\dfrac{h(b_1 + b_2)}{2}$

$a = \dfrac{2s}{t^2}$

36.   Following the example, complete the other rearrangement.

Solve for "h".

$$A = \frac{1}{2}h(b_1 + b_2)$$

$$A = \frac{h(b_1 + b_2)}{2}$$

$$2(A) = \cancel{2}\left[\frac{h(b_1 + b_2)}{\cancel{2}}\right]$$

$$2A = h(b_1 + b_2)$$

$$h = \frac{2A}{b_1 + b_2}$$

Solve for "v^2".

$$E = \frac{1}{2}(m_1 + m_2)v^2$$

$$v^2 = \frac{2E}{m_1 + m_2}$$

8-6   REARRANGING FORMULAS CONTAINING TWO FRACTIONS

In this section, we will discuss the method for rearranging formulas that contain two fractions. The formulas are limited to those with only one term on each side.

---

37.   To rearrange a formula containing two fractions, we also begin by "clearing the fractions". To do so, we multiply both sides by both denominators at the same time. That is:

For $\dfrac{P_1}{P_2} = \dfrac{V_2}{V_1}$, we multiply both sides by P_2V_1 .

For $\dfrac{I_1}{I_2} = \dfrac{(d_2)^2}{(d_1)^2}$, we multiply both sides by _____ .

---

38.   Following the example, clear the fractions in the other formula.

$$\frac{V_1}{V_2} = \frac{T_1}{T_2}$$

$$\cancel{V_2}T_2\left(\frac{V_1}{\cancel{V_2}}\right) = V_2\cancel{T_2}\left(\frac{T_1}{\cancel{T_2}}\right)$$

$$T_2V_1 = T_1V_2$$

$$\frac{P_1V_1}{T_1} = \frac{P_2V_2}{T_2}$$

$I_2(d_1)^2$

$$P_1T_2V_1 = P_2T_1V_2$$

39. Following the example, clear the fractions in the other formula.

$$\frac{I_1}{I_2} = \frac{(d_2)^2}{(d_1)^2}$$

$$\frac{a}{b} = \frac{c^2}{d^2}$$

$$I_2(d_1)^2\left(\frac{I_1}{I_2}\right) = I_2(d_1)^2\left[\frac{(d_2)^2}{(d_1)^2}\right]$$

$$(d_1)^2 I_1 = I_2(d_2)^2$$

---

40. Following the example, complete the other rearrangement.

$ad^2 = bc^2$

Solve for P_1.   Solve for T_1.

$$\frac{P_1}{P_2} = \frac{V_2}{V_1}$$

$$\frac{V_1}{V_2} = \frac{T_1}{T_2}$$

$$P_2 V_1\left(\frac{P_1}{P_2}\right) = P_2 V_1\left(\frac{V_2}{V_1}\right)$$

$$V_1 P_1 = P_2 V_2$$

$$P_1 = \frac{P_2 V_2}{V_1}$$

---

41. Following the example, complete the other rearrangement.

$T_1 = \frac{T_2 V_1}{V_2}$

Solve for I_2.   Solve for T_2.

$$\frac{I_1}{I_2} = \frac{(d_2)^2}{(d_1)^2}$$

$$\frac{P_1 V_1}{T_1} = \frac{P_2 V_2}{T_2}$$

$$I_2(d_1)^2\left(\frac{I_1}{I_2}\right) = I_2(d_1)^2\left[\frac{(d_2)^2}{(d_1)^2}\right]$$

$$I_1(d_1)^2 = I_2(d_2)^2$$

$$I_2 = \frac{I_1(d_1)^2}{(d_2)^2}$$

---

42. a) Solve for I_1.   b) Solve for V_1.

$$\frac{I_1}{I_2} = \frac{(d_2)^2}{(d_1)^2}$$

$$\frac{P_1}{P_2} = \frac{V_2}{V_1}$$

$T_2 = \frac{P_2 T_1 V_2}{P_1 V_1}$

a) $I_1 = \frac{I_2(d_2)^2}{(d_1)^2}$

b) $V_1 = \frac{P_2 V_2}{P_1}$

338   Formula Rearrangement

SELF-TEST 26 (pp. 325-338)

1. Solve for "t".

$$d = kt$$

2. Solve for "h".

$$r = h(w - v)$$

3. Solve for F_1.

$$d_1 F_1 = d_2 F_2$$

4. Solve for H.

$$L = \frac{V}{HW}$$

5. Solve for "m".

$$E = \frac{1}{2} mv^2$$

6. Solve for "f".

$$X_c = \frac{1}{2\pi fC}$$

7. Solve for R.

$$W = \frac{BR(P_2 - P_1)}{M}$$

8. Solve for "s_1".

$$\frac{t_1}{t_2} = \frac{s_1}{s_2}$$

9. Solve for A_1.

$$\frac{A_2}{A_1} = \frac{(d_1)^2}{(d_2)^2}$$

ANSWERS:

1. $t = \dfrac{d}{k}$

2. $h = \dfrac{r}{w - v}$

3. $F_1 = \dfrac{d_2 F_2}{d_1}$

4. $H = \dfrac{V}{LW}$

5. $m = \dfrac{2E}{v^2}$

6. $f = \dfrac{1}{2\pi C X_c}$

7. $R = \dfrac{MW}{B(P_2 - P_1)}$

8. $s_1 = \dfrac{t_1 s_2}{t_2}$

9. $A_1 = \dfrac{A_2 (d_2)^2}{(d_1)^2}$

8-7   THE SQUARE-ROOT PRINCIPLE AND FORMULAS CONTAINING SQUARED LETTERS

In this section, we will show how the square root principle can be used to solve for a letter that is squared in a formula.

---

43. The principal square root of the square of a number is the number itself. For example:

$$\sqrt{7^2} = \sqrt{49} = 7 \qquad\qquad \sqrt{3^2} = \sqrt{9} = 3$$

Similarly, the principal square root of the square of a letter is the letter itself. That is:

$$\sqrt{x^2} = x \qquad \sqrt{y^2} = y \qquad \text{a) } \sqrt{s^2} = \underline{\quad\quad} \qquad \text{b) } \sqrt{P^2} = \underline{\quad\quad}$$

---

44. Equations like $x^2 = 16$ have two roots, one positive and one negative. However, as we saw earlier, the negative root does not ordinarily make sense in a formula evaluation. For example:

If $s^2 = 100$ when "s" stands for "distance", "-10" does not ordinarily make sense as a value for distance.

When rearranging formulas to solve for a letter that is squared, negative square roots also do not ordinarily make sense. Therefore, to solve for "s" in the formula below, we simply take the principal (or positive) square root of each side. That is:

$$s^2 = A$$
$$\sqrt{s^2} = \sqrt{A}$$
$$s = \sqrt{A}$$

In the formula, "s" stands for the length of the side of a square. Would the solution $s = -\sqrt{A}$ make sense? _____

a) s   b) P

---

45. To solve for E below, we took the principal square root of each side. Complete the other solutions.

$$\text{If } E^2 = PR \qquad \text{a) If } v^2 = 2as \qquad \text{b) If } s^2 = \frac{a}{h}$$

$$\sqrt{E^2} = \sqrt{PR} \qquad \sqrt{v^2} = \sqrt{2as} \qquad \sqrt{s^2} = \sqrt{\frac{a}{h}}$$

$$E = \sqrt{PR} \qquad\quad v = \underline{\quad\quad} \qquad\quad s = \underline{\quad\quad}$$

No, because a length is not ordinarily negative.

---

a) $v = \sqrt{2as}$

b) $s = \sqrt{\dfrac{a}{h}}$

46.   When solving for a letter that is squared, <u>be sure to take the square root of all of the other side</u>. For example:

$$\text{If} \quad a^2 = \frac{b + c}{d}, \qquad\qquad \text{If} \quad c^2 = a^2 + b^2,$$

$$a = \sqrt{\frac{b + c}{d}} \qquad\qquad c = \sqrt{a^2 + b^2}$$

a) If $d^2 = \dfrac{I_A - I_C}{m}$ b) If $v_o{}^2 = v_f{}^2 - 2gs$,

   $d = $ _____ $v_o = $ _____

---

47.   To solve for I below, we isolated I^2 first and then took the square root of each side. Complete the other rearrangements.

   Solve for I .   a) Solve for "w".   b) Solve for "d".

   $P = I^2R$ $a = w^2r$ $A = .7854d^2$

   $I^2 = \dfrac{P}{R}$

   $I = \sqrt{\dfrac{P}{R}}$

a) $d = \sqrt{\dfrac{I_A - I_C}{m}}$

b) $v_o = \sqrt{v_f{}^2 - 2gs}$

---

48.   To isolate "v^2" below, we began by clearing the fraction. Complete the other rearrangement.

   Solve for "v".   Solve for "d".

   $F = \dfrac{mv^2}{r}$ $P = \dfrac{pL}{d^2}$

   $r(F) = \cancel{r}\left(\dfrac{mv^2}{\cancel{r}}\right)$

   $Fr = mv^2$

   $v^2 = \dfrac{Fr}{m}$

   $v = \sqrt{\dfrac{Fr}{m}}$

a) $w = \sqrt{\dfrac{a}{r}}$

b) $d = \sqrt{\dfrac{A}{.7854}}$

$d = \sqrt{\dfrac{pL}{P}}$

49.   a) Solve for D.   b) Solve for "d".   c) Solve for "t".

$$H = \frac{D^2N}{2.5}$$ $$F = \frac{m_1m_2}{rd^2}$$ $$P = \frac{rw}{2t^2}$$

50. To solve for "t" below, we began by writing $\frac{1}{2}at^2$ as $\frac{at^2}{2}$. Complete the other rearrangement.

Solve for "t".

$$s = \frac{1}{2}at^2$$

$$s = \frac{at^2}{2}$$

$$2s = at^2$$

$$t^2 = \frac{2s}{a}$$

$$t = \sqrt{\frac{2s}{a}}$$

Solve for "v".

$$E = \frac{1}{2}mv^2$$

a) $D = \sqrt{\frac{2.5H}{N}}$

b) $d = \sqrt{\frac{m_1m_2}{Fr}}$

c) $t = \sqrt{\frac{rw}{2P}}$

51. To solve for "r" below, we began by clearing both fractions. Solve for "t" in the other formula.

$$\frac{E}{d} = \frac{2r^2}{a}$$

$$ad\left(\frac{E}{d}\right) = ad\left(\frac{2r^2}{a}\right)$$

$$aE = 2dr^2$$

$$r^2 = \frac{aE}{2d}$$

$$r = \sqrt{\frac{aE}{2d}}$$

$$\frac{s}{t^2} = \frac{m}{h}$$

$v = \sqrt{\frac{2E}{m}}$

$t = \sqrt{\frac{hs}{m}}$

52.   Though neither H is squared in the formula below, we get an H^2 when the fractions are cleared. Complete the other rearrangement.

Solve for H .

$$\frac{M}{H} = \frac{H}{T}$$

$$\cancel{H}T\left(\frac{M}{\cancel{H}}\right) = H\cancel{T}\left(\frac{H}{\cancel{T}}\right)$$

$$MT = H^2$$

$$H = \sqrt{MT}$$

Solve for "p".

$$\frac{cd}{p} = \frac{pq}{a}$$

$$p = \sqrt{\frac{acd}{q}}$$

8-8   OPPOSITES AND ADDITION-SUBTRACTION CONVERSIONS

In this section, we will extend the concept of "opposites" and addition-subtraction conversions to literal expressions.

53.   <u>Two literal terms are opposites if their sum is 0 </u>. Therefore:

Since $4P + (-4P) = 0$, 4P and -4P are opposites.

Since $2ab + (-2ab) = 0$, 2ab and -2ab are opposites.

Write the opposite of each term.

a) 10t _____   b) -3s _____   c) 7TV _____   d) -5pq _____

54.   To show that the following terms are opposites, we wrote "1" and "-1" explicitly.

Since $m^2 + (-m^2) = 1m^2 + (-1m^2) = 0$, "m^2" and "$-m^2$" are opposites.

Since $AB + (-AB) = 1AB + (-1AB) = 0$, AB and -AB are opposites.

Write the opposite of each term.

a) t^2 _____   b) $-I_A$ _____   c) F_1r_1 _____   d) -pq _____

a) -10t

b) 3s

c) -7TV

d) 5pq

55.   To convert a subtraction to an addition, <u>we add the opposite of the second term</u>. That is:

$$a - b = a + (\text{the opposite of } b) = a + (-b)$$

$$F_1r_1 - F_2r_2 = F_1r_1 + (\text{the opposite of } F_2r_2) = F_1r_1 + (-F_2r_2)$$

$$c^2 - d^2 = c^2 + (\text{the opposite of } d^2) = \text{_____}$$

a) $-t^2$

b) I_A

c) $-F_1r_1$

d) pq

56. Similarly, any addition in which the second term is negative can be converted to a subtraction. To do so, <u>we subtract the opposite of the second term</u>. That is:

$$m + (-50) = m - (\text{the opposite of } -50) = m - 50$$

$$ab + (-cd) = ab - (\text{the opposite of } -cd) = ab - cd$$

Convert each addition to a subtraction.

a) $v + (-t) =$ _____

b) $V_2 + (-V_1) =$ _____

c) $F_1 r_1 + (-F_2 r_2) =$ _____

d) $c^2 + (-a^2) =$ _____

$c^2 + (-d^2)$

---

57. In the addition below, both the second and third terms are negative. We converted both to subtraction form.

$$D + (-S) + (-T) = D - S - T$$

Convert these additions to subtraction form.

a) $p + (-q) + (-r) =$ _____

b) $R_t + (-R_1) + (-R_2) =$ _____

a) $v - t$

b) $V_2 - V_1$

c) $F_1 r_1 - F_2 r_2$

d) $c^2 - a^2$

---

a) $p - q - r$ b) $R_t - R_1 - R_2$

8-9   THE ADDITION AXIOM AND FORMULA REARRANGEMENT

When a formula contains more than one term on one side, we use the addition axiom to solve for a letter on that side. We will discuss the method in this section.

---

58. To solve for A below, we added L to both sides. Solve for P in the other formula by adding D to both sides.

$$C = A - L \qquad\qquad N = P - D$$

$$L + C = A - L + L$$

$$L + C = A + \quad 0$$

$$A = L + C$$

---

59. Using the same steps, solve for the indicated letter in each formula.

a) Solve for K.

$$C = K - 273$$

b) Solve for E.

$$e = E - Ir$$

$P = D + N$

60. To solve for C below, we added -M to both sides. To solve for V below, we added -100 to both sides.

$$R = C + M \qquad\qquad F = V + 100$$

$$R + (-M) = C + \underline{M + (-M)} \qquad F + (-100) = V + \underline{100 + (-100)}$$

$$R + (-M) = C + \quad 0 \qquad\qquad F + (-100) = V + \quad 0$$

$$C = R + (-M) \qquad\qquad V = F + (-100)$$

In each solution above, the right side is an addition in which the <u>second term</u> is <u>negative</u>. Ordinarily, additions of that type are converted to subtraction form. That is:

Instead of C = R + (-M) , we write C = R - M .

Instead of V = F + (-100) , we write _____ .

a) K = C + 273

b) E = e + Ir

---

61. Solve for the indicated letter. Write each solution in subtraction form.

a) Solve for E .

$$H = E + PV$$

b) Solve for R .

$$D = R + S + T$$

V = F - 100

---

62. To solve for P below, we used the addition axiom to isolate PV . Then we used the multiplication axiom. Solve for "m" in the other formula.

$$H = E + PV \qquad\qquad I_a = I_c + md^2$$

$$H + (-E) = (-E) + E + PV$$

$$H - E = PV$$

$$P = \frac{H - E}{V}$$

a) E = H - PV

b) R = D - S - T

---

63. After clearing the fraction below, we used the addition axiom to solve for "v_2". Solve for F_i in the other formula.

$$a = \frac{v_2 - v_1}{t} \qquad\qquad w = \frac{F_o + F_i}{t}$$

$$at = v_2 - v_1$$

$$at + v_1 = v_2 - v_1 + v_1$$

$$v_2 = at + v_1$$

$$m = \frac{I_a - I_c}{d^2}$$

$$F_i = wt - F_o$$

64. To solve for "v_f" below, we used the addition axiom to isolate v_f^2 first. Solve for "c" in the other formula.

$$v_o^2 \;=\; v_f^2 - 2gs \qquad\qquad a^2 \;=\; c^2 - b^2$$

$$v_o^2 + 2gs \;=\; v_f^2 - 2gs + 2gs$$

$$v_o^2 + 2gs \;=\; v_f^2$$

$$v_f \;=\; \sqrt{v_o^2 + 2gs}$$

---

65. To solve for X below, we isolated X^2 first. Solve for "b" in the other formula.

$$Z^2 \;=\; X^2 + R^2 \qquad\qquad c^2 \;=\; a^2 + b^2$$

$$Z^2 + (-R^2) \;=\; X^2 + R^2 + (-R^2)$$

$$Z^2 - R^2 \;=\; X^2$$

$$X \;=\; \sqrt{Z^2 - R^2}$$

$$c \;=\; \sqrt{a^2 + b^2}$$

---

66. The steps needed to solve for F_1 in the formula below are described.

$$F_1 r_1 + F_2 r_2 + F_3 r_3 \;=\; 0$$

1) Add $(-F_2 r_2)$ and $(-F_3 r_3)$ to both sides and get:

$$F_1 r_1 \;=\; (-F_2 r_2) + (-F_3 r_3)$$

2) Change the addition to subtraction form.

$$F_1 r_1 \;=\; (-F_2 r_2) - F_3 r_3$$

$$\text{or}$$

$$F_1 r_1 \;=\; -F_2 r_2 - F_3 r_3$$

3) Use the multiplication axiom.

$$F_1 \;=\; \frac{-F_2 r_2 - F_3 r_3}{r_1}$$

Solve for "r_2" in the same formula.

$$F_1 r_1 + F_2 r_2 + F_3 r_3 \;=\; 0$$

$$b \;=\; \sqrt{c^2 - a^2}$$

$$r_2 \;=\; \frac{-F_1 r_1 - F_3 r_3}{F_2}$$

346   Formula Rearrangement

8-10   THE OPPOSITING PRINCIPLE AND FORMULA REARRANGEMENT

Though D and "v_1" are isolated below, they are not solved for because there is a "-" in front of each.

$$-D = N - L \qquad -v_1 = at - v_2$$

In this section, we will show how the oppositing principle can be used to complete each solution above.

67.   We have already seen that we can get the opposite of a subtraction by simply interchanging the two terms. That is:

The opposite of $(5 - 3)$ is $(3 - 5)$

The opposite of $(2x - 1)$ is $(1 - 2x)$

The same procedure can be used to get the opposite of a subtraction with two literal terms. That is:

The opposite of $(a - b)$ is $(b - a)$

The opposite of $(at - v_2)$ is $(v_2 - at)$

Write the opposite of each subtraction.

a) $4y - 1$ _____

b) $p - q$ _____

c) $a^2 - c^2$ _____

d) $v_f^2 - 2gs$ _____

68.   We have also seen the "oppositing principle for equations". That principle, which says that we can replace each side of an equation by its opposite, was used to solve for "y" below.

$$-y = 3x - 4$$
$$y = 4 - 3x$$

The same principle can be used to solve for D and v_1 below. That is:

$$-D = N - L \qquad -v_1 = at - v_2$$
$$D = L - N \qquad v_1 = v_2 - at$$

Solve for "t" in each formula below by replacing each side with its opposite.

a) $-t = a - b$

t = _____

b) $-t = cd - pq$

t = _____

a) $1 - 4y$

b) $q - p$

c) $c^2 - a^2$

d) $2gs - v_f^2$

a) $t = b - a$

b) $t = pq - cd$

69.   Use the oppositing principle for these.

   a) Solve for I .   b) Solve for "a".

       $-I = E_1 - E_2$ $-a = bF_1 - cF_2$

---

70.   In solving for L below, we had to use the oppositing principle for the final step. Solve for D in the other formula.

a) $I = E_2 - E_1$
b) $a = cF_2 - bF_1$

$$C = A - L \qquad\qquad N = P - D$$
$$C + (-A) = (-A) + A - L$$
$$C - A = -L$$
$$L = A - C$$

---

71.   In solving for T_C below, we used the oppositing principle for the final step. Solve for R_1 in the other formula.

$D = P - N$

$$T_K - T_C = 273 \qquad\qquad R_t - R_1 = R_2$$
$$(-T_K) + T_K - T_C = 273 + (-T_K)$$
$$-T_C = 273 - T_K$$
$$T_C = T_K - 273$$

---

72.   We solved for "v_1" in the formula below. Use the same steps to solve for "d" in the other formula.

$R_1 = R_t - R_2$

$$a = \frac{v_2 - v_1}{t} \qquad\qquad m = \frac{b - d}{r}$$
$$at = v_2 - v_1$$
$$at + (-v_2) = (-v_2) + v_2 - v_1$$
$$at - v_2 = -v_1$$
$$v_1 = v_2 - at$$

---

73.   We solved for I below. Use the same steps to solve for "s" in the other formula.

$d = b - mr$

$$e = E - Ir \qquad\qquad v_o^2 = v_f^2 - 2gs$$
$$e + (-E) = (-E) + E - Ir$$
$$e - E = -Ir$$
$$Ir = E - e$$
$$I = \frac{E - e}{r}$$

74.   We solved for "b" below. Use the same steps to solve for "q" in the other
formula.

$$a^2 = c^2 - b^2 \qquad\qquad p^2 - q^2 = r^2$$
$$a^2 + (-c^2) = (-c^2) + c^2 - b^2$$
$$a^2 - c^2 = -b^2$$
$$b^2 = c^2 - a^2$$
$$b = \sqrt{c^2 - a^2}$$

$$s = \frac{v_f^2 - v_o^2}{2g}$$

$$q = \sqrt{p^2 - r^2}$$

8-11   CONTRASTING COEFFICIENTS AND TERMS

Confusing a subtraction of "two terms" with a "negative factor and its coefficient" is a common error. We
will contrast expressions of those types in this section.

75. $F_1 - F_2$ and $F_1(-F_2)$ are two different algebraic expressions.

In $F_1 - F_2$, the "-" is the symbol for subtraction. Therefore,
F_1 and F_2 are <u>two</u> <u>different</u> <u>terms</u>.

In $F_1(-F_2)$, the parentheses indicate a multiplication. The "-"
is the sign of F_2 . Therefore, the whole expression
is <u>only</u> <u>one</u> <u>term</u>.

a) Which of these is a multiplication? _____

$X_L - X_C$ \qquad $X_L(-X_C)$

b) Which of these is a subtraction? _____

$E_1 - E_o$ \qquad $E_1(-E_o)$

76.   a) In $V_1 - V_2$, is V_1 a term in a subtraction or the coefficient of $-V_2$?

_____

b) In $c(-d)$, is "c" a term in a subtraction or the coefficient of $-d$?

_____

a) $X_L(-X_C)$

b) $E_1 - E_o$

77.   In which of the following is V_0 a coefficient? _____

a) $V_0 V_1$ \qquad b) $V_0 - V_1$ \qquad c) $V_0(-V_1)$

a) a term in a
subtraction

b) the coefficient of $-d$

Both (a) and (c)

78.   To solve for X_C below, we had to eliminate X_L from that side. Since X_L is a term in a subtraction, <u>we used the addition axiom</u>, not the multiplication axiom. Solve for E_0 in the other formula.

$$X = X_L - X_C$$

$$X + (-X_L) = (-X_L) + X_L - X_C$$

$$X - X_L = -X_C$$

$$X_C = X_L - X$$

$$I = E_1 - E_0$$

$$E_0 = E_1 - I$$

<u>SELF-TEST 27</u> (pp. <u>339-349</u>)

1.   Solve for "r".

$$A = \pi r^2$$

2.   Solve for "v".

$$s = \frac{v^2}{g}$$

3.   Solve for "d".

$$F = \frac{kQ_1 Q_2}{d^2}$$

4.   Solve for "t".

$$\frac{2h}{at^2} = \frac{c}{m}$$

5.   Solve for K .

$$\frac{G}{K} = \frac{K}{T}$$

6.   Solve for "d_3".

$$d_1 w_1 + d_2 w_2 + d_3 w_3 = 0$$

7.   Solve for "y_2".

$$m = \frac{y_2 - y_1}{x}$$

8.   Solve for "i".

$$p = P - ei$$

9.   Solve for N .

$$F^2 = R^2 - N^2$$

<u>ANSWERS:</u>   1. $r = \sqrt{\dfrac{A}{\pi}}$ 3. $d = \sqrt{\dfrac{kQ_1 Q_2}{F}}$ 5. $K = \sqrt{GT}$ 7. $y_2 = y_1 + mx$ 9. $N = \sqrt{R^2 - F^2}$

2. $v = \sqrt{gs}$ 4. $t = \sqrt{\dfrac{2hm}{ac}}$ 6. $d_3 = \dfrac{-d_1 w_1 - d_2 w_2}{w_3}$ 8. $i = \dfrac{P - p}{e}$

8-12   MULTIPLYING BY THE DISTRIBUTIVE PRINCIPLE IN REARRANGEMENTS

In this section, we will discuss formula rearrangements that involve multiplying by the distributive principle.

---

79.   When letters are involved, we multiply by the distributive principle in the usual way. That is:

$$P(Q + R) = PQ + PR \qquad\qquad C(T_1 - T_2) = CT_1 - CT_2$$

Following the examples, complete these.

a) $h(b_1 + b_2)$ = _____   b) $3.14(d_2 - d_1)$ = _____

---

80.   In $D = P(Q + R)$, the letter "Q" appears within the parentheses. To solve for Q , we can begin by multiplying by the distributive principle as we have done below. Use the same method to solve for "t" in the other formula.

a) $hb_1 + hb_2$
b) $3.14d_2 - 3.14d_1$

$$D = P(Q + R)$$
$$D = PQ + PR$$
$$D + (-PR) = PQ + PR + (-PR)$$
$$D - PR = PQ$$
$$Q = \frac{D - PR}{P}$$

$$m = b(d + t)$$

---

81.   In the formula below, T_1 is within the parentheses. Therefore to solve for it, we began by multiplying by the distributive principle. Solve for L in the other formula.

$t = \dfrac{m - bd}{b}$

$$Q = C(T_1 - T_2)$$
$$Q = CT_1 - CT_2$$
$$Q + CT_2 = CT_1 - CT_2 + CT_2$$
$$Q + CT_2 = CT_1$$
$$T_1 = \frac{Q + CT_2}{C}$$

$$M = P(L - X)$$

---

82.   To solve for V_2 below, we began by multiplying by the distributive principle. Notice that we had to use the oppositing principle. Solve for "t_1" in the other formula.

$L = \dfrac{M + PX}{P}$

$$W_S = P(V_1 - V_2)$$
$$W_S = PV_1 - PV_2$$
$$W_S + (-PV_1) = (-PV_1) + PV_1 - PV_2$$
$$W_S - PV_1 = -PV_2$$
$$PV_1 - W_S = PV_2$$
$$V_2 = \frac{PV_1 - W_S}{P}$$

$$H = ms(t_2 - t_1)$$

83.   To solve for either "b_1" or "b_2" below, we have to get them out of the parentheses.
Notice the steps.

$$t_1 = \frac{mst_2 - H}{ms}$$

$$A = \frac{1}{2}h(b_1 + b_2)$$

$$A = \frac{h(b_1 + b_2)}{2}$$

$$2A = h(b_1 + b_2)$$

$$2A = hb_1 + hb_2$$

Beginning with the last formula, solve for both "b_1" and "b_2".

a) Solve for "b_1".   b) Solve for "b_2".

---

84.   To solve for R below, we began by multiplying by the distributive principle.
Solve for "d" in the other formula.

$$A = \pi(R^2 - r^2)$$ $$m = (d^2 - 1)t$$

$$A = \pi R^2 - \pi r^2$$

$$A + \pi r^2 = \pi R^2 - \pi r^2 + \pi r^2$$

$$A + \pi r^2 = \pi R^2$$

$$R^2 = \frac{A + \pi r^2}{\pi}$$

$$R = \sqrt{\frac{A + \pi r^2}{\pi}}$$

a) $b_1 = \dfrac{2A - hb_2}{h}$

b) $b_2 = \dfrac{2A - hb_1}{h}$

---

85.   To solve for C , we also began by multiplying by the distributive principle.
Notice that we had to use the oppositing principle. Solve for "r" in the other
formula.

$$d = \sqrt{\frac{m + t}{t}}$$

$$H = (1 - C^2)h$$ $$A = \pi(R^2 - r^2)$$

$$H = h - C^2h$$

$$H + (-h) = (-h) + h - C^2h$$

$$H - h = -C^2h$$

$$h - H = C^2h$$

$$C^2 = \frac{h - H}{h}$$

$$C = \sqrt{\frac{h - H}{h}}$$

86.   To solve for "t_2" below, we began by clearing the fraction. Solve for "t" in the other formula.

$$r = \sqrt{\frac{\pi R^2 - A}{\pi}}$$

$$H = \frac{AKT(t_2 - t_1)}{L}$$

$$p = \frac{k(t^2 - b^2)}{ar}$$

$$HL = AKT(t_2 - t_1)$$

$$HL = AKTt_2 - AKTt_1$$

$$HL + AKTt_1 = AKTt_2$$

$$t_2 = \frac{HL + AKTt_1}{AKT}$$

87.   To solve for V_2 below, we also began by clearing the fraction. Notice that we had to use the oppositing principle. Solve for F_2 in the other formula.

$$t = \sqrt{\frac{apr + b^2k}{k}}$$

$$H = \frac{M(V_1^2 - V_2^2)}{1,100gt}$$

$$H = \frac{\pi dR(F_1 - F_2)}{33,000}$$

$$1,100gtH = M(V_1^2 - V_2^2)$$

$$1,100gtH = MV_1^2 - MV_2^2$$

$$1,100gtH - MV_1^2 = -MV_2^2$$

$$MV_1^2 - 1,100gtH = MV_2^2$$

$$V_2^2 = \frac{MV_1^2 - 1,100gtH}{M}$$

$$V_2 = \sqrt{\frac{MV_1^2 - 1,100gtH}{M}}$$

$$F_2 = \frac{\pi dRF_1 - 33,000H}{\pi dR}$$

8-13   ANOTHER METHOD AND EQUIVALENT FORMS OF SOLUTIONS

There is another method that can be used for the rearrangements in the last section. The other method leads to a solution that looks different but is really equivalent. We will discuss the other method in this section and show that the two forms of the solution are equivalent.

88.   To solve for L at the left below, we began by multiplying by the distributive principle. At the right below, we used another method to solve for L . In the second method, we began by isolating the binomial.

$$M = P(L - X)$$

$$M = PL - PX$$

$$M + PX = PL - PX + PX$$

$$M + PX = PL$$

$$L = \frac{M + PX}{P}$$

$$M = P(L - X)$$

$$\frac{M}{P} = L - X$$

$$\frac{M}{P} + X = L - X + X$$

$$L = \frac{M}{P} + X$$

Though the two solutions look different, they are equivalent. We have shown that fact below.

To show that the left solution is equivalent to the right solution, we can break up the sum of fractions and reduce.

$$L = \frac{M + PX}{P} = \frac{M}{P} + \frac{\not{P}X}{\not{P}} = \frac{M}{P} + X$$

To show that the right solution is equivalent to the left solution, we can add the fraction and non-fraction.

$$L = \frac{M}{P} + X = \frac{M}{P} + \frac{PX}{P} = \underline{\hspace{3cm}}$$

89.   To solve for "r" below, we began by multiplying by the distributive principle. At the right below, we used another method in which we began by isolating the binomial.

$$\frac{M + PX}{P}$$

$$A = \pi(R^2 - r^2)$$

$$A = \pi R^2 - \pi r^2$$

$$A - \pi R^2 = -\pi r^2$$

$$\pi R^2 - A = \pi r^2$$

$$r^2 = \frac{\pi R^2 - A}{\pi}$$

$$r = \sqrt{\frac{\pi R^2 - A}{\pi}}$$

$$A = \pi(R^2 - r^2)$$

$$\frac{A}{\pi} = R^2 - r^2$$

$$\frac{A}{\pi} - R^2 = -r^2$$

$$R^2 - \frac{A}{\pi} = r^2$$

$$r = \sqrt{R^2 - \frac{A}{\pi}}$$

Continued on following page.

89.   Continued

Though the two solutions look different, they are also equivalent. We have shown that fact below.

    a) To show that the left radicand is equivalent to the right radicand, we can break up the sum of fractions and reduce.

$$r = \sqrt{\frac{\pi R^2 - A}{\pi}} = \sqrt{\frac{\pi R^2}{\pi} - \frac{A}{\pi}} = \underline{\hspace{3cm}}$$

    b) To show that the right radicand is equivalent to the left radicand, we can add the non-fraction and fraction.

$$r = \sqrt{R^2 - \frac{A}{\pi}} = \sqrt{\frac{\pi R^2}{\pi} - \frac{A}{\pi}} = \underline{\hspace{3cm}}$$

---

90.   To solve for "t_2" by the second method below, we cleared the fraction and then isolated the binomial. Use the same method to solve for F_1 in the other formula.

$$H = \frac{AKT(t_2 - t_1)}{L} \qquad\qquad H = \frac{\pi dR(F_1 - F_2)}{33,000}$$

$$HL = AKT(t_2 - t_1)$$

$$\frac{HL}{AKT} = t_2 - t_1$$

$$t_2 = \frac{HL}{AKT} + t_1$$

a) $\sqrt{R^2 - \dfrac{A}{\pi}}$

b) $\sqrt{\dfrac{\pi R^2 - A}{\pi}}$

---

91.   To solve for V_2 by the second method below, we isolated the binomial after clearing the fraction. Notice how we used the oppositing principle. Use the same method to solve for "b" in the other formula.

$$F_1 = \frac{33,000H}{\pi dR} + F_2$$

$$H = \frac{M(V_1^2 - V_2^2)}{1,100gt} \qquad\qquad p = \frac{k(t^2 - b^2)}{ar}$$

$$1,100gtH = M(V_1^2 - V_2^2)$$

$$\frac{1,100gtH}{M} = V_1^2 - V_2^2$$

$$\frac{1,100gtH}{M} - V_1^2 = -V_2^2$$

$$V_1^2 - \frac{1,100gtH}{M} = V_2^2$$

$$V_2 = \sqrt{V_1^2 - \frac{1,100gtH}{M}}$$

$$b = \sqrt{t^2 - \frac{apr}{k}}$$

92.   Two methods are possible for the rearrangements below. Use the one that you prefer.

   a) Solve for V_1 .

$$W_S = P(V_1 - V_2)$$

   b) Solve for "t_1".

$$H = \frac{AKT(t_2 - t_1)}{L}$$

---

93.   Use either method for the rearrangements below.

   a) Solve for "m".

$$b = c(d^2 - m^2)$$

   b) Solve for G .

$$V = \frac{R(G^2 - H^2)}{100}$$

a) $V_1 = \dfrac{W_S}{P} + V_2$

   or

$V_1 = \dfrac{W_S + PV_2}{P}$

b) $t_1 = t_2 - \dfrac{LH}{AKT}$

   or

$t_1 = \dfrac{AKTt_2 - LH}{AKT}$

---

94.   We solved for R_1 below. Solve for R_3 in the same formula

$$E = IR_1 + IR_2 + IR_3$$

$$IR_1 = E - IR_2 - IR_3$$

$$R_1 = \frac{E - IR_2 - IR_3}{I}$$

   or

$$R_1 = \frac{E}{I} - R_2 - R_3$$

$$E = IR_1 + IR_2 + IR_3$$

a) $m = \sqrt{d^2 - \dfrac{b}{c}}$

   or

$m = \sqrt{\dfrac{cd^2 - b}{c}}$

b) $G = \sqrt{\dfrac{100V}{R} + H^2}$

   or

$G = \sqrt{\dfrac{100V + RH^2}{R}}$

$R_3 = \dfrac{E - IR_1 - IR_2}{I}$

   or

$R_3 = \dfrac{E}{I} - R_1 - R_2$

8-14   FACTORING BY THE DISTRIBUTIVE PRINCIPLE IN REARRANGEMENTS

To solve for a variable that appears in two terms in a formula, we must factor by the distributive principle. We will discuss rearrangements of that type in this section.

95. In a solution for a letter, <u>the same letter cannot appear in the expression on the other side</u>. For example, the following is not a solution for H because H appears on the right side.

$$H = \frac{A + BH}{C}$$

Which of the following is a solution for "v"? _____

a) $v = \dfrac{p + qr}{t}$ b) $v = \dfrac{p + qv}{t}$

> Only (a). In (b), "v" appears on the right side.

96. In $\boxed{am = b + cm}$, "m" appears in one term on each side. Let's try to solve for "m" by isolating "am" alone or "cm" alone.

<u>Isolating "am" alone</u>.   <u>Isolating "cm" alone</u>.

$$am = b + cm$$ $$cm = am - b$$

$$m = \frac{b + cm}{a}$$ $$m = \frac{am - b}{c}$$

Neither expression above is a solution for "m". Why not? _____

> Because in each, "m" also appears on the right side.

97. To solve for "m" in $\boxed{am = b + cm}$, we must get both "m" terms on one side, <u>factor by the distributive principle</u>, and then complete the solution as we have done below.

$$am = b + cm$$
$$am - cm = b$$
$$m(a - c) = b$$
$$m = \frac{b}{a - c}$$

Before rearranging formulas like those above, let's review the procedure for factoring by the distributive principle. Two examples are shown.

$$bd + bt = b(d + t)$$
$$AV_1 - CV_1 = V_1(A - C)$$

Following the examples, factor these.

a) $mx + my = $ _____   c) $HR_1 + HR_2 = $ _____

b) $aR - bR = $ _____   d) $c_1T_1 - c_2T_1 = $ _____

98.   To solve for "d" below, we factored by the distributive principle and then used the multiplication axiom. Solve for I in the other formula.

$$m \ = \ dp + dq$$

$$m \ = \ d(p + q)$$

$$d \ = \ \frac{m}{p + q}$$

$$E \ = \ IR_1 + IR_2$$

a) $m(x + y)$

b) $R(a - b)$

c) $H(R_1 + R_2)$

d) $T_1(c_1 - c_2)$

99.   To solve for S below, we isolated both "S" terms on one side before factoring. Solve for P_2 in the other formula.

$$MS \ = \ R - QS$$

$$MS + QS \ = \ R$$

$$S(M + Q) \ = \ R$$

$$S \ = \ \frac{R}{M + Q}$$

$$P_1P_2 \ = \ I - RP_2$$

$$I \ = \ \frac{E}{R_1 + R_2}$$

100.   To solve for "t" below, we isolated both "t" terms on one side before factoring. Solve for T in the other formula.

$$bt + cm \ = \ dt$$

$$cm \ = \ dt - bt$$

$$cm \ = \ t(d - b)$$

$$t \ = \ \frac{cm}{d - b}$$

$$RV + ST \ = \ QT$$

$$P_2 \ = \ \frac{I}{P_1 + R}$$

101.   To clear the fractions in the formula below, we multiplied both sides by "abc". Notice how we multiplied by the distributive principle on the left. Clear the fractions in the other formula by multiplying both sides by "pqr".

$$\frac{1}{a} + \frac{1}{b} \ = \ \frac{1}{c}$$

$$abc\left(\frac{1}{a} + \frac{1}{b}\right) \ = \ abc\left(\frac{1}{c}\right)$$

$$abc\left(\frac{1}{a}\right) + abc\left(\frac{1}{b}\right) \ = \ abc\left(\frac{1}{c}\right)$$

$$bc + ac \ = \ ab$$

$$\frac{1}{p} + \frac{1}{q} \ = \ \frac{1}{r}$$

$$T \ = \ \frac{RV}{Q - S}$$

$$qr + pr \ = \ pq$$

102. To clear the fractions below, we multiplied both sides by $R_t R_1 R_2$. Clear the fractions in the other formula by multiplying both sides by $C_1 C_t C_2$.

$$\frac{1}{R_t} = \frac{1}{R_1} + \frac{1}{R_2}$$

$$\frac{1}{C_1} = \frac{1}{C_t} - \frac{1}{C_2}$$

$$R_t R_1 R_2\left(\frac{1}{R_t}\right) = R_t R_1 R_2\left(\frac{1}{R_1} + \frac{1}{R_2}\right)$$

$$R_t R_1 R_2\left(\frac{1}{R_t}\right) = R_t R_1 R_2\left(\frac{1}{R_1}\right) + R_t R_1 R_2\left(\frac{1}{R_2}\right)$$

$$R_1 R_2 = R_t R_2 + R_t R_1$$

---

103. To solve for "c" below, we cleared the fractions and then factored by the distributive principle. Solve for "f" in the other formula.

$$\boxed{C_t C_2 = C_1 C_2 - C_1 C_t}$$

$$\frac{1}{a} + \frac{1}{b} = \frac{1}{c}$$

$$\frac{1}{D} + \frac{1}{d} = \frac{1}{f}$$

$$abc\left(\frac{1}{a} + \frac{1}{b}\right) = abc\left(\frac{1}{c}\right)$$

$$abc\left(\frac{1}{a}\right) + abc\left(\frac{1}{b}\right) = abc\left(\frac{1}{c}\right)$$

$$bc + ac = ab$$

$$c(b + a) = ab$$

$$c = \frac{ab}{b + a}$$

---

104. Using the same steps, complete these:

a) Solve for R_t .

$$\frac{1}{R_t} = \frac{1}{R_1} + \frac{1}{R_2}$$

b) Solve for C_1 .

$$\frac{1}{C_1} = \frac{1}{C_t} - \frac{1}{C_2}$$

$$\boxed{f = \frac{Dd}{d + D}}$$

$$\boxed{\text{a) } R_t = \frac{R_1 R_2}{R_2 + R_1} \qquad \text{b) } C_1 = \frac{C_t C_2}{C_2 - C_t}}$$

105. To solve for W below, we cleared the fractions by multiplying both sides by 2. Solve for A in the other formula.

$$M = \frac{WLX}{2} - \frac{WX^2}{2}$$

$$2(M) = 2\left(\frac{WLX}{2} - \frac{WX^2}{2}\right)$$

$$2M = \cancel{2}\left(\frac{WLX}{\cancel{2}}\right) - \cancel{2}\left(\frac{WX^2}{\cancel{2}}\right)$$

$$2M = WLX - WX^2$$

$$2M = W(LX - X^2)$$

$$W = \frac{2M}{LX - X^2}$$

$$T = \frac{AFG}{3} - \frac{AG^2}{3}$$

---

106. To solve for "d" below, we had to isolate the "d" terms before factoring. Solve for C_t in the other formula.

$$\frac{1}{D} + \frac{1}{d} = \frac{1}{f}$$

$$Ddf\left(\frac{1}{D} + \frac{1}{d}\right) = Ddf\left(\frac{1}{f}\right)$$

$$\cancel{D}df\left(\frac{1}{\cancel{D}}\right) + D\cancel{d}f\left(\frac{1}{\cancel{d}}\right) = Dd\cancel{f}\left(\frac{1}{\cancel{f}}\right)$$

$$df + Df = Dd$$

$$Df = Dd - df$$

$$Df = d(D - f)$$

$$d = \frac{Df}{D - f}$$

$$\frac{1}{C_1} = \frac{1}{C_t} - \frac{1}{C_2}$$

$$\boxed{A = \frac{3T}{FG - G^2}}$$

---

107. We solved for T_R below. Notice how we factored $T - ET$ on the right side of the solution. Solve for R in the other formula.

$$E = \frac{T - T_R}{T}$$

$$ET = T - T_R$$

$$ET - T = -T_R$$

$$T - ET = T_R$$

$$T_R = T(1 - E)$$

$$V = \frac{S - R}{S}$$

$$\boxed{C_t = \frac{C_1 C_2}{C_1 + C_2}}$$

$$\boxed{\begin{array}{l} R = S(1 - V) \text{, from:} \\ R = S - SV \end{array}}$$

108.   To solve for M below, we had to isolate the "M" terms after multiplying by the distributive principle. Solve for β in the other formula.

$$\mu = \frac{Mm}{M + m}$$

$$\alpha = \frac{\beta}{\beta + 1}$$

$$\mu(M + m) = (\cancel{M + m})\left(\frac{Mm}{\cancel{M + m}}\right)$$

$$\mu M + \mu m = Mm$$

$$\mu m = Mm - \mu M$$

$$\mu m = M(m - \mu)$$

$$M = \frac{\mu m}{m - \mu}$$

109.   Using the same steps, complete these.

$$\beta = \frac{\alpha}{1 - \alpha}$$

a) Solve for "m".

$$\mu = \frac{Mm}{M + m}$$

b) Solve for A .

$$A_f = \frac{A}{1 - BA}$$

110.   To solve for "a" below, we factored by the distributive principle. Solve for I in the other formula.

$$m = ab + ac + ad$$

$$m = a(b + c + d)$$

$$a = \frac{m}{b + c + d}$$

$$E = IR_1 + IR_2 + IR_3$$

a) $m = \dfrac{\mu M}{M - \mu}$

b) $A = \dfrac{A_f}{1 + A_f B}$

$$I = \frac{E}{R_1 + R_2 + R_3}$$

111.   Two methods of solving for "v" in the formula below are shown. At the left, we did not clear the fraction first. At the right, we did clear the fraction first.

$$P = \frac{mv^2}{r} - mg$$

$$P + mg = \frac{mv^2}{r}$$

$$r(P + mg) = mv^2$$

$$v^2 = \frac{r(P + mg)}{m}$$

$$v = \sqrt{\frac{r(P + mg)}{m}}$$

$$P = \frac{mv^2}{r} - mg$$

$$r(P) = r\left(\frac{mv^2}{r}\right) - r(mg)$$

$$Pr = mv^2 - mgr$$

$$Pr + mgr = mv^2$$

$$v^2 = \frac{Pr + mgr}{m}$$

$$v = \sqrt{\frac{r(P + mg)}{m}}$$

Using either method above, do these:

a) Solve for "a".

$$a^2 + 1 = \frac{b}{c}$$

b) Solve for "d".

$$T = \frac{bd^2}{p} - S$$

a) $a = \sqrt{\frac{b}{c} - 1}$ or $a = \sqrt{\frac{b - c}{c}}$

b) $d = \sqrt{\frac{p(T + S)}{b}}$

<u>SELF</u>-<u>TEST</u> <u>28</u> (pp. <u>350</u>-<u>362</u>)

1.   Solve for "t_1".

$$d = r(t_1 + t_2)$$

2.   Solve for "p".

$$H = \frac{1}{2}k(P^2 - p^2)$$

3.   Solve for R .

$$V = \frac{AH(R - r)}{M}$$

4.   Solve for "i".

$$e_1 i = p - e_2 i$$

5.   Solve for F .

$$\frac{1}{F} = \frac{1}{P} + \frac{1}{Q}$$

6.   Solve for "h_2".

$$\frac{1}{h_1} + \frac{1}{h_2} = \frac{1}{H}$$

7.   Solve for "d".

$$v = \frac{ad}{a - d}$$

<u>ANSWERS</u>:

1. $t_1 = \dfrac{d - rt_2}{r}$

2. $p = \sqrt{\dfrac{kP^2 - 2H}{k}}$

3. $R = \dfrac{AHr + MV}{AH}$

4. $i = \dfrac{p}{e_1 + e_2}$

5. $F = \dfrac{PQ}{P + Q}$

6. $h_2 = \dfrac{h_1 H}{h_1 - H}$

7. $d = \dfrac{av}{a + v}$

8-15   A PREFERRED FORM FOR FRACTIONAL SOLUTIONS

When rearranging formulas, fractional solutions are usually written so that they contain as few negative signs as possible. We will discuss that preferred form for solutions in this section.

112. Two fractions are equivalent <u>if</u> <u>their</u> <u>numerators</u> <u>are</u> <u>opposites</u> <u>and</u> <u>their</u> <u>denominators</u> <u>are</u> <u>opposites</u>. That is:

$\frac{6}{3}$ and $\frac{-6}{-3}$ are equivalent, because both equal +2 .

$\frac{8}{2}$ and $\frac{-8}{-2}$ are equivalent, because both equal +4 .

Write each of these in an equivalent form by replacing each term with its opposite.

a) $\frac{7}{4}$ = _____   b) $\frac{-2}{-9}$ = _____   c) $\frac{a}{b}$ = _____   d) $\frac{-P}{-Q}$ = _____

---

113. Two fractions with binomial denominators are also equivalent <u>if</u> <u>their</u> <u>numerators</u> <u>are</u> <u>opposites</u> <u>and</u> <u>their</u> <u>denominators</u> <u>are</u> <u>opposites</u>. That is:

$\frac{10}{6-4}$ and $\frac{-10}{4-6}$ are equivalent, because both equal +5 .

Write each of these in an equivalent form by replacing each term with its opposite.

a) $\frac{12}{7-4}$ = _____   c) $\frac{P_1}{R_1 - R_2}$ = _____

b) $\frac{-15}{3-8}$ = _____   d) $\frac{-ab}{c-d}$ = _____

a) $\frac{-7}{-4}$

b) $\frac{2}{9}$

c) $\frac{-a}{-b}$

d) $\frac{P}{Q}$

---

114. Depending on the method used, you can get either of the following <u>equivalent</u> solutions when solving $dm = cm - b$ for "m".

$$m = \frac{b}{c-d} \qquad m = \frac{-b}{d-c}$$

The solution on the left is preferred because it contains fewer negative signs. If you get the solution on the right, you can convert it to the preferred form <u>by</u> <u>replacing</u> <u>each</u> <u>term</u> <u>with</u> <u>its</u> <u>opposite</u>. That is:

$$m = \frac{-b}{d-c} = \frac{b}{c-d}$$

Write each solution in the preferred form.

a) $P = \frac{-S}{V-T}$ = _____   b) $R = \frac{-wr}{2F-w}$ = _____

a) $\frac{-12}{4-7}$

b) $\frac{15}{8-3}$

c) $\frac{-P_1}{R_2 - R_1}$

d) $\frac{ab}{d-c}$

115.   At the left below, we used the oppositing principle when solving for T .
At the right below, we did not use the oppositing principle when solving
for T .

a) $P = \dfrac{S}{T - V}$

b) $R = \dfrac{wr}{w - 2F}$

$$E = \frac{T - T_R}{T}$$

$$ET = T - T_R$$

$$ET - T = -T_R$$

$$T - ET = T_R$$

$$T(1 - E) = T_R$$

$$T = \frac{T_R}{1 - E}$$

$$E = \frac{T - T_R}{T}$$

$$ET = T - T_R$$

$$ET - T = -T_R$$

$$T(E - 1) = -T_R$$

$$T = \frac{-T_R}{E - 1}$$

The solution at the left is preferred. If you get the solution at the right,
you can convert it to the preferred form by replacing each term with its
opposite. That is:

$$T = \frac{-T_R}{E - 1} = \underline{\hspace{2cm}}$$

$$T = \frac{T_R}{1 - E}$$

116.   Do these. Write each solution in the preferred form.

a) Solve for "q".

$$\frac{1}{p} - \frac{1}{q} = \frac{1}{r}$$

b) Solve for C_2 .

$$\frac{1}{C_1} = \frac{1}{C_t} - \frac{1}{C_2}$$

a) $q = \dfrac{pr}{r - p}$

b) $C_2 = \dfrac{C_1 C_t}{C_1 - C_t}$

117. In the formula below, "b" appears on both sides. To solve for "b", we multiplied by the distributive principle and then isolated the "b" terms. <u>Notice</u> <u>how</u> <u>we</u> <u>wrote</u> <u>the</u> <u>solution</u> <u>in</u> <u>the</u> <u>preferred</u> <u>form</u>. Solve for R in the other formula.

$$ab = c(b - t) \qquad\qquad 2RF = w(R - r)$$

$$ab = bc - ct$$

$$ab - bc = -ct$$

$$b(a - c) = -ct$$

$$b = \frac{-ct}{a - c}$$

$$b = \frac{ct}{c - a}$$

---

118. Two fractions with binomial numerators are also equivalent <u>if</u> <u>their</u> <u>numerators</u> <u>are</u> <u>opposites</u> <u>and</u> <u>their</u> <u>denominators</u> <u>are</u> <u>opposites</u>. That is:

$$\frac{8 - 2}{3} \text{ and } \frac{2 - 8}{-3} \text{ are equivalent because both equal } +2 .$$

Write each of these in an equivalent form by replacing each term with its opposite.

a) $\dfrac{c - d}{-b}$ = _____   b) $\dfrac{R - V}{-ST}$ = _____

$$R = \frac{wr}{w - 2F}$$

---

119. Depending on the method used, you can get either of the following <u>equivalent</u> solutions when solving $Q = R - MP$ for P .

$$P = \frac{R - Q}{M} \qquad\qquad P = \frac{Q - R}{-M}$$

The solution on the left is preferred because it contains fewer negative signs. If you get the solution on the right, you can convert it to the preferred form <u>by</u> <u>replacing</u> <u>each</u> <u>term</u> <u>with</u> <u>its</u> <u>opposite</u>. That is:

$$P = \frac{Q - R}{-M} = \underline{\qquad\qquad}$$

a) $\dfrac{d - c}{b}$ b) $\dfrac{V - R}{ST}$

$$P = \frac{R - Q}{M}$$

120.   We solved for "t" below. Notice how we wrote the solution in the preferred form. Solve for B in the other formula.

$$a = \frac{d}{1 - dt}$$
$$a(1 - dt) = d$$
$$a - adt = d$$
$$-adt = d - a$$
$$t = \frac{d - a}{-ad}$$
$$t = \frac{a - d}{ad}$$

$$A_f = \frac{A}{1 - BA}$$

$$B = \frac{A_f - A}{A_f A}$$

8-16   OPPOSITES OF INSTANCES OF THE DISTRIBUTIVE PRINCIPLE

In this section, we will discuss the meaning of "opposites" of instances of the distributive principle. Then we will use that concept to rearrange formulas in the next section.

121.   Since (2)(3) = 6 , the opposite of (2)(3) must equal -6 . To get the opposite of (2)(3) , <u>we replace one (not both) of its factors by its opposite</u>. That is:

$$\left.\begin{array}{c} (-2)(3) \\ \text{and} \\ (2)(-3) \end{array}\right\}$$ are opposites of (2)(3) , since both equal -6 .

If we replace both factors by their opposites we get (-2)(-3) . Is (-2)(-3) the opposite of (2)(3) ? _____

122.   Which of the following are opposites of (6)(5) ? _____

     a) (-6)(5)   b) (-6)(-5)   c) (6)(-5)

No, since (-2)(-3) = 6 , and the opposite of (2)(3) must equal -6 .

123.   Since 2(3 + 4) = 14 , the opposite of 2(3 + 4) must equal -14 . To get the opposite of 2(3 + 4) , we also replace one factor by its opposite. Usually we replace the monomial factor "2" by its opposite. That is:

     The opposite of 2(3 + 4) is (-2)(3 + 4) .

(-2)(3 + 4) is the opposite of 2(3 + 4) because (-2)(3 + 4) = _____

(a) and (c) , <u>not</u> (b)

-14

124. Since $3(8 - 4) = 12$, the opposite of $3(8 - 4)$ must equal -12 . To get the opposite of $3(8 - 4)$, we also replace the monomial factor by its opposite. That is:

The opposite of $3(8 - 4)$ is $(-3)(8 - 4)$.

$(-3)(8 - 4)$ is the opposite of $3(8 - 4)$ because $(-3)(8 - 4) =$ _____

---

125. By replacing the monomial factor by its opposite, we can get the opposite of $4(y + 5)$. That is:

The opposite of $4(y + 5)$ is $(-4)(y + 5)$.

To show that $4(y + 5)$ and $(-4)(y + 5)$ are opposites, we added them below to show that their sum is "0".

$$4(y + 5) + (-4)(y + 5) = 4y + 20 + (-4y) + (-20)$$
$$= 4y + (-4y) + 20 + (-20)$$
$$= 0 \quad + \quad 0 \quad = 0$$

Write the opposite of each expression by replacing the monomial factor by its opposite.

a) $7(x + 3)$ _____   b) $R(S + T)$ _____

-12

---

126. By replacing the monomial factor by its opposite, we can get the opposite of $2(x - 5)$. That is:

The opposite of $2(x - 5)$ is $(-2)(x - 5)$.

To show that $2(x - 5)$ and $(-2)(x - 5)$ are opposites, we added them below to show that their sum is "0".

$$2(x - 5) + (-2)(x - 5) = 2x - 10 + (-2x) + 10$$
$$= 2x + (-2x) - 10 + 10$$
$$= 0 \quad + \quad 0 \quad = 0$$

Write the opposite of each expression by replacing the monomial factor by its opposite.

a) $8(m - 1)$ _____   b) $a(b - c)$ _____

a) $(-7)(x + 3)$

b) $(-R)(S + T)$

---

a) $(-8)(m - 1)$

b) $(-a)(b - c)$

127.   To convert each subtraction below to addition, we add the opposite of the instance of the distributive principle.

$$7 - 6(y + 3) = 7 + [\text{the opposite of } 6(y + 3)]$$
$$= 7 + (-6)(y + 3)$$

$$V - M(T - S) = V + [\text{the opposite of } M(T - S)]$$
$$= V + \underline{\hspace{2cm}}$$

---

128.   To convert each addition below to subtraction form, we subtract the opposite of the instance of the distributive principle.

$$4 + (-3)(t + 7) = 4 - [\text{the opposite of } (-3)(t + 7)]$$
$$= 4 - 3(t + 7)$$

$$H + (-d)(b - c) = H - [\text{the opposite of } (-d)(b - c)]$$
$$= H - \underline{\hspace{2cm}}$$

> $V + \underline{(-M)(T - S)}$

> $H - \underline{d(b - c)}$

8-17   THE ADDITION AXIOM AND INSTANCES OF THE DISTRIBUTIVE PRINCIPLE

In this section, we will show how the addition axiom is used in formula rearrangements when one of the terms is an instance of the distributive principle.

---

129.   To solve for "m" below, we add the opposite of $a(c - d)$ to both sides. We get:

$$V = m + a(c - d)$$
$$V + (-a)(c - d) = m + \underbrace{a(c - d) + (-a)(c - d)}$$
$$V + (-a)(c - d) = m \quad + \quad 0$$
$$V + (-a)(c - d) = m$$

To complete the solution, we convert the addition on the left side to subtraction form. Do so.

$$m = \underline{\hspace{4cm}}$$

> $m = V - a(c - d)$

130.   Using the same steps, complete these.

    a) Solve for "p_1".   b) Solve for T .

$$P = p_1 + w(h - h_1)$$ $$T + k(a + b) = S$$

---

131.   To solve for D below, we used the addition axiom to isolate CD . Solve for I_b in the other formula.

$$B = CD + (v + 1)F \qquad I_c = \beta I_b + (\beta + 1)I_{co}$$

$$B - (v + 1)F = CD$$

$$D = \frac{B - (v + 1)F}{C}$$

a) $p_1 = P - w(h - h_1)$

b) $T = S - k(a + b)$

---

132.   To solve for "a" in the formula below, we began by using the addition axiom to isolate the instance of the distributive principle. Solve for "w" in the other formula.

$$m = t + a(b + c) \qquad\qquad P = p_1 + w(h - h_1)$$

$$m - t = a(b + c)$$

$$a = \frac{m - t}{b + c}$$

$$I_b = \frac{I_c - (\beta + 1)I_{co}}{\beta}$$

$$w = \frac{P - p_1}{h - h_1}$$

133. To solve for R in the formula below, we isolated the instance of the distributive principle and then multiplied to get R out of the parentheses. Solve for "h" in the other formula.

$$b = v + m(R - 1) \qquad\qquad P = p_1 + w(h - h_1)$$

$$b - v = m(R - 1)$$

$$b - v = mR - m$$

$$b - v + m = mR$$

$$R = \frac{b - v + m}{m}$$

or

$$R = \frac{b - v}{m} + 1$$

---

134. a) Solve for I_{co}.

$$I_c = \beta I_b + (\beta + 1)I_{co}$$

b) Solve for "p".

$$R = ab + m(p + q)$$

$$h = \frac{P - p_1 + wh_1}{w}$$

or

$$h = \frac{P - p_1}{w} + h_1$$

---

135. To solve for "v" below, we began by multiplying by the distributive principle to get "v" out of the parentheses. Then we isolated the "v" terms. Solve for β in the other formula.

$$R = vt + (v + 1)d \qquad\qquad I_c = \beta I_b + (\beta + 1)I_{co}$$

$$R = vt + vd + d$$

$$R - d = vt + vd$$

$$R - d = v(t + d)$$

$$v = \frac{R - d}{t + d}$$

a) $I_{co} = \dfrac{I_c - \beta I_b}{\beta + 1}$

b) $p = \dfrac{R - ab - mq}{m}$

or

$p = \dfrac{R - ab}{m} - q$

$$\beta = \frac{I_c - I_{co}}{I_b + I_{co}}$$

136.   At the left below, we used the oppositing principle when solving for "h_1". At the right below, we did not use the oppositing principle.

$$P = p_1 + w(h - h_1)$$

$$P - p_1 = w(h - h_1)$$

$$P - p_1 = wh - wh_1$$

$$P - p_1 - wh = -wh_1$$

$$p_1 + wh - P = wh_1$$

$$h_1 = \frac{p_1 + wh - P}{w}$$

$$P = p_1 + w(h - h_1)$$

$$P - p_1 = w(h - h_1)$$

$$P - p_1 = wh - wh_1$$

$$P - p_1 - wh = -wh_1$$

$$h_1 = \frac{P - p_1 - wh}{-w}$$

The form of the solution at the left is preferred because it contains fewer negative signs. If you get the solution on the right, you can convert it to the preferred form by replacing each term of the fraction with its opposite. That is:

$$h_1 = \frac{P - p_1 - wh}{-w} = \underline{\hspace{3cm}}$$

$$h_1 = \frac{p_1 + wh - P}{w}$$

8-18   FORMULAS WITH DISTRIBUTIVE-PRINCIPLE DENOMINATORS

In this section, we will rearrange formulas containing one fraction whose denominator is an instance of the distributive principle.

137.   The formula below contains one fraction whose denominator is an instance of the distributive principle. To clear the fraction, we multiplied both sides by $p(q - r)$. Clear the fraction in the other formula.

$$m = \frac{a - t}{p(q - r)} \qquad T = \frac{V_1 - V_2}{S(R_1 - R_2)}$$

$$mp(q - r) = a - t$$

$$ST(R_1 - R_2) = V_1 - V_2$$

138.   To solve for "a" below, we cleared the fraction and then used the addition axiom to isolate "a". Notice how we were able to factor the solution. Solve for L_2 in the other formula.

$$m = \frac{a - t}{t(p - r)} \qquad \alpha = \frac{L_2 - L_1}{L_1(t_2 - t_1)}$$

$$mt(p - r) = a - t$$

$$a = mt(p - r) + t$$

or

$$a = t[m(p - r) + 1]$$

139.   We solved for "b" below. Solve for T in the other formula.

$$H = \frac{b(t_2 - t_1)}{c(d_2 - d_1)} \qquad\qquad B = \frac{T(V_1 - V_2)}{V_1(P_2 - P_1)}$$

$$cH(d_2 - d_1) = b(t_2 - t_1)$$

$$b = \frac{cH(d_2 - d_1)}{t_2 - t_1}$$

$$L_2 = \alpha L_1(t_2 - t_1) + L_1$$

or

$$L_2 = L_1[\alpha(t_2 - t_1) + 1]$$

140.   We solved for "k" below. Solve for V_1 in the other formula.

$$t = \frac{a(b - c)}{k(p - q)} \qquad\qquad B = \frac{T(R_1 - R_2)}{V_1(P_2 - P_1)}$$

$$kt(p - q) = a(b - c)$$

$$k = \frac{a(b - c)}{t(p - q)}$$

$$T = \frac{BV_1(P_2 - P_1)}{V_1 - V_2}$$

141.   To solve for "r" below, we had to multiply by the distributive principle on the left side to get "r" out of the parentheses. Notice how we were able to write the solution in an equivalent form. Solve for P_2 in the other formula.

$$V_1 = \frac{T(R_1 - R_2)}{B(P_2 - P_1)}$$

$$b = \frac{m(d_1 - d_2)}{d_1(r - s)} \qquad\qquad B = \frac{T(V_1 - V_2)}{V_1(P_2 - P_1)}$$

$$bd_1(r - s) = m(d_1 - d_2)$$

$$bd_1 r - bd_1 s = m(d_1 - d_2)$$

$$bd_1 r = m(d_1 - d_2) + bd_1 s$$

$$r = \frac{m(d_1 - d_2) + bd_1 s}{bd_1}$$

or

$$r = \frac{m(d_1 - d_2)}{bd_1} + s$$

$$P_2 = \frac{T(V_1 - V_2) + BV_1 P_1}{BV_1}$$

or

$$P_2 = \frac{T(V_1 - V_2)}{BV_1} + P_1$$

142. To solve for "b_2" below, we also had to multiply by the distributive principle on the left side to get "b_2" out of the parentheses. Notice how we were able to write the solution in an equivalent form. Solve for "t_2" in the other formula.

$$R = \frac{P - Q}{Q(b_2 - b_1)} \qquad\qquad \alpha = \frac{L_2 - L_1}{L_1(t_2 - t_1)}$$

$$QR(b_2 - b_1) = P - Q$$

$$QRb_2 - QRb_1 = P - Q$$

$$QRb_2 = P - Q + QRb_1$$

$$b_2 = \frac{P - Q + QRb_1}{QR}$$

or

$$b_2 = \frac{P - Q}{QR} + b_1$$

---

143. Notice how we used the oppositing principle in solving for "t_1" below. Complete the solution.

$$\alpha = \frac{L_2 - L_1}{L_1(t_2 - t_1)}$$

$$\alpha L_1(t_2 - t_1) = L_2 - L_1$$

$$\alpha L_1 t_2 - \alpha L_1 t_1 = L_2 - L_1$$

$$-\alpha L_1 t_1 = L_2 - L_1 - \alpha L_1 t_2$$

$$\alpha L_1 t_1 = L_1 - L_2 + \alpha L_1 t_2$$

$$t_1 = \rule{3cm}{0.4pt}$$

$$t_2 = \frac{L_2 - L_1 + \alpha L_1 t_1}{\alpha L_1}$$

or

$$t_2 = \frac{L_2 - L_1}{\alpha L_1} + t_1$$

---

144. Some of the steps needed to solve for L_1 below are shown.

$$\alpha = \frac{L_2 - L_1}{L_1(t_2 - t_1)}$$

$$\alpha L_1(t_2 - t_1) = L_2 - L_1$$

$$\alpha L_1 t_2 - \alpha L_1 t_1 = L_2 - L_1$$

$$t_1 = \frac{L_1 - L_2 + \alpha L_1 t_2}{\alpha L_1}$$

or

$$t_1 = \frac{L_1 - L_2}{\alpha L_1} + t_2$$

As you can see, L_1 appears in three terms. Isolate those three terms on the left side, factor by the distributive principle, and complete the solution.

$$\alpha L_1 t_2 - \alpha L_1 t_1 + L_1 = L_2$$

$$L_1(\alpha t_2 - \alpha t_1 + 1) = L_2$$

$$L_1 = \frac{L_2}{\alpha t_2 - \alpha t_1 + 1}$$

or $$L_1 = \frac{L_2}{\alpha(t_2 - t_1) + 1}$$

SELF-TEST 29 (pp. 363-374)

1.   Solve for H .

$$P = \frac{H - F}{2H}$$

2.   Solve for R_2 .

$$\frac{1}{R_1} = \frac{1}{R_t} - \frac{1}{R_2}$$

3.   Solve for P .

$$T = \frac{V}{1 - PV}$$

4.   Solve for "s".

$$w = s + k(d - b)$$

5.   Solve for "r".

$$p = rt + a(r - 1)$$

6.   Solve for V_2 .

$$F = \frac{P_1 + P_2}{C(V_1 + V_2)}$$

ANSWERS:

1. $H = \dfrac{F}{1 - 2P}$

3. $P = \dfrac{T - V}{TV}$

5. $r = \dfrac{a + p}{a + t}$

2. $R_2 = \dfrac{R_1 R_t}{R_1 - R_t}$

4. $s = w - k(d - b)$

6. $V_2 = \dfrac{P_1 + P_2 - CFV_1}{CF}$

or $V_2 = \dfrac{P_1 + P_2}{CF} - V_1$

SUPPLEMENTARY PROBLEMS - CHAPTER 8

Assignment 26

1. Solve for "s".

$$P = 4s$$

2. Solve for L .

$$A = LW$$

3. Solve for "r".

$$C = 2\pi r$$

4. Solve for R.

$$E^2 = PR$$

5. Solve for "h".

$$V = \pi r^2 h$$

6. Solve for "v".

$$d = v(t_2 - t_1)$$

7. Solve for M .

$$F(p - a) = 2Mm$$

8. Solve for V_2 .

$$T_1 V_2 = T_2 V_1$$

9. Solve for W .

$$F = \frac{W}{d}$$

10. Solve for "f".

$$t = \frac{1}{f}$$

11. Solve for "c".

$$T = \frac{Q}{cm}$$

12. Solve for G .

$$W = \frac{GKN}{R^2}$$

13. Solve for R .

$$I = \frac{E + e}{R}$$

14. Solve for C .

$$s = \frac{1}{2}Cgt^2$$

15. Solve for "h".

$$p = \frac{bh(1 - a)}{2d}$$

16. Solve for L .

$$\frac{AL}{H} = K(b_1 + b_2)$$

17. Solve for A_2 .

$$\frac{P_1}{P_2} = \frac{A_2}{A_1}$$

18. Solve for R_2 .

$$\frac{(E_1)^2}{(E_2)^2} = \frac{R_1}{R_2}$$

19. Solve for "d".

$$\frac{bd}{h} = \frac{pt}{w}$$

20. Solve for "a".

$$\frac{r}{a} = \frac{m}{r}$$

Assignment 27

1. Solve for H .

$$H^2 = D + T$$

2. Solve for "c".

$$E = mc^2$$

3. Solve for "v".

$$a = \frac{v^2}{2s}$$

4. Solve for "t".

$$2k = \frac{d}{t^2}$$

5. Solve for "h".

$$M = \frac{1}{2}bh^2$$

6. Solve for P .

$$\frac{P}{N} = \frac{G}{P}$$

7. Solve for "r".

$$\frac{2c}{r} = \frac{ar}{w}$$

8. Solve for V .

$$\frac{F}{W} = \frac{V^2}{gR}$$

9. Solve for F_t .

$$F_1 = F_t - F_2$$

10. Solve for W .

$$P = 2L + 2W$$

11. Solve for Z .

$$X^2 = Z^2 - R^2$$

12. Solve for E .

$$i = \frac{E - e}{r}$$

13. Solve for "s_2".

$$s_1 f_1 + s_2 f_2 = 0$$

14. Solve for "h".

$$r^2 - h^2 = v^2$$

15. Solve for P_1 .

$$P_2 - P_1 = EI$$

16. Solve for "k".

$$w = ah - ks$$

17. Solve for "d_2".

$$v = \frac{d_1 - d_2}{t}$$

18. Solve for "r".

$$hN = R^2 - r^2$$

19. Solve for G_1 .

$$W = \frac{G_1 + G_2}{P}$$

20. Solve for "c".

$$d = \frac{1 - cm}{2a}$$

Assignment 28

1. Solve for "w".

$$b = h(d + w)$$

2. Solve for "t_1".

$$v = a(t_2 - t_1)$$

3. Solve for "d".

$$A = \frac{\pi(D^2 - d^2)}{4}$$

4. Solve for "p".

$$w = k(p^2 - 1)$$

5. Solve for "d_2".

$$V = \frac{1}{2}bh(d_1 + d_2)$$

6. Solve for P_1.

$$W = \frac{FT(P_1 - P_2)}{S}$$

7. Solve for "v".

$$t = \frac{a(v^2 + h^2)}{cr}$$

8. Solve for P_2.

$$F = \frac{dH(P_1 - P_2)}{3ms}$$

9. Solve for P.

$$F = A_1P + A_2P + A_3P$$

10. Solve for "x".

$$mx = b + x$$

11. Solve for "f_1".

$$w = df_1 - df_2$$

12. Solve for T.

$$AK + 2T = PT$$

13. Solve for "r".

$$hr = a - rv$$

14. Solve for A.

$$A + G = AD$$

15. Solve for "t".

$$\frac{1}{t} = \frac{1}{s} + \frac{1}{v}$$

16. Solve for G_2.

$$\frac{1}{G_1} + \frac{1}{G_2} = \frac{1}{G_t}$$

17. Solve for "d".

$$s = \frac{b - d}{d}$$

18. Solve for H.

$$V = \frac{FH}{F + H}$$

19. Solve for N.

$$L = \frac{N}{1 - N}$$

20. Solve for "v".

$$e = \frac{mv^2}{p} + k$$

Assignment 29

1. Solve for R.

$$H = \frac{R - A}{R}$$

2. Solve for "a".

$$\frac{1}{c} - \frac{1}{a} = \frac{1}{b}$$

3. Solve for F_1.

$$\frac{1}{F_2} = \frac{1}{F_t} - \frac{1}{F_1}$$

4. Solve for "k".

$$hk = m(k - t)$$

5. Solve for P.

$$BP + T = M(P - 1)$$

6. Solve for "r".

$$N = \frac{r}{1 - hr}$$

7. Solve for M.

$$s = \frac{1 + M}{1 - M}$$

8. Solve for "t_1".

$$T = t_1 + b(p - p_1)$$

9. Solve for "a".

$$a + m(d_1 + d_2) = t$$

10. Solve for "k".

$$v = h + k(w - 1)$$

11. Solve for T.

$$p_1 = p_2 + b(T - t)$$

12. Solve for R_2.

$$R_t = dR_1 + (d + 1)R_2$$

13. Solve for F.

$$h = Ft + (F - 1)w$$

14. Solve for "r_1".

$$G = \frac{r_1 + r_2}{d(p_1 - p_2)}$$

15. Solve for H.

$$S = \frac{H(V_2 - V_1)}{A(B_1 + B_2)}$$

16. Solve for "m".

$$p = \frac{w(h - t)}{m(a - k)}$$

17. Solve for "e_2".

$$Q = \frac{C_1 + C_2}{C_1(e_1 + e_2)}$$

18. Solve for "d_1".

$$A = \frac{W - P}{P(d_2 - d_1)}$$

19. Solve for "r_1".

$$f = \frac{r_2 - r_1}{r_1(w_1 + w_2)}$$

20. Solve for "d".

$$h = \frac{r - d}{a(r + d)}$$

Chapter 9 AN INTRODUCTION TO SYSTEMS OF EQUATIONS

In this chapter, we will define what is meant by a "system" of two equations and then solve systems of that type by both the graphing method and the addition method. Some applied problems are included.

9-1 THE MEANING OF A SYSTEM OF EQUATIONS

In this section, we will define what is meant by a system of equations and the solution of a system of equations.

---

1. It is sometimes easier to solve a problem if we use two variables (or letters) and two equations. An example is shown.

    The sum of two numbers is 21. The difference between the larger number and the smaller number is 5. Find the two numbers.

    Using "x" for the larger number and "y" for the smaller number, we can set up two different equations. That is:

    The sum of two numbers is 21.

    $$x + y = 21$$

    The difference between the larger and smaller is 5.

    $$x - y = 5$$

    Therefore, the problem has been translated to the following pair of equations which is called a "system" of equations.

    $$\boxed{\begin{array}{l} x + y = 21 \\ x - y = 5 \end{array}}$$

    The solution of a system of equations is a pair of values that satisfies both equations. For example, for the system above:

    (x = 15, y = 6) is not a solution because it only satisfies the top equation.

    (x = 10, y = 5) is not a solution because it only satisfies the bottom equation.

    (x = 13, y = 8) is the solution because it satisfies both equations.

    Show that (x = 13, y = 8) satisfies both equations in the system.

    $$x + y = 21 \qquad\qquad x - y = 5$$

2.   Remember that a pair of values is a solution of a system of equations
     <u>only</u> <u>if</u> <u>it</u> <u>satisfies</u> <u>both</u> <u>equations</u>.

     Which pair of values below is the solution
     of the system at the right? _____

     $$p = q + 1$$
     $$p + q = 5$$

     a) $p = 1$, $q = 0$ b) $p = 3$, $q = 2$ c) $p = 4$, $q = 1$

     $$x + y = 21$$
     $$13 + 8 = 21$$
     $$21 = 21$$

     $$x - y = 5$$
     $$13 - 8 = 5$$
     $$5 = 5$$

---

3.   We made up a solution-table at
     the right for each equation in
     the system below.

     $$x + y = 5$$
     $$y = 4x$$

     By examining the tables, you can
     see that the equations have one
     common solution. The common
     solution is the solution of the system.

     It is: $x =$ _____ , $y =$ _____

     $x + y = 5$

x	y
0	5
1	4
2	3
3	2
4	1

     $y = 4x$

x	y
0	0
1	4
2	8
3	12
4	16

     b) $p = 3$, $q = 2$

     $x = 1$, $y = 4$

9-2   THE GRAPHING METHOD FOR SOLVING SYSTEMS

Various methods can be used to solve systems of equations. In this section, we will discuss the graphing
method.

---

4.   We have graphed both equations
     in the system below at the right.

     $$x - y = 1$$
     $$x + y = 5$$

     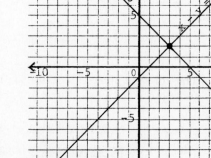

     The point where the two graphed lines cross is called
     the "<u>point of intersection</u>". Since that point lies on
     both lines, its coordinates satisfy both equations.
     Therefore, <u>its coordinates are the solution of the
     system</u>. Check that fact by answering the questions
     below.

     a) The coordinates of the point of intersection are:

         $x =$ _____ , $y =$ _____

     b) Do those coordinates satisfy both equations? _____

     c) Therefore, the solution of the system is:

         $x =$ _____ , $y =$ _____

5.   When the graphs of both equations are straight lines and the lines intersect, there is only one point of intersection. Therefore, there is only one solution of the system of equations.

Two systems of equations have been graphed below. Using the coordinates of the point of intersection, write the one solution of each system.

a) x = 3, y = 2

b) Yes

c) x = 3, y = 2

a) $\boxed{\begin{array}{l} x + y = 3 \\ \quad\quad y = 2x \end{array}}$

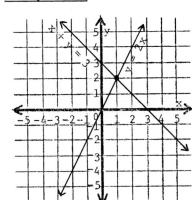

b) $\boxed{\begin{array}{l} 2x - y = 3 \\ x + y = 0 \end{array}}$

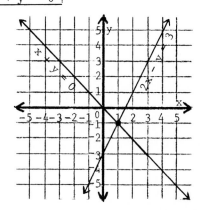

The solution is: x = _____ , y = _____   The solution is: x = _____ , y = _____

---

6.   To solve a system of equations by the graphing method, we use these steps:

a) x = 1, y = 2

b) x = 1, y = -1

    1. Graph both equations on the same set of axes.

    2. Use the coordinates of the point of intersection as the solution of the system.

    3. Check the solution in both equations of the system.

Let's use the graphing method to solve the system at the right.

$\boxed{\begin{array}{l} \quad\quad y = 3x \\ x + y = 4 \end{array}}$

a) Using the tables provided, graph each equation on the same set of axes.

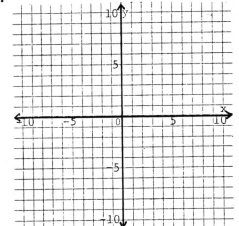

$y = 3x$			$x + y = 4$	
x	y		x	y
2	6		5	-1
1	3		3	1
0	0		0	4
-1	-3		-3	7
-2	-6		-5	9

b) The solution of the system is:

         x = _____ , y = _____

c) Check the solution in both original equations.

         y = 3x   x + y = 4

a)

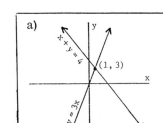

b) x = 1, y = 3

c) 3 = 3(1)

     3 = 3

     1 + 3 = 4

         4 = 4

7.   Let's use the graphing method to solve the system below.

$$y = 2x$$
$$y - x = 3$$

a) Complete each table and graph each equation.

y = 2x	
x	y
5	
3	
0	
-3	
-5	

y - x = 3	
x	y
4	
2	
0	
-2	
-4	

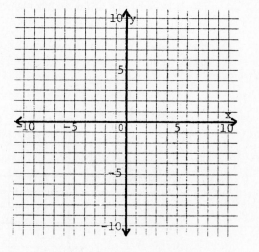

b) The solution of the system is: x = _____ , y = _____

c) Check the solution in both original equations.

$$y = 2x \qquad\qquad y - x = 3$$

---

8.   Using the tables provided, graph the following system on the axes below.

$$2x - y = 6$$
$$x + 2y = 13$$

2x - y = 6	
x	y

x + 2y = 13	
x	y

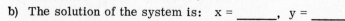

The solution of the system is: x = _____ , y = _____

a)
x	y		x	y
5	10		4	7
3	6		2	5
0	0		0	3
-3	-6		-2	1
-5	-10		-4	-1

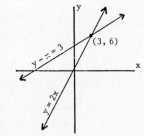

b) x = 3, y = 6

c) 6 = 2(3)   6 - 3 = 3
    6 = 6   3 = 3

The solution is:

x = 5, y = 4

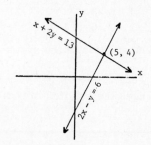

9.   A system has one solution if its graph is two straight lines that intersect at one point. A system can have no solution or an infinite number of solutions. An example of each type is discussed below.

SYSTEM WITH NO SOLUTION

The system below is graphed at the right. Notice that the straight lines are parallel.

$$y = 3x + 2$$
$$y = 3x - 1$$

Since the parallel lines do not intersect, the equations have no common solution. Therefore, the system has no solution.

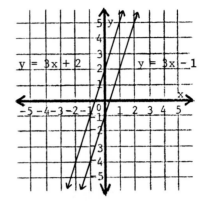

SYSTEM WITH AN INFINITE NUMBER OF SOLUTIONS

The system below is graphed at the right. Notice that each equation has the same graph.

$$x + y = 2$$
$$2x + 2y = 4$$

Since the lines are identical, the equations have an infinite number of common solutions. Therefore, the system has an infinite number of solutions.

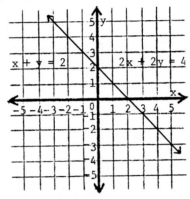

Note: Systems with no solution or an infinite number of solutions are extremely rare in applied problems. Therefore, we will not include any more systems of those types in this book.

9-3   THE ADDITION METHOD FOR SOLVING SYSTEMS

Because the graphing method for solving systems of equations is time-consuming and only approximate when the solutions are not whole numbers, an algebraic method is ordinarily used. In this section, we will discuss an algebraic method called the "addition" method for solving systems.

10.   In the addition method for solving systems, the key step is adding the two equations. To do so, we add the like terms in columns as we have done below.

$$
\begin{aligned}
2x + 5y &= 7 \\
3x - 2y &= 4 \\
\hline
5x + 3y &= 11
\end{aligned}
$$

Note: To get 5x, we added 2x and 3x.

To get 3y, we added 5y and -2y.

To get 11, we added 7 and 4.

Continued on following page.

10.   Continued

Use the same method to add each pair of equations below.

a)   4x + 7y = 5
      3x - 4y = 10

b)   9s + 4t = 3
      5s - 2t = 7

---

11.   When adding a pair of equations, remember that the coefficient of a letter is "1" or "-1" if the coefficient is not explicitly shown.   That is:

x + 3y = 5   Note: To get 3x, we added 1x and 2x.
2x - y = 7   To get 2y, we added 3y and -1y.
3x + 2y = 12

Use the same method to add each pair of equations below.

a)   3x + 4y = 3
       x - y = 1

b)   m + 7f = 9
      m - f = 0

a) 7x + 3y = 15

b) 14s + 2t = 10

---

12.   When the "y" terms were added below, we got "0y".   Since "0y" can be dropped, we eliminated "y" by performing the addition.   Complete the other addition.

x + 2y = 3
x - 2y = 5
2x + 0y = 8

    or

2x = 8

x + 4y = 0
2x - 4y = 7

a) 4x + 3y = 4

b) 2m + 6f = 9

---

13.   In the addition method for solving a system, we add the equations <u>in order to eliminate a letter</u>.   An example is shown.   Complete the other addition.

2x + 3y = 0
4x - 3y = 6
6x   = 6

p - q = 4
p + q = 6

3x + 0y = 7

  or 3x = 7

---

14.   We used the addition method below to solve the system at the right.   The two steps are described.

x + y = 5
x - y = 1

1. <u>Finding the value of "x"</u>. By adding the two equations, we can eliminate "y" and solve for "x".

x + y = 5
x - y = 1
2x   = 6
      x = 3

2p = 10

Continued on following page.

14.   Continued

2. Finding the value of "y". We can now find the corresponding value of "y" by substituting "3" for "x" in either of the original equations.

$$x + y = 5$$
$$3 + y = 5$$
$$y = 2$$

$$x - y = 1$$
$$3 - y = 1$$
$$y = 2$$

The obtained solution is (x = 3, y = 2). Check that solution in each original equation below.

a) $x + y = 5$

b) $x - y = 1$

---

15.   We used the addition method below to solve the system at the right.

$$s + 2t = 5$$
$$3s - 2t = 7$$

1. Finding the value of "s".

By adding the equations, we can eliminate "t" and solve for "s".

$$s + 2t = 5$$
$$3s - 2t = 7$$
$$4s = 12$$
$$s = 3$$

2. Finding the value of "t".

To find the corresponding value of "t", we substituted "3" for "s" in the top equation.

$$s + 2t = 5$$
$$3 + 2t = 5$$
$$2t = 2$$
$$t = 1$$

Note: We could also have substituted "3" for "s" in the bottom equation.

The obtained solution is (s = 3, t = 1). Show that the solution satisfies each of the original equations.

a) $s + 2t = 5$

b) $3s - 2t = 7$

a) 3 + 2 = 5
    5 = 5

b) 3 - 2 = 1
    1 = 1

---

16.   Use the addition method to solve each system below.

a) $2x + y = 10$
      $2x - y = 6$

b) $p - 4q = 16$
      $p + 4q = 0$

a) 3 + 2(1) = 5
    3 + 2 = 5
        5 = 5

b) 3(3) - 2(1) = 7
    9 - 2 = 7
        7 = 7

17.   The solution of a system can contain one or two decimal numbers. An example is shown. Solve the other system.

$$2x + y = 5.54$$
$$x - y = 1.18$$

$$p + 2q = 4.8$$
$$3p - 2q = 3.2$$

Finding the value of "x"

$$2x + y = 5.54$$
$$\underline{x - y = 1.18}$$
$$3x \qquad = 6.72$$
$$x = 2.24$$

Finding the value of "y"

$$2x + y = 5.54$$
$$2(2.24) + y = 5.54$$
$$4.48 + y = 5.54$$
$$y = 1.06$$

a) x = 4, y = 2

b) p = 8, q = -2

p = 2, q = 1.4

SELF-TEST 30 (pages 377-384)

1. Solve this system of equations by the graphing method.

$$x - 2y = 8$$
$$2x + y = 6$$

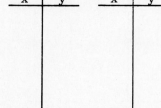

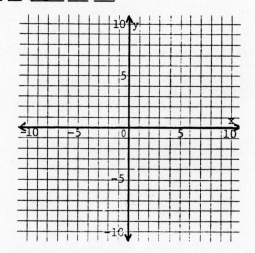

Solve each system of equations by the addition method.

2.
$$4x - 3y = 15$$
$$5x + 3y = 12$$

3.
$$2t + v = 6.5$$
$$3t - v = 3.5$$

ANSWERS: 1.   x = 4, y = -2   2.   x = 3, y = -1   3.   t = 2, v = 2.5

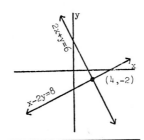

9-4   MULTIPLYING AN EQUATION BY A NUMBER

When using the addition method to solve a system, we sometimes have to multiply one or both equations by a number. We will discuss that multiplication process in this section.

---

18.   We multiplied $x + 4y = 5$ by 2 below. To do so, we used the distributive principle on the left side.

$$\begin{aligned}
\underline{\text{Original:}} \quad x \;+\; 4y &= 5 \\
2(x \;+\; 4y) &= 2(5) \\
2(x) \;+\; 2(4y) &= 2(5) \\
\underline{\text{New:}} \quad 2x \;+\; 8y &= 10
\end{aligned}$$

Notice that each term in the new equation is 2 times the corresponding term in the original equation.

$$\begin{aligned}
2x &= 2(x) \\
8y &= 2(4y) \\
10 &= 2(5)
\end{aligned}$$

Complete these by simply multiplying each term by the number.

a) Multiply this equation by 3.

$$3x + y = 6$$

b) Multiply this equation by 4.

$$p + 2q = 0$$

---

19.   We multiplied $2p - q = 4$ by 3 below. To do so, we again used the distributive principle on the left side.

$$\begin{aligned}
\underline{\text{Original:}} \quad 2p \;-\; q &= 4 \\
3(2p \;-\; q) &= 3(4) \\
3(2p) \;-\; 3(q) &= 3(4) \\
\underline{\text{New:}} \quad 6p \;-\; 3q &= 12
\end{aligned}$$

Notice again that each term in the new equation is 3 times the corresponding term in the original equation.

a) $9x + 3y = 18$

b) $4p + 8q = 0$

Continued on following page.

19.   Continued

Complete these by simply multiplying each term by the number.

a) Multiply this equation by 2.   b) Multiply this equation by 5.

    x − 3y = 7   a − b = 3

---

20.   We multiplied 2x + 4y = 1 by −3 below. Notice that we converted the addition to subtraction in the last step.

$$
\begin{aligned}
\underline{\text{Original:}} \qquad 2x + 4y &= 1 \\
-3(2x + 4y) &= -3(1) \\
-3(2x) + (-3)(4y) &= -3(1) \\
-6x + (-12y) &= -3 \\
\underline{\text{New:}} \qquad -6x - 12y &= -3
\end{aligned}
$$

Notice that each term in the new equation is −3 times the corresponding term in the original equation. That is:

$$
\begin{aligned}
-6x &= -3(2x) \\
-12y &= -3(4y) \\
-3 &= -3(1)
\end{aligned}
$$

Complete these by simply multiplying each term by the number.

a) Multiply this equation by −2.   b) Multiply this equation by −4.

    3x + y = 0   2s + 5t = 3

a) 2x − 6y = 14

b) 5a − 5b = 15

---

21.   We multiplied 3a − b = 5 by −2 below. Notice that we converted the subtraction to addition in the last step.

$$
\begin{aligned}
\underline{\text{Original:}} \qquad 3a - b &= 5 \\
-2(3a - b) &= -2(5) \\
-2(3a) - (-2)(b) &= -2(5) \\
-6a - (-2b) &= -10 \\
\underline{\text{New:}} \qquad -6a + 2b &= -10
\end{aligned}
$$

Notice again that each term in the new equation is −2 times the corresponding term in the original equation.

a) −6x − 2y = 0

b) −8s −20t = −12

Continued on following page.

21.   Continued

Complete these by simply multiplying each term by the number.

a) Multiply this equation by -5.   b) Multiply this equation by -3.

$m - 4t = 2$ $2x - 3y = 7$

---

22.   Multiply each equation below by the circled number at its left.

a)   ⑤ $x + y = 3$ c)   ④̶ $a + b = 1$

b)   ② $4p - 3q = 7$ d)   ③̶ $2m - 5f = 4$

a) $-5m + 20t = -10$

b) $-6x + 9y = -21$

---

23.   We multiplied both $3x + y = 2$ and $p - 2q = 3$ by -1 below.

Original: $3x + y = 2$

$-1(3x + y) = -1(2)$

$-1(3x) + (-1)(y) = -1(2)$

$-3x + (-y) = -2$

New: $-3x - y = -2$

Original: $p - 2q = 3$

$-1(p - 2q) = -1(3)$

$-1(p) - (-1)(2q) = -1(3)$

$-p - (-2q) = -3$

New: $-p + 2q = -3$

Notice that each new equation can be obtained by changing the sign of each
term in the original equation. Use that method to multiply each equation
below by -1.

a) $5x + 4y = 9$ c) $p + q = 1$

b) $2x - 3y = 7$ d) $a - b = 0$

a) $5x + 5y = 15$

b) $8p - 6q = 14$

c) $-4a - 4b = -4$

d) $-6m + 15f = -12$

24. Multiply each equation below by the circled number at its left.

a) (-1) $4t - 3s = 5$

b) (-2) $3a - 2b = 1$

c) (-1) $p - q = 5$

d) (-3) $x - y = 0$

a) $-5x - 4y = -9$

b) $-2x + 3y = -7$

c) $-p - q = -1$

d) $-a + b = 0$

25. When an equation is multiplied by a number, the new equation is <u>equivalent</u> to the original equation. That is, <u>both equations have the same solutions</u>. An example is discussed below.

We multiplied the equation below by 3.

<u>Original</u>: $x + y = 2$

<u>New</u>: $3x + 3y = 6$

A table of solutions for each equation is given below.

$x + y = 2$			$3x + 3y = 6$	
x	y		x	y
2	0		2	0
1	1		1	1
0	2		0	2
-1	3		-1	3
-2	4		-2	4

Since the solution-tables are identical, both equations have the same solutions. Therefore, we say that they are _____ equations.

a) $-4t + 3s = -5$

b) $-6a + 4b = -2$

c) $-p + q = -5$

d) $-3x + 3y = 0$

equivalent

9-5   ONE MULTIPLICATION IN THE ADDITION METHOD

When using the addition method, we sometimes have to multiply one equation by a number before adding the equations. We will discuss systems of that type in this section.

26. If we added the equations in the system at the right, neither letter would be eliminated.

However, if the "-y" in the bottom equation were a "-2y", we could eliminate "y". We can get a "-2y" in that equation by multiplying it by 2. We did so at the right.

$x + 2y = 7$
$2x - y = 4$

$2(2x - y) = 2(4)$
$4x - 2y = 8$

Continued on following page.

26.   Continued

Substituting the new equation for
$2x - y = 4$, we get the new system at
the right. We added the equations and
solved for "x".

$$x + 2y = 7$$
$$\underline{4x - 2y = 8}$$
$$5x \qquad = 15$$
$$x = 3$$

Substituting 3 for "x" in one of the
original equations, we can find the
corresponding value for "y".

$$x + 2y = 7$$
$$3 + 2y = 7$$
$$2y = 4$$
$$y = 2$$

Check $(x = 3, y = 2)$ in each original equation below.

   a) $x + 2y = 7$ b) $2x - y = 4$

---

27.   We could eliminate "b" in the system
at the right if the "2b" in the top equation were a "–6b".

$$\boxed{\begin{array}{l} a + 2b = 0 \\ 5a + 6b = 4 \end{array}}$$

To get a "–6b" in the top equation, we
multiplied it by –3 at the right and then
solved for "a".

$$-3a - 6b = 0$$
$$\underline{5a + 6b = 4}$$
$$2a \qquad = 4$$
$$a = 2$$

Substituting "2" for "a" in the bottom
equation, we found the corresponding
value of "b" at the right.

$$5a + 6b = 4$$
$$5(2) + 6b = 4$$
$$10 + 6b = 4$$
$$6b = -6$$
$$b = -1$$

Check $(a = 2, b = -1)$ in each original equation below.

   a) $a + 2b = 0$ b) $5a + 6b = 4$

a) $3 + 2(2) = 7$
    $3 + 4 = 7$
    $7 = 7$

b) $2(3) - 2 = 4$
    $6 - 2 = 4$
    $4 = 4$

---

28.   We could eliminate "y" in the system
at the right if the "3y" were "–3y"
in either equation.

$$\boxed{\begin{array}{l} 2x + 3y = 8 \\ x + 3y = 7 \end{array}}$$

To get a "–3y" in the bottom equation,
we multiplied it by "–1" at the right
and then solved for "x".

$$2x + 3y = 8$$
$$\underline{-x - 3y = -7}$$
$$x \qquad = 1$$

Substituting "1" for "x" in the original
bottom equation, we found the corresponding value of "y" at the right.

$$x + 3y = 7$$
$$1 + 3y = 7$$
$$3y = 6$$
$$y = 2$$

a) $2 + 2(-1) = 0$
    $2 + (-2) = 0$
    $0 = 0$

b) $5(2) + 6(-1) = 4$
    $10 + (-6) = 4$
    $4 = 4$

Continued on following page.

28.   Continued

Check (x = 1, y = 2) in each original equation below.

a) 2x + 3y = 8   b) x + 3y = 7

---

29.   Let's solve the system at the right by eliminating "t". To do so, we must multiply the top equation by 4.

2s − t = 7
s + 4t = 8

a) Write the new system obtained if the top equation is multiplied by 4.

b) Solve the system.

a) 2(1) + 3(2) = 8
     2 + 6 = 8
         8 = 8

b) 1 + 3(2) = 7
     1 + 6 = 7
         7 = 7

---

30.   Let's solve the system at the right by eliminating "x". To do so, we must multiply the bottom equation by −2.

4x + y = 1
2x + 3y = 13

a) Write the new system obtained if the bottom equation is multiplied by −2.

b) Solve the system.

a) 8s − 4t = 28
     s + 4t = 8

b) s = 4, t = 1

---

a) 4x + y = 1
    −4x − 6y = −26

b) x = −1, y = 5

31. Let's solve the system at the right by eliminating "p". To do so, we can multiply either equation by -1. Let's multiply the bottom equation by -1.

$$\boxed{\begin{array}{rcr} p + q &=& 7 \\ p + 3q &=& 15 \end{array}}$$

  a) Write the new system obtained if the bottom equation is multiplied by -1.

  b) Solve the system.

---

32. We multiply one equation by a number to make the coefficients of one of the letters a pair of opposites. When doing so, it is important that you know what number to use. For example:

  a) To eliminate "t" in the system at the right, we multiply the bottom equation by _____.

$$\begin{array}{rcr} s + 5t &=& 9 \\ 2s - t &=& 0 \end{array}$$

  b) To eliminate "y" in the system at the right, we multiply the top equation by _____.

$$\begin{array}{rcr} x + 2y &=& 0 \\ 3x - 4y &=& 5 \end{array}$$

  c) To eliminate "a" in the system at the right, we multiply the bottom equation by _____.

$$\begin{array}{rcr} 3a + 2b &=& 5 \\ a + 3b &=& 9 \end{array}$$

  d) To eliminate "p" in the system at the right, we can multiple either the top equation or the bottom equation by _____.

$$\begin{array}{rcr} 4p + 3q &=& 7 \\ 4p + q &=& 1 \end{array}$$

a) $p + q = 7$
    $-p - 3q = -15$

b) $p = 3, q = 4$

a) 5   c) -3

b) 2   d) -1

---

33. When solving a system, we can eliminate either letter and still get the same solution. As an example, we solved the system below in two different ways.

$$\boxed{\begin{array}{rcr} 2x + y &=& 2 \\ x + 4y &=& 1 \end{array}}$$

To eliminate "x", we multiplied the bottom equation by -2 and then completed the solution.

$$\begin{array}{rcr} 2x + y &=& 2 \\ -2x - 8y &=& -2 \\ \hline -7y &=& 0 \\ y &=& 0 \end{array}$$

$$\begin{array}{rcl} x + 4y &=& 1 \\ x + 4(0) &=& 1 \\ x + 0 &=& 1 \\ x &=& 1 \end{array}$$

To eliminate "y", we multiplied the top equation by -4 and then completed the solution.

$$\begin{array}{rcr} -8x - 4y &=& -8 \\ x + 4y &=& 1 \\ \hline -7x &=& -7 \\ x &=& 1 \end{array}$$

$$\begin{array}{rcl} x + 4y &=& 1 \\ 1 + 4y &=& 1 \\ 4y &=& 0 \\ y &=& 0 \end{array}$$

Continued on following page.

33.   Continued

We obtained the same solution $(x = 1, y = 0)$ both ways. Check that solution in each original equation below.

a) $2x + y = 2$ b) $x + 4y = 1$

---

34.   To solve the system at the right:

$$x + 4y = 0$$
$$3x + 2y = 5$$

a) we can eliminate "x" by multiplying the top equation by _____.

b) we can eliminate "y" by multiplying the bottom equation by _____.

To solve the system at the right:

$$2a - b = 1$$
$$a + 5b = 9$$

c) we can eliminate "a" by multiplying the bottom equation by _____.

d) we can eliminate "b" by multiplying the top equation by _____.

To solve the system at the right:

$$2s + 3t = 7$$
$$2s - t = 3$$

e) we can eliminate "t" by multiplying the bottom equation by _____.

f) we can eliminate "s" by multiplying either equation by _____.

a) $2(1) + 0 = 2$
    $2 + 0 = 2$
    $2 = 2$

b) $1 + 4(0) = 1$
    $1 + 0 = 1$
    $1 = 1$

---

35.   By eliminating either letter, solve each system below.

a) $$\boxed{\begin{array}{l} x + 4y = 10 \\ 3x - 2y = 16 \end{array}}$$

b) $$\boxed{\begin{array}{l} 2k + m = 0 \\ 2k + 3m = 8 \end{array}}$$

a) −3
b) −2

c) −2
d) 5

e) 3
f) −1

a) $x = 6, y = 1$

b) $k = -2, m = 4$

9-6   TWO MULTIPLICATIONS IN THE ADDITION METHOD

When using the addition method, we sometimes have to multiply both equations by a number before adding the equations. We will discuss systems of that type in this section.

36. We could eliminate "y" in the system at the right if the "-3y" were a "-6y" and the "2y" were a "6y".

$$4x - 3y = 5$$
$$3x + 2y = 8$$

To get a "-6y" in the top equation, we multiplied it by 2 at the right. To get a "6y" in the bottom equation, we multiplied it by 3 at the right. Then we eliminated "y" and solved for "x".

$$8x - 6y = 10$$
$$9x + 6y = 24$$
$$17x = 34$$
$$x = 2$$

Substituting 2 for "x" in the original bottom equation, we found the corresponding value of "y" at the right.

$$3x + 2y = 8$$
$$3(2) + 2y = 8$$
$$6 + 2y = 8$$
$$2y = 2$$
$$y = 1$$

Check (x = 2, y = 1) in each original equation below.

a) 4x - 3y = 5   b) 3x + 2y = 8

37. We could eliminate "q" in the system at the right if the "3q" were a "15q" and the "5q" were a "-15q".

$$5p + 3q = 2$$
$$3p + 5q = -2$$

To get a "15q" in the top equation, we multiplied it by 5 at the right. To get a "-15q" in the bottom equation, we multiplied it by -3 at the right. Then we eliminated "q" and solved for "p".

$$25p + 15q = 10$$
$$-9p - 15q = 6$$
$$16p = 16$$
$$p = 1$$

Substituting "1" for "p" in the original top equation, we found the corresponding value of "q" at the right.

$$5p + 3q = 2$$
$$5(1) + 3q = 2$$
$$5 + 3q = 2$$
$$3q = -3$$
$$q = -1$$

Check (p = 1, q = -1) in each original equation.

a) 5p + 3q = 2   b) 3p + 5q = -2

a) 4(2) - 3(1) = 5
8 - 3 = 5
5 = 5

b) 3(2) + 2(1) = 8
6 + 2 = 8
8 = 8

38. We could eliminate "x" in the system at the right if the "6x" were a "30x" and the "10x" were a "-30x".

$$6x + 2y = 4$$
$$10x + 7y = -8$$

a) $5(1) + 3(-1) = 2$
    $5 + (-3) = 2$
    $2 = 2$

To get a "30x" in the top equation, we multiplied it by 5 at the right. To get a "-30x" in the bottom equation, we multiplied it by -3 at the right. Then we eliminated "x" and solved for "y".

$$30x + 10y = 20$$
$$-30x - 21y = 24$$
$$-11y = 44$$
$$y = -4$$

b) $3(1) + 5(-1) = -2$
    $3 + (-5) = -2$
    $-2 = -2$

Substituting "-4" for "y" in the original top equation, we found the corresponding value of "x" at the right.

$$6x + 2y = 4$$
$$6x + 2(-4) = 4$$
$$6x + (-8) = 4$$
$$6x = 12$$
$$x = 2$$

Check (x = 2, y = -4) in each original equation.

a) $6x + 2y = 4$ b) $10x + 7y = -8$

---

39. Let's solve the system at the right by eliminating "t". We can do so if we get a "-12t" in the top equation and a "12t" in the bottom equation.

$$2v - 3t = -1$$
$$3v + 4t = 24$$

a) $6(2) + 2(-4) = 4$
    $12 + (-8) = 4$
    $4 = 4$

a) Write the new system obtained if we multiply the top equation by 4 and the bottom equation by 3.

b) $10(2) + 7(-4) = -8$
    $20 + (-28) = -8$
    $-8 = -8$

b) Solve the system.

---

40. Let's solve the system at the right by eliminating "a". We can do so if we get a "10a" in the top equation and a "-10a" in the bottom equation.

$$5a - 2b = 0$$
$$2a - 3b = -11$$

a) $8v - 12t = -4$
    $9v + 12t = 72$

a) Write the new system obtained if the top equation is multiplied by 2 and the bottom equation is multiplied by -5.

b) v = 4, t = 3

Continued on following page.

40.   Continued

　　b) Solve the system.

---

41.   Sometimes we have to multiply both equations by a number to make the coefficients of one of the letters a pair of opposites. When doing so, it is important that you know what numbers to use.   For example:

　　　To eliminate "y" at the right, we
　　　can multiply the top equation by 7
　　　and the bottom equation by 2.

$$3x + 2y = 7$$
$$2x - 7y = 0$$

　　　To eliminate "q" at the right, we
　　　can multiply the top equation by 3
　　　and the bottom equation by 2.

$$5p - 4q = 9$$
$$2p + 6q = 1$$

　　a) To eliminate "t" at the right, we
　　　　can multiply the top equation by
　　　　_____ and the bottom equation
　　　　by _____.

$$3v - 5t = 10$$
$$5v + 2t = 15$$

　　b) To eliminate "b" at the right, we
　　　　can multiply the top equation by
　　　　_____ and the bottom equation
　　　　by _____.

$$7a + 6b = 14$$
$$3a - 8b = 10$$

a) $10a - 4b = 0$
　　$-10a + 15b = 55$

b) $a = 2, b = 5$

---

42.   To eliminate "y" at the right, we
　can multiply the equations by either
　pair of numbers below.

　　　top by 4, bottom by -5

　　　top by -4, bottom by 5

$$6x + 5y = 20$$
$$5x + 4y = 15$$

　　a) to eliminate "a" at the right, we
　　　　can multiply the equations by
　　　　either of two pairs of numbers.
　　　　They are:

　　　　　top by _____, bottom by _____

　　　　　top by _____, bottom by _____

$$3a - 4b = 7$$
$$7a - 2b = 10$$

a) top by 2, bottom by 5

b) top by 4, bottom by 3

Continued on following page.

42.   Continued

b) To eliminate "t" at the right, we
can multiply the equations by
either of two pairs of numbers.
They are:

$$5m + 4t = 9$$
$$2m + 10t = 7$$

top by _____, bottom by _____

top by _____, bottom by _____

---

43.   To eliminate "y" at the right, we can
multiply the equations by either pair
of numbers below.

$$3x - 5y = 11$$
$$7x - 3y = 10$$

top by 3, bottom by -5

top by -3, bottom by 5

a) To eliminate "q" at the right, we
can multiply the equations by
either of two pairs of numbers.
They are:

$$2p - 7q = 0$$
$$3p - 4q = 9$$

top by _____, bottom by _____

top by _____, bottom by _____

b) To eliminate "m" at the right, we
can multiply the equations by either
of two pairs of numbers. They
are:

$$5d - 8m = 40$$
$$3d - 10m = 25$$

top by _____, bottom by _____

top by _____, bottom by _____

a) top by 7, bottom by -3
   top by -7, bottom by 3

b) top by 5, bottom by -2
   top by -5, bottom by 2

---

44.   By eliminating either letter, solve each system below.

a)
| $2p + 5q = 9$ |
| $3p - 2q = 4$ |

b)
| $3x + 4y = 6$ |
| $2x + 3y = 5$ |

a) top by 4, bottom by -7
   top by -4, bottom by 7

b) top by 5, bottom by -4
   top by -5, bottom by 4

45. It is usually easier to solve a system when we only have to multiply <u>one</u> equation by a number. Therefore, we would eliminate "x" in the system at the left below. What letter would we eliminate to solve each of the other equations?

$$2x - 4y = 7$$
$$x + 3y = 10$$

a) $4a - t = 0$
$$6a + 4t = 5$$

b) $3m + 6d = 9$
$$6m + 4d = 7$$

a) p = 2, q = 1

b) x = -2, y = 3

---

46. By eliminating either letter, solve each system.

a) $2x + 3y = 2$
$$x - 2y = 8$$

b) $5a - 4b = 7$
$$3a + 2b = 13$$

a) t

b) m

---

a) x = 4, y = -2   b) a = 3, b = 2

SELF-TEST 31 (pages 385-398)

Solve each system.

1. $3x + y = 13$
$$2x - 3y = 5$$

2. $h + 3m = 12$
$$h + 2m = 7$$

Continued on following page.

<u>SELF-TEST</u> <u>31</u> (Continued)

| 3. | 2t - 3w = 7 |
| | 5t + 2w = 8 |

4. 3a - 5d = 0
   5a - 4d = 13

<u>ANSWERS</u>:   1. x = 4, y = 1   3. t = 2, w = -1

2. h = -3, m = 5   4. a = 5, d = 3

9-7   SOLVING SYSTEMS BY CONVERTING TO STANDARD FORM

To solve a system by the addition method, both equations must be in standard form. When an equation is not in standard form, we must convert it to standard form first. We will discuss solutions of that type in this section.

47.   In the two systems below, the equations are <u>in standard form</u> because both letter terms are on the left side and the number term is on the right side.

$$x + y = 5 \qquad 3a + b = 7$$
$$x - y = 0 \qquad a + 5b = 9$$

An equation in a system <u>is not in standard form</u> if it has a letter term on the right side or a number term on the left side. For example, in the systems below, $x + y - 8 = 0$ and $t = 5s - 6$ are not in standard form.

$$x + y - 8 = 0 \qquad s - 2t = 0$$
$$5x + y = 4 \qquad t = 5s - 6$$

In each system below, underline the equation that <u>is not in standard form</u>.

a)   x + y = 10   c)   3p + 2q = 5
         y = 3x - 1   2p + 5q + 1 = 0

b)   m = k   d)   5d = 7 - 4h
      3k - m = 7   d + 3h = 9

48.   When an equation in a system is not in standard form, we must convert it to standard form before using the addition method to solve the system. An example is shown. Notice that we used the addition axiom to get the number term on the right side.

Original System

2x - 3y - 10 = 0
x + y = 5

Conversion

2x - 3y - 10 = 0
2x - 3y - 10 + 10 = 10
2x - 3y = 10

Standard System

2x - 3y = 10
x + y = 5

Convert each system below to one with two standard-form equations.

a)   m - p = 0
3m + p - 5 = 0

b)   5d + 6k + 7 = 0
2d + 5k = 8

a) y = 3x - 1

b) m = k

c) 2p + 5q + 1 = 0

d) 5d = 7 - 4h

49.   Two more examples of conversions to standard form are shown below. Notice these two points.

1. We used the addition axiom to get the letter-term to the left side.

2. We lined up the terms with the same letters on the left side.

Original System

x + y = 10
y = 2x + 3

Conversion

y = 2x + 3
(-2x) + y = 2x + (-2x) + 3
-2x + y = 3

Standard System

x + y = 10
-2x + y = 3

Original System

a = b + 7
2a + b = 5

Conversion

a = b + 7
a + (-b) = b + (-b) + 7
a - b = 7

Standard System

a - b = 7
2a + b = 5

a) m - p = 0
3m + p = 5

b) 5d + 6k = -7
2d + 5k = 8

Continued on following page.

49.   Continued

Convert each system below to one with two standard-form equations.

a) $2p + q = 9$
$q = 4p + 5$

b) $t = 2d - 1$
$t + d = 10$

---

50.   Two more examples of conversions to standard form are shown below. Notice in each that we lined up the terms with the same letters on the left side.

Original System	Original System
$p - q = 7$	$x = 5 - y$
$q = 2p$	$2x + y = 3$

Conversion

$q = 2p$
$(-2p) + q = 2p + (-2p)$
$-2p + q = 0$

Conversion

$x = 5 - y$
$x + y = 5 - y + y$
$x + y = 5$

Standard System	Standard System
$p - q = 7$	$x + y = 5$
$-2p + q = 0$	$2x + y = 3$

Convert each system below to one with two standard-form equations.

a) $b = a$
$2a - 3b = 8$

b) $t + 6m = 1$
$t = 4 - 3m$

a) $2p + q = 9$
$-4p + q = 5$

b) $t - 2d = -1$
$t + d = 10$

a) $-a + b = 0$
$2a - 3b = 8$

b) $t + 6m = 1$
$t + 3m = 4$

51. Sometimes we have to use the addition axiom twice to convert to standard form.   An example is shown.

Original System

$$x + 1 = 2y$$
$$3x - 5y = 9$$

Conversion

$$x + 1 = 2y$$
$$x + 1 + (-1) = 2y + (-1)$$
$$x = 2y - 1$$
$$x + (-2y) = (-2y) + 2y - 1$$
$$x - 2y = -1$$

Standard System

$$x - 2y = -1$$
$$3x - 5y = 9$$

Convert each system below to one with two standard-form equations.

a) $t - 5 = p$
     $t + 2p = 0$

b) $3a - 4b = 9$
     $2a + 7 = 5b$

---

52. In the system at the right, $y = 3x - 8$ is not in standard form.

$$x - y = 2$$
$$y = 3x - 8$$

To solve the system, we begin by converting $y = 3x - 8$ to standard form.

$$(-3x) + y = 3x + (-3x) - 8$$
$$-3x + y = -8$$

Then we can use the addition method to solve the standard system at the right.

$$x - y = 2$$
$$-3x + y = -8$$

a) Solve the standard system.

b) Check your solution in each original equation.

$$x - y = 2 \qquad\qquad y = 3x - 8$$

a) $t - p = 5$
   $t + 2p = 0$

b) $3a - 4b = 9$
   $2a - 5b = -7$

53.   Following the steps in the last frame, solve each system.

a)   2p + q = 30

p = 2q

b)   3a - 4 = 2b

2a + 3b = 7

a) x = 3, y = 1

b) 3 - 1 = 2

2 = 2

1 = 3(3) - 8

1 = 9 - 8

1 = 1

a) p = 12, q = 6   b) a = 2, b = 1

9-8   APPLIED PROBLEMS

In this section, we will discuss some applied problems that can be solved by setting up and solving a system of equations.

54.   The following problem can be solved by setting up and solving a system of equations.

The sum of two numbers is 50. The difference between the larger and smaller numbers is 14. Find the two numbers.

Using "x" for the larger number and "y" for the smaller number, we can set up two equations.

The sum of two numbers is 50.

x + y = 50

The difference between the larger and smaller numbers is 14.

x - y = 14

Therefore, the problem can be solved by solving the system at the right.

x + y = 50

x - y = 14

a) Solve the system.

b) Therefore, the larger number is _____ and the smaller number is _____.

55. Here is another problem that can be solved by setting up and solving a system of equations.

> A portable radio costs twice as much as a digital clock. If the total cost of the radio and clock is $72, how much does each cost?

The two statements in the problem lead to two equations involving the cost of the radio (r) and the cost of the clock (c).   That is:

A radio costs twice as much as a clock.

$$r = 2c \quad \text{or} \quad r - 2c = 0$$

The total cost of the radio and clock is $72.

$$r + c = \$72$$

Therefore, the problem can be solved by solving the system at the right.

$$r - 2c = 0$$
$$r + c = 72$$

a) Solve the system.

b) Therefore, the radio costs _____ and the clock costs _____.

---

56. A system of equations can be used to solve the problem below.

> A 10-foot board is cut into two parts.   The larger part is 2 feet longer than the smaller.   How long is each part?

Using "x" for the larger part and "y" for the smaller part, we can set up two equations.

The sum of the two parts is 10 feet.

$$x + y = 10$$

The larger part is 2 feet longer than the smaller part.

$$x = y + 2$$

or

$$x - y = 2$$

Continued on following page.

a) x = 32, y = 18

b) larger number is 32 smaller number is 18

a) r = $48, c = $24

b) Radio costs $48 and clock costs $24

56.   Continued

Therefore, we can solve the problem $x + y = 10$
by solving the system at the right. $x - y = 2$

   a) Solve the system.

   b) Therefore, the larger part is _____ feet and the smaller
      part is _____ feet.

---

57.   The geometric problem below can be solved by means of a system of
equations.

     The perimeter of a rectangle is 180
     centimeters. If its length is 20
     centimeters longer than its width,
     what are its length and width?

Using "L" for length and "W" for width, we can set up two equations.

     The perimeter of a rectangle is 180 centimeters.

$$2L + 2W = 180$$

     The length is 20 centimeters longer than the width.

$$L = W + 20 \quad \text{or} \quad L - W = 20$$

Therefore, we can solve the problem by $2L + 2W = 180$
solving the system at the right. $L - W = 20$

   a) Solve the system.

   b) Therefore, its length is _____ centimeters and its width
     is _____ centimeters.

---

a) $x = 6$, $y = 4$

b) Larger is 6 feet,
   smaller is 4 feet

---

a) $L = 55$, $W = 35$

b) Length is 55 centi-
   meters, width is 35
   centimeters.

58. When using a system of equations to solve an applied problem, the key step is setting up the two equations in the system. Therefore, we will give practice in setting up systems in the following frames.

Without solving, set up the system for each problem.

a) There are 36 adults at a party. If there are 3 times as many non-smokers (n) as smokers (s), how many non-smokers and smokers are there?

b) A husband (h) weighs 60 pounds more than his wife (w). Together they weigh 300 pounds. Find the weight of each.

---

59. Without solving, set up the system for each problem.

a) The difference between two numbers is 6. The larger number (x) is four times the smaller number (y). Find the two numbers.

b) The sum of two numbers is 40. If the larger (x) is 20 more than the smaller (y), find the two numbers.

a) $n + s = 36$
$n = 3s$

b) $h = w + 60$
$h + w = 300$
or
$h - w = 60$
$h + w = 300$

---

60. Without solving, set up the system for each problem.

a) A piece of wire 28 centimeters long is cut into two parts. If the larger part (x) is three times the smaller part (y), how long is each part?

b) A metal rod 12 feet long is cut into two parts. If the larger part (x) is 3 feet longer than the smaller part (y), how long is each part?

a) $x - y = 6$
$x = 4y$

b) $x + y = 40$
$x = y + 20$
or
$x + y = 40$
$x - y = 20$

---

61. Without solving, set up the system for each problem.

a) The perimeter of a rectangle is 48 inches. If the length (L) is five times the width (W), find the length and width.

b) The perimeter of a rectangle is 100 centimeters. If the length (L) is 10 centimeters longer than the width (W), find the length and width.

a) $x + y = 28$
$x = 3y$

b) $x + y = 12$
$x = y + 3$
or
$x + y = 12$
$x - y = 3$

62.   Solve each problem.

a) The sum of two numbers is 64. The larger (x) is 18 more than the smaller (y). Find the two numbers.

b) The perimeter of a rectangular lot is 360 meters. If the length (L) is twice the width (W), find the length and width.

a) 2L + 2W = 48
       L = 5W

b) 2L + 2W = 100
       L = W + 10
        or
   2L + 2W = 100
     L - W = 10

a) x = 41, y = 23   b) L = 120 meters, W = 60 meters

9-9   DECIMAL COEFFICIENTS IN SYSTEMS

When solving applied problems, we sometimes set up systems that contain equations with decimal coefficients. We will discuss systems of that type in this section.

63.   Numbers like 10, 100, and 1,000 are called powers of ten. Multiplying a decimal number by a power of ten is the same as shifting the decimal point to the right. The number of places the decimal point is shifted depends on the number of 0's in the power of ten. That is:

Since 10 has one "0", we shift it one place.

$$10 \times 1.6 = 1.6 = 16. \text{ or } 16$$

Since 100 has two 0's, we shift it two places.

$$100 \times 3.24 = 3.24 = 324. \text{ or } 324$$

Since 1,000 has three 0's, we shift it three places.

$$1,000 \times .525 = .525 = 525. \text{ or } 525$$

Use the decimal-point-shift method to complete these.

a) 10 x 2.5 = _____

c) 100 x .75 = _____

b) 10 x .4 = _____

d) 1,000 x .125 = _____

a) 25   c) 75

b) 4   d) 125

64. By multiplying both sides of the equations below by 10, we obtained an equation that contains whole numbers instead of decimals.

$$2.3x = 4.6$$
$$10(2.3x) = 10(4.6)$$
$$23x = 46$$

The root of $23x = 46$ is 2. Is 2 the root of $2.3x = 4.6$? _____

---

65. By multiplying both sides of the equation below by 100, we obtained an equation that contains whole numbers instead of decimals.

$$1.12y = 3.36$$
$$100(1.12y) = 100(3.36)$$
$$112y = 336$$

a) The root of $112y = 336$ is _____.

b) The root of $1.12y = 3.36$ is _____.

c) Do both equations have the same root? _____

Yes

---

66. When we multiply both sides of an equation by a power of ten to clear the decimals, the new equation is <u>equivalent</u> to the original equation because it has the same root. For example, we multiplied both sides of the equation below by 1,000 to eliminate the decimals.

$$.025x = .125$$
$$1,000(.025x) = 1,000(.125)$$
$$25x = 125$$

Is $25x = 125$ equivalent to $.025x = .125$? _____

a) 3

b) 3

c) Yes

---

67. When multiplying by a power of ten to clear the decimals in an equation, the power of ten is determined by the number <u>with the most decimal places</u>. If its last digit is:

    in the <u>tenths</u> place, we multiply by 10.

    in the <u>hundredths</u> place, we multiply by 100.

    in the <u>thousandths</u> place, we multiply by 1,000.

What power of ten should be used to clear the decimals in each of these?

a) $2.4t = 6.48$ _____   c) $.3D = .369$ _____

b) $.6m = 1.2$ _____   d) $1.25y = 2.5$ _____

Yes. The root of each is 5.

---

a) 100   c) 1,000

b) 10   d) 100

68.   The same method can be used to clear the decimals in a two-variable
       equation. For example, we cleared the decimals below by multiplying
       both sides by 100. Notice that we multiplied by the distributive principle
       on the left side.

$$.12x + .48y = .96$$
$$100(.12x + .48y) = 100(.96)$$
$$100(.12x) + 100(.48y) = 100(.96)$$
$$12x + 48y = 96$$

Using the same method, clear the decimals in these.

   a) $.4a - .8b = 1.6$ b) $.025m + .075p = .125$

---

69.   What power of ten should be used to clear the decimals in each of these?

   a) $.25s + .5t = .75$ _____   b) $.8v - .6t = .248$ _____

| a) $4a - 8b = 16$ |
| b) $25m + 75p = 125$ |

---

70.   Clear the decimals in each of these.

   a) $.7x + .14y = 2.1$ b) $.9m - .027t = .81$

| a) 100 |
| b) 1,000 |

---

71.   When an equation contains a whole number, don't forget to multiply the
       whole number by the power of ten. For example:

$$5x + 1.5y = 7.5$$
$$10(5x + 1.5y) = 10(7.5)$$
$$10(5x) + 10(1.5y) = 10(7.5)$$
$$50x + 15y = 75$$

Clear the decimals in these.

   a) $1.25m + 4t = 2.5$ b) $3a - 1.5b = 4.5$

| a) $70x + 14y = 210$ (multiply by 100) |
| b) $900m - 27t = 810$ (multiply by 1,000) |

| a) $125m + 400t = 250$ |
| b) $30a - 15b = 45$ |

72. Since it is easier to solve a system with whole-number coefficients, we usually clear any decimal coefficients before using the addition method. For example, we cleared the decimal coefficients in the system below.

Original System

$$.2x + .5y = 1.6$$
$$3x - 1.5y = 1.2$$

Equivalent System

$$2x + 5y = 16$$
$$30x - 15y = 12$$

Following the example, write the equivalent system obtained after clearing the decimals in these:

a) $2a + 1.14b = 6.8$
    $.57a - .9b = 3$

b) $7m - .4t = 1.9$
    $m + 1.8t = 5$

---

73. Let's solve the system at the right.

$$1.5x + y = 8$$
$$3x - .5y = 3.5$$

a) Write the equivalent system obtained by clearing the decimal coefficients.

b) Solve the equivalent system.

c) Show that the solution satisfies both original equations.

$$1.5x + y = 8 \qquad\qquad 3x - .5y = 3.5$$

---

a) $200a + 114b = 680$
    $57a - 90b = 300$

b) $70m - 4t = 19$
    $10m + 18t = 50$

---

a) $15x + 10y = 80$ b) $x = 2,\ y = 5$ c) $1.5(2) + 5 = 8$ $3(2) - .5(5) = 3.5$
   $30x - 5y = 35$ $3 + 5 = 8$ $6 - 2.5 = 3.5$
                                                        $8 = 8$ $3.5 = 3.5$

9-10   MIXTURE PROBLEMS

Systems of equations can be used to solve various mixture problems. We will discuss problems of that type in this section.

74.   A system of equations can be used to solve various mixture problems. An example is discussed below.

A chemist has one solution that is 70% acid and a second solution that is 20% acid. She wants to mix them to get 100 liters of a solution that is 50% acid. How many liters of each solution should she use?

She must mix $\underline{x}$ liters of the first solution and $\underline{y}$ liters of the second solution to get 100 liters. Therefore:

$$x + y = 100$$

The amount of acid in the new solution must be 50% of 100 liters. This amount will be made up by the acid in the two solutions mixed. These amounts of acid are 70% of x and 20% of y, or 70%x and 20%y. Therefore:

$$70\%x + 20\%y = 50\%(100)$$

or

$$.7x + .2y = 50$$

a) We can solve the problem by solving the system below. Do so.

$$x + \quad y = 100$$
$$.7x + .2y = \quad 50$$

b) She should use _____ liters of the 70% solution and _____ liters of the 20% solution.

---

75.   Here is another problem that can be solved by a system of equations.

One alloy contains 55% silver; a second alloy contains 80% silver. By mixing them, we want to get 40 grams of an alloy that contains 75% silver. How many grams of each of the two original alloys should we use?

We must mix $\underline{x}$ grams of the 55% alloy and $\underline{y}$ grams of the 80% alloy to get 40 grams. Therefore:

$$x + y = 40$$

Continued on following page.

a) x = 60, y = 40

b) <u>60</u> liters of the 70% solution and <u>40</u> liters of the 20% solution.

75.   Continued

The amount of silver in the new alloy must be 75% of 40 grams.
This amount will be made up by the silver in the two alloys
mixed. These amounts of silver are 55%x and 80%y. Therefore:

$$55\%x + 80\%y = 75\%(40)$$

or

$$.55x + .8y = 30$$

a) We can solve the problem by solving the system below. Do so.

$$x + y = 40$$
$$.55x + .8y = 30$$

b) Therefore, we should mix _____ grams of the 55% alloy and
_____ grams of the 80% alloy.

---

76.   Let's use a system of equations to solve the problem below.

Solution A is 30% alcohol and solution B is 80% alcohol.
How much of each should be used to make 100 milliliters
of a solution that is 60% alcohol?

a) Set up the system needed to solve the problem.

b) Solve the system.

c) Therefore, we must use _____ milliliters of solution A and
_____ milliliters of solution B.

a) x = 8, y = 32

b) <u>8</u> grams of the 55%
    alloy and <u>32</u> grams
    of the 80% alloy.

77. Here is another type of problem that can be solved by a system of equations. Since one of the equations in the system contains only one variable, we do not have to use the addition method because there is a simpler method.

Brine is a solution of salt and water. We want to mix pure water and a solution of brine that is 30% salt to get 100 gallons of a brine solution that is 24% salt. How many gallons of each should be mixed?

We must mix $\underline{x}$ gallons of the 30% brine and $\underline{y}$ gallons of water to get 100 gallons. Therefore:

$$x + y = 100$$

The amount of salt in the new solution must be 24% of 100 gallons. This amount will be made up by the salt in the 30% brine and the water. These amounts are 30%x and 0%y. Therefore:

$$30\%x + 0\%y = 24\%(100)$$

or

$$.3x + 0 = 24$$

a) We can solve the problem by solving the system below. Do so by solving for $\underline{x}$ immediately in the bottom equation and then substituting in the top equation to find $\underline{y}$.

$$x + \quad y = 100$$
$$.3x = \quad 24$$

b) Therefore, we should mix _____ gallons of the 30% brine and _____ gallons of water.

────────────────

78. Let's use a system of equations to solve the problem below.

We want to mix some milk containing 6% butterfat with skimmed milk (containing 0% butterfat) to get 100 gallons of milk containing 4.5% butterfat. How many gallons of the 6% milk and the skimmed milk should we use?

a) Set up the system needed to solve the problem.

Continued on following page.

Answers (right column):

77.
a)   A +   B = 100
    .3A + .8B = 60

b) A = 40, B = 60

c) <u>40</u> milliliters of solution A and <u>60</u> milliliters of solution B

78.
a) x = 80, y = 20

b) <u>80</u> gallons of the 30% brine and <u>20</u> gallons of water.

78.   Continued

    b) Solve the system.

    c) Therefore, we should mix _____ gallons of the 6% milk and _____ gallons of the skimmed milk.

---

a) $x + y = 100$
    $.06x = 4.5$

b) $x = 75$, $y = 25$

c) 75 gallons of the 6% milk and 25 gallons of the skimmed milk

---

<u>SELF-TEST 32</u> (pages 398-413)

Solve each system of equations by the addition method.

1.
$$6x + 6 = 5y$$
$$2y = 3x$$

2.
$$1.5t + 2w = 6.5$$
$$t - 2.5 = 0.5w$$

3. The sum of two numbers is 131. The larger is 15 greater than the smaller. Find the two numbers.

4. We want to add water to a 60% solution of radiator coolant to get 15 liters of a 40% solution of radiator coolant. How many liters of each should be mixed?

ANSWERS: 1. $x = 4$, $y = 6$ 3. The numbers are 73 and 58.

            2. $t = 3$, $w = 1$ 4. 10 liters of 60% coolant and 5 liters of water

SUPPLEMENTARY PROBLEMS - CHAPTER 9

Assignment 30

Solve each system of equations by the graphing method.

1. $x - y = 2$
   $x + y = 8$

2. $x - 2y = 10$
   $4x + y = 4$

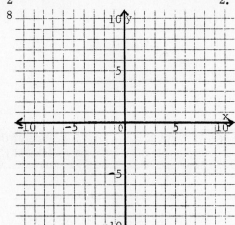
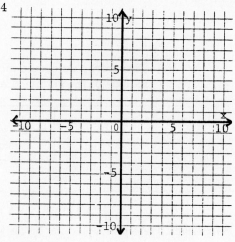

Solve each system of equations by the addition method.

3. $x - y = 5$
   $x + y = 3$

4. $p + d = 6$
   $p - d = 4$

5. $3h + 2m = 24$
   $h - 2m = 0$

6. $5x - 4y = 23$
   $3x + 4y = 1$

7. $t + w = 0$
   $t - w = 8$

8. $4x - 3y = 1$
   $2x + 3y = 5$

9. $8a + 5b = 1$
   $3a - 5b = 21$

10. $7r - 10v = 40$
    $3r + 10v = 60$

11. $5h - 2k = 35$
    $h + 2k = 7$

12. $8s + 7t = 26$
    $2s - 7t = 24$

13. $3b - m = 20$
    $2b + m = 0$

14. $p + 9r = 65$
    $4p - 9r = 35$

15. $p - v = 2.8$
    $p + v = 5.2$

16. $2m + 5n = 11$
    $4m - 5n = 4$

17. $x - 7y = 4.4$
    $2x + 7y = 4.6$

18. $7r + 2w = 5.82$
    $3r - 2w = 8.38$

Assignment 31

Solve each system of equations by the addition method.

1. $x - 3y = 2$
   $2x + y = 11$

2. $x + y = 5$
   $3x - 2y = 5$

3. $5x + 4y = 8$
   $3x + y = 9$

4. $4x + y = 3$
   $7x + 3y = 4$

5. $t - w = 1$
   $4t + 3w = 11$

6. $3h + 2k = 28$
   $2h - k = 7$

7. $4a + 3b = 3$
   $2a + b = 3$

8. $3r + 6s = 21$
   $5r + 2s = 3$

9. $7m + 6n = 14$
   $5m + 2n = 10$

10. $3c - 4d = 11$
    $7c - 2d = 11$

11. $p + 2t = 16$
    $p + 3t = 26$

12. $6v - w = 19$
    $2v - w = 7$

13. $7x + 3y = 26$
    $9x - 2y = 10$

14. $8a + 3b = 15$
    $5a - 4b = 27$

15. $9p - 2r = 11$
    $7p + 5r = 2$

16. $5v - 4w = 6$
    $4v + 3w = -20$

17. $5h + 7k = 32$
    $4h + 5k = 25$

18. $2d + 5h = 5$
    $7d + 9h = 43$

19. $8x - 3y = 10$
    $5x - 2y = 7$

20. $4s - 9t = 32$
    $3s - 8t = 24$

21. $11m - 6n = 8$
    $12m - 5n = 1$

22. $4x - 9y = 17$
    $7x - 8y = 22$

23. $10b + 3d = 15$
    $9b + 5d = 2$

24. $7t + 5v = 17$
    $2t + 9v = 20$

SUPPLEMENTARY PROBLEMS - CHAPTER 9 (Continued)

Assignment 32

Convert each system to standard form; then solve by the addition method.

1. $2x + y = 13$
   $x - 2 = y$

2. $b = 3a$
   $2a = b - 3$

3. $t - w = 5$
   $3w + 9 = 2t$

4. $r + 2v - 2 = 0$
   $v + 5 - r = 0$

5. $h - 4m = 0$
   $2m + 50 = 3h$

6. $p + 3 = s$
   $2s - 5 = 3p$

7. $2d = 6 - 5h$
   $2h = 4d - 36$

8. $3k + 13 = 4m$
   $5m - 11 = 2k$

Clear the decimal coefficients in each system; then solve by the addition method.

9. $.3r - .1t = 1.1$
   $.2r + .1t = .9$

10. $.21p + .08w = .5$
    $.17p + .08w = .42$

11. $1.5x - y = 1$
    $3x - 1.5y = 3$

12. $2d = 1.25p + 1$
    $d = 4.4 - .35p$

Use a system of equations to solve each problem.

13. The sum of two numbers is 218 and their difference is 82. Find the two numbers.

14. The sum of two numbers is 64. The first number is three times the second number. Find the two numbers.

15. A metal rod 100 centimeters long is to be sawed into two pieces so that one piece is 10 centimeters longer than the other. Find the length of each piece.

16. The total cost of a camera and a calculator is $420. The camera costs $80 more than the calculator. Find the cost of each.

17. We want to construct a rectangle whose perimeter is 60 centimeters and whose width is 6 centimeters less than its length. Find the length and the width.

18. A wire 20 meters long is to be cut into two parts so that the longer part is four times the smaller part. How long is each part?

19. Solution A is 30% acid and solution B is 60% acid. How much of each solution should be mixed to get 60 liters of a solution that is 40% acid?

20. How much of a 10% solution of alcohol and how much of a 50% solution of alcohol should be mixed to make 200 milliliters of a 35% solution?

21. An alloy contains 8% chromium; another alloy contains 5% chromium. How much of each alloy should be mixed to get 300 pounds of an alloy containing 6% chromium?

22. How much pure zinc (100% zinc) and how much alloy containing 40% zinc must be melted together to get 15 kilograms of an alloy containing 60% zinc?

Chapter 10 SYSTEMS OF TWO EQUATIONS AND FORMULAS

In the last chapter, we discussed the graphing and addition methods for solving systems of two equations. In this chapter, we will discuss the equivalence and substitution methods for solving the same type of systems. We will also use the equivalence and substitution methods to eliminate a variable (or variables) from a system of two formulas to derive a new formula.

10-1 THE EQUIVALENCE METHOD FOR SOLVING SYSTEMS OF EQUATIONS

In this section, we will discuss the equivalence method for solving systems of two equations. To emphasize the procedure, only very simple systems are solved.

---

1. The equivalence principle says this: <u>If the same quantity equals two other quantities, those two quantities are equal.</u> That is:

   If $y = 2x$ If $V = R + 5$

   and $y = 5x - 1$,   and $V = 2R - 7$,

   then $2x = 5x - 1$.   then _____ = _____

---

2. The equivalence principle can be used to eliminate a variable in a system of equations <u>when the same variable is solved for in both equations.</u> For example, we eliminated "t" at the left below. Eliminate "a" in the other system.

   $$t = s + 5 \qquad\qquad a = 4b$$
   $$t = 3s - 1 \qquad\qquad a = b - 9$$
   $$s + 5 = 3s - 1 \qquad\quad \text{____} = \text{_____}$$

   $R + 5 = 2R - 7$

---

3. In the system at the right, "y"
   is solved for in both equations.
   Let's use the equivalence method
   to solve it. Three steps are needed:

   $$\boxed{\begin{aligned} y &= 2x \\ y &= 4x - 6 \end{aligned}}$$

   4b = b - 9

   1. Use the equivalence principle to
      eliminate "y" and obtain an
      equation that contains only "x".

      $$2x = 4x - 6$$

   2. Solve that equation to find the
      value of "x".

      $$2x = 4x - 6$$
      $$-2x = -6$$
      $$x = 3$$

   3. Substitute that value of "x" in one
      of the original equations to find
      the corresponding value of "y".
      (We used the original top equation.)

      $$y = 2x$$
      $$y = 2(3)$$
      $$y = 6$$

   Therefore, the solution of the system is x = _____ , y = _____

4.   Let's use the steps in the last frame | p = q + 2 | x = 3, y = 6
        to solve the system at the right. | p = 2q - 3 |

        a) Use the equivalence principle to
            eliminate "p" and write the
            equation that contains only "q".   _____ = _____

        b) Solve that equation to find the
            value of "q".

        c) Substitute in one of the original
            equations to find the correspond-
            ing value of "p".

        d) The solution is: p = _____ , q = _____

---

5.   Use the equivalence method to solve each system below.   a) q + 2 = 2q - 3

        a)   a = 4b   b)   t = v - 1   b) q = 5
              a = b + 3   t = 4v + 5
                                                                     c) p = 7

                                                                     d) p = 7, q = 5

---

a) a = 4, b = 1   b) t = -3, v = -2

10-2   REARRANGEMENTS NEEDED IN THE EQUIVALENCE METHOD

Before using the equivalence principle to eliminate a variable in a system, that variable must be solved for in both equations. Frequently, we have to rearrange one or both equations to solve for the same variable in each. We will review rearrangements of that type in this section.

---

6.   To solve for "y" below, we added "x" to both sides. Solve for "a" in the
        other equation.

                    y - x = 1   a - 3b = 5

                y - x + x = 1 + x

                        y = 1 + x

                or   y = x + 1

7. When solving for "y" below, we converted the addition of a negative term to a subtraction in the last step. Solve for "p" in the other equation.

$$x + y = 10 \qquad\qquad p + 5q = 12$$

$$(-x) + x + y = 10 + (-x)$$

$$y = 10 + (-x)$$

$$y = 10 - x$$

| $a = 5 + 3b$
| or
| $a = 3b + 5$

8. To solve for "y" below, we divided the entire right side of the equation by "2". Solve for "m" in the other equation.

$$2y = 3x + 5 \qquad\qquad 7m = 1 - 3v$$

$$y = \frac{3x + 5}{2}$$

$p = 12 - 5q$

9. We solved for "x" below. Solve for "a" in the other equation.

$$3x - 4y = 5 \qquad\qquad 7a - 8b = 10$$

$$3x - 4y + 4y = 5 + 4y$$

$$3x = 4y + 5$$

$$x = \frac{4y + 5}{3}$$

$m = \dfrac{1 - 3v}{7}$

10. We solved for "y" below. Solve for D in the other equation.

$$2x + 5y = 10 \qquad\qquad 4D + 3H = 11$$

$$(-2x) + 2x + 5y = 10 + (-2x)$$

$$5y = 10 - 2x$$

$$y = \frac{10 - 2x}{5}$$

$a = \dfrac{8b + 10}{7}$

11. As we saw earlier, we get the opposite of a subtraction by interchanging the terms. That is:

The opposite of $2y - 5$ is $5 - 2y$.

The opposite of $1 - 7d$ is $7d - 1$.

Furthermore, to solve for "y" in each equation below, we used the "oppositing principle for equations." That is, we replaced each side by its opposite.

$$-y = 10 - 2x \qquad\qquad -y = 5x - 1$$

$$y = 2x - 10 \qquad\qquad\quad y = 1 - 5x$$

Solve for V in each equation below.

a) $-V = T - 9$ b) $-V = 8 - 5S$

    $V = $ _____ $V = $ _____

$D = \dfrac{11 - 3H}{4}$

12.   We used the oppositing principle to solve for "y" below. Solve for "b" in the other equation.

$$x - y = 1 \qquad\qquad 3a - b = 10$$

$$(-x) + x - y = 1 + (-x)$$

$$-y = 1 - x$$

$$y = x - 1$$

a) V = 9 – T

b) V = 5S – 8

---

13.   To solve for "y" below, we used the oppositing principle <u>before</u> dividing by 4. Solve for "q" in the other equation.

$$3x - 4y = 5 \qquad\qquad 7p - 9q = 1$$

$$(-3x) + 3x - 4y = 5 + (-3x)$$

$$-4y = 5 - 3x$$

$$4y = 3x - 5$$

$$y = \frac{3x - 5}{4}$$

b = 3a – 10

---

14.   Solve for "t" in each equation.

    a) 7v – t = 12   b) 10 – 5t = s

$$q = \frac{7p - 1}{9}$$

---

15.   Neither variable is solved for in the equations in the system below. Solve for "y" in each. Write the new equations in the lower box.

$$x + y = 7$$
$$3y = 9x - 5$$

y =

y =

a) t = 7v – 12

b) $t = \dfrac{10 - s}{5}$

---

16.   Solve for "m" in each equation below. Write the new equations in the lower box.

$$5d - m = 1$$
$$7m - 8d = 3$$

m =

m =

y = 7 – x

$$y = \frac{9x - 5}{3}$$

17. Solve for "i_2" in each equation below. Write the new equations in the lower box.

$$m = 5d - 1$$

$$m = \frac{8d + 3}{7}$$

$3i_1 + 2i_2 = 15$
$7i_1 - 5i_2 = 20$

$i_2 =$
$i_2 =$

$$i_2 = \frac{15 - 3i_1}{2}$$

$$i_2 = \frac{7i_1 - 20}{5}$$

10-3   FRACTIONAL EQUATIONS ENCOUNTERED IN THE EQUIVALENCE METHOD

When using the equivalence method, we frequently have to solve a fractional equation to find the value of the first variable. We will discuss fractional equations of that type in this section.

18. To clear the fraction in the equation below, we multiplied both sides by 3. Since $3(4x - 1)$ is an instance of the distributive principle, we multiplied both 4x and "1" by 3. <u>Without solving</u>, clear the fraction in the other equation.

$$4x - 1 = \frac{x + 9}{3} \qquad\qquad \frac{1 - 3y}{5} = y + 6$$

$$3(4x - 1) = \cancel{3}\left(\frac{x + 9}{\cancel{3}}\right)$$

$$12x - 3 = x + 9$$

19. Solve each equation. Begin by clearing the fraction.

$$1 - 3y = 5y + 30$$

a) $b - 1 = \dfrac{2b + 5}{9}$

  b) $\dfrac{3R - 1}{5} = 4R + 3$

20. To clear the fractions below, we multiplied both sides by 7. Without solving, clear the fractions in the other equation.

a) $b = 2$

b) $R = -\dfrac{16}{17}$

$$\frac{x - 5}{7} = \frac{3x + 9}{7} \qquad\qquad \frac{t + 10}{3} = \frac{5t - 1}{3}$$

$$\cancel{7}\left(\frac{x - 5}{\cancel{7}}\right) = \cancel{7}\left(\frac{3x + 9}{\cancel{7}}\right)$$

$$x - 5 = 3x + 9$$

21. To clear the fractions below, we multiplied both sides by 8. Notice that we multiplied by the distributive principle on the right side. Without solving, clear the fractions in the other equation by multiplying both sides by 12.

$$\frac{3m - 1}{8} = \frac{m + 10}{4}$$

$$8\left(\frac{3m - 1}{8}\right) = \overset{2}{8}\left(\frac{m + 10}{4}\right)$$

$$3m - 1 = 2m + 20$$

$$\frac{y + 2}{3} = \frac{5y - 7}{12}$$

| $t + 10 = 5t - 1$ |

---

22. To clear the fractions below, we multiplied both sides by 6. Notice that we multiplied by the distributive principle on both sides. Without solving, clear the fractions in the other equation by multiplying both sides by 12.

$$\frac{y + 7}{2} = \frac{2y - 5}{3}$$

$$\overset{3}{6}\left(\frac{y + 7}{2}\right) = \overset{2}{6}\left(\frac{2y - 5}{3}\right)$$

$$3y + 21 = 4y - 10$$

$$\frac{3m - 1}{6} = \frac{m + 10}{4}$$

| $4y + 8 = 5y - 7$ |

---

23. When a fractional equation contains two fractions, we clear the fractions by multiplying both sides <u>by the smallest common multiple of the two denominators</u>. That is:

        If both denominators are 2's, we multiply by 2.

        If the denominators are 3 and 9, we multiply by 9.

        If the denominators are 5 and 4, we multiply by 20.

        If the denominators are 6 and 8, we multiply by 24.

What multiplier would we use to clear the fractions in each of these?

a) $\dfrac{m + 7}{5} = \dfrac{m - 9}{2}$ b) $\dfrac{2y - 3}{15} = \dfrac{y + 11}{5}$ c) $\dfrac{R - 10}{4} = \dfrac{7R + 1}{10}$

_____   _____   _____

| $6m - 2 = 3m + 30$ |

---

24. Without solving, clear the fractions in these.

a) $\dfrac{3x + 1}{5} = \dfrac{2x - 7}{5}$ b) $\dfrac{d - 9}{10} = \dfrac{d + 5}{2}$

| a) 10 |
| b) 15 |
| c) 20 |

| a) $3x + 1 = 2x - 7$ |
| b) $d - 9 = 5d + 25$ |

25.   Without solving, clear the fractions in these.

a) $\dfrac{1 - 4V}{3} = \dfrac{2V + 3}{5}$ b) $\dfrac{F_1 - 3}{12} = \dfrac{5 - 3F_1}{8}$

26.   Solve each equation.

a) $\dfrac{y - 2}{3} = \dfrac{5y - 1}{6}$ b) $\dfrac{3 - 2P}{4} = \dfrac{P - 4}{3}$

a) $5 - 20V = 6V + 9$

b) $2F_1 - 6 = 15 - 9F_1$

a) $y = -1$

b) $P = \dfrac{5}{2}$

10-4   REARRANGEMENTS AND THE EQUIVALENCE METHOD

Before using the equivalence principle to eliminate a variable in a system, the same variable must be solved for in both equations. In this section, we will solve systems in which we must begin by rearranging one or both equations to solve for the same variable in each.

27.   Neither variable is solved for in the equations in the system at the right. Four steps are needed to solve the system by the equivalent method. They are:

$a + 3b = 10$
$a + b = 6$

1. Solve for the same variable in each equation. (We solved for "a".)

$a = 10 - 3b$
$a = 6 - b$

2. Use the equivalence principle to eliminate "a".

$10 - 3b = 6 - b$

3. Find the value of "b".

$4 = 2b$
$b = 2$

4. Find the corresponding value of "a".
    Note: It is easier to substitute into one of the equations in which "a" is solved for.

$a = 6 - b$
$a = 6 - 2$
$a = 4$

Check $(a = 4, b = 2)$ in each of the original equations.

a) $a + 3b = 10$ b) $a + b = 6$

28. Here is a general summary of the equivalence method. We will use the system at the right as an example.

$$V - 3T = 16$$
$$V + T = 8$$

a) $4 + 3(2) = 10$
$4 + 6 = 10$
$10 = 10$

1. Solve for the same variable in each equation. (We solved for V.)

$$V = 3T + 16$$
$$V = 8 - T$$

b) $4 + 2 = 6$
$6 = 6$

2. Use the equivalence principle to eliminate the solved-for variable V.

$$3T + 16 = 8 - T$$

3. Find the value of T.

$$4T = -8$$
$$T = -2$$

4. Find the corresponding value of V. (We substituted -2 for T in one of the equations in which V is solved for.)

$$V = 3T + 16$$
$$V = 3(-2) + 16$$
$$V = (-6) + 16$$
$$V = 10$$

5. Check the solution (T = -2, V = 10) in each of the original equations.

$$V - 3T = 16$$
$$10 - 3(-2) = 16$$
$$10 - (-6) = 16$$
$$10 + 6 = 16$$
$$16 = 16$$

$$V + T = 8$$
$$10 + (-2) = 8$$
$$8 = 8$$

---

29. Let's use the equivalence method to solve the system at the right.

$$3p = q - 7$$
$$2p + q = 2$$

a) Solve for "p" in each equation.

p =
p =

b) Use the equivalence principle to eliminate "p".

_____ = _____

c) Find the value of "q".

d) Find the corresponding value of p".

e) The solution of the system is: p = _____, q = _____

30.   Let's use the same method to solve
the system at the right.

$$3R - T = 4$$
$$3T - R = 12$$

a) Solve for R in each equation.

R =

R =

b) Use the equivalence principle
to eliminate R.

_____ = _____

c) Find the value of T.

d) Find the corresponding value of R.

e) The solution is: R = _____ , T = _____

a) $p = \dfrac{q - 7}{3}$

$p = \dfrac{2 - q}{2}$

b) $\dfrac{q - 7}{3} = \dfrac{2 - q}{2}$

c) $q = 4$

d) $p = -1$

e) $p = -1,\ q = 4$

31.   Let's use the same method to solve
this system.

$$4F_1 = 15 - 3F_2$$
$$8F_2 = 29 - 7F_1$$

a) Solve for F_1 in each system.

$F_1 =$

$F_1 =$

b) Use the equivalence principle
to eliminate F_1.

_____ = _____

c) Find the value of F_2.

d) Find the corresponding value of F_1.

e) The solution is: $F_1 =$ _____ , $F_2 =$ _____

a) $R = \dfrac{T + 4}{3}$

$R = 3T - 12$

b) $\dfrac{T + 4}{3} = 3T - 12$

c) $T = 5$

d) $R = 3$

e) $R = 3,\ T = 5$

a) $F_1 = \dfrac{15 - 3F_2}{4}$

$F_1 = \dfrac{29 - 8F_2}{7}$

b) $\dfrac{15 - 3F_2}{4} = \dfrac{29 - 8F_2}{7}$

c) $F_2 = 1$

d) $F_1 = 3$

e) $F_1 = 3,\ F_2 = 1$

SELF-TEST 33   (pages 416–425)

Solve each system by the equivalence method.

1. $P = 1 - 4R$
      $P = 3R + 8$

2. $x + 2y = 2$
      $3y - 2x = 10$

3. $2t_1 + 3t_2 = 44$
      $5t_2 + 20 = 6t_1$

4. $3A = 27 - 4B$
      $6B = 5A - 26$

ANSWERS: 1. $P = 5$, $R = -1$ 2. $x = -2$, $y = 2$ 3. $t_1 = 10$, $t_2 = 8$ 4. $A = 7$, $B = \frac{3}{2}$

10-5   THE SUBSTITUTION METHOD FOR SOLVING SYSTEMS OF EQUATIONS

We have discussed both the addition method and the equivalence method for solving systems of equations. In this section, we will discuss a third algebraic method for solving systems. The third method is called the "substitution method".

32.   Instead of adding equations or using the equivalence principle to eliminate a variable, we can substitute to eliminate a variable. An example is discussed below.

In the system at the right, "y" is solved for in the top equation but not in the bottom equation. Since "y" equals 2x, we can eliminate "y" by substituting 2x for it in the bottom equation. We get:

$y = 2x$
$x + 4y = 45$

Continued on following page.

32.   Continued

$$x + 4y = 45$$

$$x + 4(2x) = 45$$

$$x + 8x = 45$$

$$9x = 45$$

"t" is solved for in one equation in each system below. Eliminate "t" by substituting 3d for "t" in the other equation in each system. <u>Simplify the resulting equation</u>.

a)
$t = 3d$
$t + 2d = 5$

b)
$5d - 3t = 10$
$t = 3d$

---

33.   Let's use the substitution method to solve the system at the right. The steps are:

$a = 4b$
$2a + b = 9$

1.   Eliminate "a" by substituting 4b for "a" in the bottom equation.

$$2(4b) + b = 9$$
$$8b + b = 9$$

2.   Find the value of "b".

$$9b = 9$$
$$b = 1$$

3.   Find the corresponding value of "a".
     (<u>Note</u>: We substituted in the top equation.)

$$a = 4b$$
$$a = 4(1)$$
$$a = 4$$

Check (a = 4, b = 1) in each original equation below.

a) $a = 4b$

b) $2a + b = 9$

a) $3d + 2d = 5$
     $5d = 5$

b) $5d - 3(3d) = 10$
    $5d - 9d = 10$
       $-4d = 10$

---

34.   The only new step in the substitution method is eliminating a variable by substitution. We will discuss more examples of the substitution process in the next few frames.

"y" is solved for in the top equation below. We eliminated "y" by substituting (x − 3) for it in the bottom equation. Eliminate "q" in the other system.

$y = x - 3$
$4y - 3x = 7$

$q = p + 4$
$2p + q = 9$

$$4(x - 3) - 3x = 7$$

$$4x - 12 - 3x = 7$$

$$x - 12 = 7$$

a) $4 = 4(1)$
   $4 = 4$

b) $2(4) + 1 = 9$
    $8 + 1 = 9$
       $9 = 9$

35. "t" is solved for in the bottom equation below. We eliminated "t" by sub-
     stituting (v + 2) for it in the top equation. Eliminate "b" in the other
     system.

$$\boxed{\begin{array}{rcl} 2v - 3t & = & 11 \\ t & = & v + 2 \end{array}}$$

$$\boxed{\begin{array}{rcl} 4a - b & = & 7 \\ b & = & a + 5 \end{array}}$$

$$2v - 3(v + 2) = 11$$

$$2v - (3v + 6) = 11$$

$$2v + (-3v) + (-6) = 11$$

$$-v - 6 = 11$$

$$2p + (p + 4) = 9$$
$$2p + p + 4 = 9$$
$$3p + 4 = 9$$

---

36. By substituting (2x - 1) for "y" in the bottom equation, we eliminated "y"
     from the system below. Eliminate R from the other system.

$$\boxed{\begin{array}{rcl} y & = & 2x - 1 \\ 5x - y & = & 4 \end{array}}$$

$$\boxed{\begin{array}{rcl} S - 2R & = & 10 \\ R & = & 3S - 7 \end{array}}$$

$$5x - (2x - 1) = 4$$

$$5x + (-2x) + 1 = 4$$

$$3x + 1 = 4$$

$$4a - (a + 5) = 7$$
$$4a + (-a) + (-5) = 7$$
$$3a - 5 = 7$$

---

37. Neither variable is solved for in the
     equations in the system at the right.

$$\boxed{\begin{array}{rcl} p + q & = & 7 \\ 3p - q & = & 2 \end{array}}$$

$$S - 2(3S - 7) = 10$$
$$S - (6S - 14) = 10$$
$$S + (-6S) + 14 = 10$$
$$-5S + 14 = 10$$

To use the substitution method, we must solve for one variable in one
equation. If we solve for "p" in the top equation, we get: p = 7 - q.
Substituting that solution for "p" in the bottom equation, we get:

$$3(7 - q) - q = 2$$

$$21 - 3q - q = 2$$

$$21 - 4q = 2$$

In each system below, solve for "v" in the top equation and then substitute
that solution for "v" in the bottom equation.

a) $$\boxed{\begin{array}{rcl} t + v & = & 9 \\ 2t - v & = & 7 \end{array}}$$

b) $$\boxed{\begin{array}{rcl} v - t & = & 5 \\ 2t - 3v & = & 1 \end{array}}$$

---

a) v = 9 - t

$$2t - (9 - t) = 7$$
$$2t - 9 + t = 7$$
$$3t - 9 = 7$$

b) v = t + 5

$$2t - 3(t + 5) = 1$$
$$2t - (3t + 15) = 1$$
$$2t - 3t - 15 = 1$$
$$-t - 15 = 1$$

38. Neither variable is solved for in the equations in the system at the right.

$$2x - 3y = 0$$
$$3x + y = 4$$

Solving for "x" in the top equation, we get:

$$2x = 3y$$
$$x = \frac{3y}{2}$$

Substituting that solution for "x" in the bottom equation, we get:

$$3\left(\frac{3y}{2}\right) + y = 4$$
$$\frac{9y}{2} + y = 4$$

In each system below, solve for "m" in the top equation and then substitute that solution for "m" in the bottom equation.

a)
$$3m = 4d$$
$$5d + 2m = 8$$

b)
$$5m - 4d = 0$$
$$3d - 2m = 5$$

---

39. After using the substitution process to eliminate a variable, the steps in the substitution method are identical to the steps in the addition and equivalence methods.

Let's use the substitution method to solve this system:

$$3x - 4y = 0$$
$$5y - 2x = 7$$

1. Solve for "x" in the top equation.

$$3x = 4y$$
$$x = \frac{4y}{3}$$

2. Substitute that solution for "x" in the bottom equation.

$$5y - 2\left(\frac{4y}{3}\right) = 7$$
$$5y - \frac{8y}{3} = 7$$

3. Clear the fraction and solve for "y".

$$3\left(5y - \frac{8y}{3}\right) = 3(7)$$
$$3(5y) - 3\left(\frac{8y}{3}\right) = 21$$
$$15y - 8y = 21$$
$$7y = 21$$
$$y = 3$$

4. Substitute 3 for "y" in the top equation to find the corresponding value of "x".

$$3x - 4y = 0$$
$$3x - 4(3) = 0$$
$$3x = 12$$
$$x = 4$$

The solution of the system is: x = _____, y = _____

a) $m = \frac{4d}{3}$

$$5d + 2\left(\frac{4d}{3}\right) = 8$$
$$5d + \frac{8d}{3} = 8$$

b) $m = \frac{4d}{5}$

$$3d - 2\left(\frac{4d}{5}\right) = 5$$
$$3d - \frac{8d}{5} = 5$$

40. Let's solve this system using the substitution method.

$$\boxed{\begin{array}{l} a + 4b = 6 \\ 5a - 3b = 7 \end{array}}$$

a) Solve for "a" in the top equation.

b) Substitute that solution for "a" in the bottom equation.

c) Find the value of "b".

d) Find the corresponding value of "a" by substituting in one of the original equations.

e) The solution is: a = _____ , b = _____

x = 4, y = 3

---

41. When using the substitution method, we sometimes encounter a complex fraction. An example is discussed.

Let's use the substitution method to solve this system.

$$\boxed{\begin{array}{l} 2x - 3y = 9 \\ 4x + 3y = 9 \end{array}}$$

1. We solved for "x" in the top equation.

$$2x = 3y + 9$$
$$x = \frac{3y + 9}{2}$$

2. We substituted that solution for "x" in the bottom equation.

$$\cancel{4}\left(\frac{3y + 9}{\cancel{2}}\right) + 3y = 9$$
$$6y + 18 + 3y = 9$$
$$9y + 18 = 9$$

Complete the solution. It is: x = _____ , y = _____

a) a = 6 − 4b

b) 5(6 − 4b) − 3b = 7
$$30 - 20b - 3b = 7$$
$$30 - 23b = 7$$

c) b = 1
   (from −23b = −23)

d) a = 2

e) a = 2, b = 1

x = 3, y = −1

10-6   A STRATEGY FOR USING THE VARIOUS METHODS

We have discussed the graphing method and three algebraic methods for solving systems of equations. Ordinarily, one of the algebraic methods is used. In this section, we will discuss a strategy for deciding whether to use the addition, equivalence, or substitution method.

42. The addition method is a good method to use when the coefficients are small and the coefficients of one variable are opposites or can easily be made into opposites. For example, it is a good method to use with the systems below.

$$7x + 5y = 9 \qquad 4a + 3b = 7 \qquad 3m - 2t = 5$$
$$3x - 5y = 10 \qquad 5a - b = 1 \qquad 4m - 3t = 8$$

The addition method is not a good method to use when the coefficients are large and it is difficult to make the coefficients of one variable into opposites. For example, it is <u>not</u> a good method to use with the systems below.

$$14x + 17y = 25 \qquad 18R - 27S = 50$$
$$12x - 11y = 19 \qquad 16R - 24S = 75$$

The addition method would be a good method to use with which systems below? _____

$$\text{a)} \quad 12m - 13v = 19 \qquad \text{b)}\ 8T - 9V = 11 \qquad \text{c)}\ 25x + 50y = 100$$
$$10m + 17v = 21 \qquad\quad 7T + 9V = 19 \qquad\quad 35x + 75y = 150$$

43. When a system is not in standard form, we must put it in standard form first before deciding whether to use the addition method. For example, the addition method is not a good method for the system at the left below. Put the other system in standard form and decide whether the addition method is a good method to use.

Only (b)

Original System

10x = 13y + 11
9x - 15 = 14y

Standard System

10x - 13y = 11
9x - 14y = 15

Original System

9p = 7q + 10
7q - 24 = 8p

Standard System

44. The substitution method is easy to use with some systems. For example, it is easy to use with the systems below.

$$y = 2x \qquad p = 2q - 7 \qquad a + b = 8$$
$$x + y = 9 \qquad p + q = 5 \qquad 2a - 3b = 5$$

However, the substitution method is more difficult to use when the substitution leads to a complex fraction. For example, no matter what variable we solve for in either equation below, we get a complex fraction to substitute.

$$7x + 13y = 15$$
$$9x - 6y = 11$$

9p - 7q = 10
-8p + 7q = 24

The addition method is a good one to use.

Continued on following page.

44. Continued

For which systems below would it be more difficult to use the substitution method? _____

a) $m = t - 3$
$\quad m + t = 9$

b) $R = 5S$
$\quad 2R - S = 10$

c) $11a - 12b = 25$
$\quad 10a - 11b = 50$

---

45. Though it requires solving for the same variable in both equations, the equivalence method is a good method to use with any system. In fact, for a system like the one below which would be difficult to solve by either the addition method or the substitution method, the equivalence method is the best to use.

$$11x - 13y = 15$$
$$17x + 19y = 21$$

For which systems below would it be best to use the equivalence method? _____

a) $12d = 17 - 14t$
$\quad 19t - 13 = 15d$

b) $y = 4x$
$\quad x - y = 6$

c) $25m - 16p = 50$
$\quad 35m + 13p = 99$

Only (c)

Both (a) and (c)

10-7   SYSTEMS CONTAINING MORE-COMPLICATED EQUATIONS

The typical equation in a system is a non-fractional equation containing two letter terms and a number term. However, some systems contain more-complicated equations. We will discuss systems of that type in this section.

---

46. The system at the right contains two fractional equations. To solve a system of that type, we begin by clearing the fractions to get two non-fractional equations. To do so, we multiply both sides of the top equation by 15 and both sides of the bottom equation by 4. We get:

$$\frac{x}{5} + \frac{y}{3} = 1$$
$$\frac{x}{4} - \frac{y}{2} = 3$$

$$15\left(\frac{x}{5} + \frac{y}{3}\right) = 15(1) \qquad 4\left(\frac{x}{4} - \frac{y}{2}\right) = 4(3)$$

$$\overset{3}{\cancel{15}}\left(\frac{x}{\cancel{5}}\right) + \overset{5}{\cancel{15}}\left(\frac{y}{\cancel{3}}\right) = 15 \qquad \cancel{4}\left(\frac{x}{\cancel{4}}\right) - \overset{2}{\cancel{4}}\left(\frac{y}{\cancel{2}}\right) = 12$$

$$3x + 5y = 15 \qquad\qquad x - 2y = 12$$

Using the two non-fractional equations, we get the simpler system below.

$$3x + 5y = 15$$
$$x - 2y = 12$$

Since the simpler system is equivalent to the original system, we can solve it to find the solution of the original system.

47.   Whenever a system contains more-complicated equations, we begin by
        simplifying the equations to get a simpler but equivalent system. We
        will discuss examples of that simplification process in the following frames.

        Clear the fractions in the equations
        in the system at the right. Write
        the simpler system in the box provided.

$$\frac{a}{3} - \frac{b}{2} = 1$$

$$\frac{a}{2} + 1 = \frac{b}{4}$$

---

48.   Each equation in the system at the
        right contains an instance of the
        distributive principle.

$$5x - 3(x - y) = 12$$
$$7y - 5(2x + y) = 13$$

$2a - 3b = 6$
$2a + 4 = b$

        To simplify the top equation, we multiplied by the distributive principle
        and then combined like terms. Simplify the bottom equation.

$$5x - 3(x - y) = 12 \qquad\qquad 7y - 5(2x + y) = 13$$

$$5x - (3x - 3y) = 12$$

$$5x - 3x + 3y = 12$$

$$2x + 3y = 12$$

---

49.   Simplify each equation in the system
        at the right. Write the simpler system
        in the box provided.

$$V + \frac{3R}{2} = 5$$

$$2V + 5(R - V) = 19$$

$2y - 10x = 13$
or
$-10x + 2y = 13$

---

50.   Simplify each equation in the
        system at the right. Write the
        simpler system in the box provided.

$$7(a - b) - 3a = 0$$
$$8a - 3(a - 2b) = 10$$

$2V + 3R = 10$
$-3V + 5R = 19$

$4a - 7b = 0$
$5a + 6b = 10$

51. Simplify each equation in the system at the right. Write the simpler system in the box provided.

$$T = 6.4P + 8.7$$
$$T = 4(1 - P)$$

   Note: Eliminate the decimals in the top equation.

   [box]

---

52. Let's solve the system at the right.

$$3m - (m + t) = 0$$
$$5m - 2(m - t) = 7$$

   a) Write the simpler system in the box provided.

   [box]

   b) Use any method to complete the solution.

$$10T = 64P + 87$$
$$T = 4 - 4P$$

a) $2m - t = 0$
    $3m + 2t = 7$

b) $m = 1$, $t = 2$

10-8   APPLIED PROBLEMS

The purpose of this section is to show various applied problems based on scientific principles that require solving a system of equations. Since setting up the original equations depends on knowing the scientific principles involved, we will set up the original equations for each problem. Don't worry if you cannot understand how we got the equations.

---

53. A steel beam 8 meters long weighs 800 kilograms and is supported at each end. A load of 1,200 kilograms is applied to the beam 2 meters from its left end. To find the forces at the ends of the beam (called F_1 and F_2), we can use equilibrium principles to set up the following system of equations.

$$F_1 + F_2 = 2,000$$
$$4F_2 + 2,400 = 4F_1$$

   Use any method to solve for F_1 and F_2.

54.   A voltage of 32 volts is applied to a circuit consisting of two parallel resistors of 10 ohms and 40 ohms connected in series with a 30-ohm resistor. We want to find the currents in the 10-ohm and 40-ohm resistors. If "i_1" is the current in the 10-ohm resistor and "i_2" is the current in the 40-ohm resistor, we can use basic circuit principles to set up the following system of equations.

$$30(i_1 + i_2) + 10i_2 = 32$$
$$10i_1 - 40i_2 = 0$$

Using any method, solve for "i_1" and "i_2".

$F_1 = 1,300$ kilograms
$F_2 = 700$ kilograms

---

55.   The formula $\boxed{V = a(t_2 - t_1)}$ is used in the study of "motion". It contains four variables. If we know the value of the same two variables in two situations, we can use a system of equations to find the value of the other two variables. For example, if $a = 10$ and $t_2 = 40$ is one situation and $a = 20$ and $t_2 = 30$ is a second situation, we can set up the following system.

$$V = 10(40 - t_1)$$
$$V = 20(30 - t_1)$$

Using any method, solve for V and t_1.

$i_1 = 0.8$ ampere

$i_2 = 0.2$ ampere

$t_1 = 20$

$V = 200$

56. To determine the tensile forces (F_1 and F_2) in two steel cables holding a 3,400 kilogram weight, we can use equilibrium principles to set up the following system of equations.

$$\boxed{\begin{array}{l} 0.8\,F_1 - 0.5\,F_2 = 0 \\ 0.4\,F_1 + 0.6\,F_2 = 3{,}400 \end{array}}$$

Using any method, solve for F_1 and F_2.

$F_1 = 2{,}500$ kilograms

$F_2 = 4{,}000$ kilograms

57. To find two currents (I_1 and I_2) in a complicated circuit, we can use basic circuit principles to set up the following system.

$$\boxed{\begin{array}{l} 10 - 12I_2 - 16I_2 + 8(I_1 - I_2) = 0 \\ 20 - 8(I_1 - I_2) - 24I_1 - 12I_1 = 0 \end{array}}$$

a) Simplify each equation in the system and write the simplified equations below.

b) Use any method to solve for I_1 and I_2. Report the answers as decimal numbers <u>rounded to thousandths</u>.

a) $\begin{array}{l} 8I_1 - 36I_2 = -10 \\ -44I_1 + 8I_2 = -20 \end{array}$ or $\begin{array}{l} 4I_1 - 18I_2 = -5 \\ -11I_1 + 2I_2 = -5 \end{array}$

b) $I_1 = 0.526$ ampere (from $\frac{10}{19}$)

$I_2 = 0.395$ ampere (from $\frac{15}{38}$)

436   Systems of Two Equations and Formulas

SELF-TEST 34 (pages 425-436)

Solve each system by the substitution method.

1.
$$8t - 3p = 1$$
$$p = 2 - 2t$$

2.
$$3x - y = 13$$
$$x + 3y = 1$$

Solve each system by any method.

3.
$$\frac{r}{2} + \frac{s}{3} = 6$$
$$\frac{r}{4} - 1 = \frac{s}{6}$$

4.
$$2G - 3(2H - G) = 80$$
$$G - \frac{2H}{5} = 12$$

5. The following system was set up to calculate the pitch diameters of two gears. Find d_1 and d_2.

$$d_1 + d_2 = 8.3$$
$$d_2 - d_1 = 3.9$$

ANSWERS: 1. $p = 1$, $t = \frac{1}{2}$ 3. $r = 8$, $s = 6$ 5. $d_1 = 2.2$, $d_2 = 6.1$

2. $x = 4$, $y = -1$ 4. $G = 10$, $H = -5$

10-9   SYSTEMS OF FORMULAS AND FORMULA DERIVATION

By eliminating a variable or variables from a system of formulas, we can derive a new formula. This process is called <u>formula derivation</u>. In this section, we will define a system of formulas, show an example of formula derivation, and show how formula derivation is used in evaluations.

---

58. Two formulas form a system only if they contain one or more common variables. For example:

$$\boxed{\begin{array}{l} Pt = W \\ W\ \ = Fs \end{array}}$$

form a system. The common variable is W.

$$\boxed{\begin{array}{l} E = IR \\ P = EI \end{array}}$$

form a system. The common variables are _____ and _____.

---

59. State whether each pair of formulas below forms a system or not.

a) $\boxed{\begin{array}{l} H = \dfrac{M}{t} \\ at = V \end{array}}$ ____

b) $\boxed{\begin{array}{l} S = 0.26DN \\ v = gt \end{array}}$ ____

c) $\boxed{\begin{array}{l} F = ma \\ F = \dfrac{GMm}{r^2} \end{array}}$ ____

> E and I

---

60. By eliminating a variable or variables from a system of formulas, we can derive a new formula. An example of formula derivation is discussed below.

In the system of formulas at the right, W is solved for in each formula. Therefore, we can use the equivalence principle to eliminate W and obtain the new formula below.

$$\boxed{\begin{array}{l} W = Pt \\ W = Fs \end{array}}$$

$$Pt = Fs$$

In the new formula, no variable is solved for. However, a formula is ordinarily written with one of the variables solved for. Therefore, after deriving a new formula, we solve for one of the variables. Any of the variables can be solved for.

In Pt = Fs: a) Solve for "t".   b) Solve for F.

> a) Yes. The common variable is "t".
>
> b) No. There is no common variable.
>
> c) Yes. The common variables are F and "m".

> a) $t = \dfrac{Fs}{P}$
>
> b) $F = \dfrac{Pt}{s}$

61.   Formula derivations can be used in evaluations.   An example is discussed below.

Suppose we are given the system of formulas at the right and we are asked to find the value of P when I = 5 and R = 10.

$$E = IR$$
$$P = EI$$

One way to find P is to substitute for I and R in the system and then solve for P. We get:

$$E = 5(10)$$
$$P = E(5)$$
    or
$$E = 50$$
$$P = 5E$$

Since E = 50, we can substitute for E in the bottom equation and find P. That is:

$$P = 5E = 5(50) = 250$$

A second way to find P is to derive a new formula in which P is solved for in terms of I and R. By eliminating E from the system, we can get the needed formula below.

$$P = I^2R$$

If we substitute 5 for I and 10 for R, do we also get 250 for P using this new formula? _____

---

62.   Suppose we are given the system of formulas at the right and we want to find "a" when G = 8, M = 3, and r = 2.

$$F = ma$$
$$F = \frac{GMm}{r^2}$$

Answer column:
Yes, since:
$$P = (5^2)(10)$$
$$= (25)(10) = 250$$

One way to find P is to substitute for G, M, and "r" in the system and then solve for "a". We get:

$$F = ma$$
$$F = \frac{(8)(3)m}{2^2}$$
    or
$$F = ma$$
$$F = 6m$$

Since F is solved for in each equation, we can use the equivalence principle to eliminate F. Notice how "m" is also eliminated when we reduce to lowest terms.

$$ma = 6m$$
$$a = \frac{6m}{m} = 6$$

A second way to find "a" is to derive a new formula in which "a" is solved for in terms of G, M, and "r". By eliminating F and "m" from the system, we can get the needed formula below.

$$a = \frac{GM}{r^2}$$

If we substitute 8 for G, 3 for M, and 2 for "r", do we also get 6 for "a" using this new formula? _____

Note: When we are given a system of formulas and have
to perform an evaluation, we usually derive a new
formula when needed rather than substitute directly
into the system.

Yes, since:

$$a = \frac{8(3)}{2^2} = \frac{24}{4} = 6$$

10-10   THE EQUIVALENCE METHOD AND FORMULA DERIVATION

In this section, we will show how the equivalence method can be used to perform formula derivations.

---

63.   The two major steps in a formula derivation are:

    1.   Eliminating a common variable or variables from the system.
    2.   Solving for one of the variables in the new formula.

The common variable in the system at
the right is P. Let's eliminate P and then
solve for M in the new formula.

$$T = 0.15DP$$
$$M = \frac{AL}{P}$$

    1.   Solve for P in each formula.

$$P = \frac{T}{0.15D}$$
$$P = \frac{AL}{M}$$

    2.   Use the equivalence principle to
eliminate P.

$$\frac{T}{0.15D} = \frac{AL}{M}$$

    3.   Now solve for M in the new formula.

---

64.   The common variable in the system at
the right is B. Let's eliminate B and then
solve for C.

$$H = \frac{B}{KT}$$
$$A = BC$$

    a)   Solve for B in each formula.

$$B = $$

$$B = $$

    b)   Use the equivalence principle to
eliminate B.

_____ = _____

    c)   Now solve for C.

$$M = \frac{0.15ADL}{T}$$

65. There are two common variables in the system at the right. At the left below, we eliminated D and then solved for W. Using the same steps, eliminate H and then solve for T.

$$D = HT \quad \text{and} \quad D = \frac{W}{H}$$

$$HT = \frac{W}{H}$$

$$W = H^2T$$

$$\boxed{\begin{array}{l} D = HT \\ W = DH \end{array}}$$

a) $B = HKT$

   $B = \dfrac{A}{C}$

b) $HKT = \dfrac{A}{C}$

c) $C = \dfrac{A}{HKT}$

---

66. The common variable in the system at the right is "a". Let's eliminate "a" and then solve for V.

   a) Solve for "a" in each formula.

   b) Use the equivalence principle.

   c) Solve for V.

$$\boxed{\begin{array}{l} F = ma \\ a = \dfrac{V^2}{r} \end{array}}$$

$$\boxed{\begin{array}{l} a = \\ \\ a = \end{array}}$$

_____ = _____

$H = \dfrac{D}{T}$ and $H = \dfrac{W}{D}$

$$\dfrac{D}{T} = \dfrac{W}{D}$$

$$D^2 = TW$$

$$T = \dfrac{D^2}{W}$$

---

67. At the left below, we eliminated "t" and then solved for "r". Using the same steps, eliminate "m" from the other system and then solve for "v".

$$\boxed{\begin{array}{l} E = dt \\ t = \dfrac{2r^2}{a} \end{array}}$$

$$t = \frac{E}{d} \quad \text{and} \quad t = \frac{2r^2}{a}$$

$$\frac{E}{d} = \frac{2r^2}{a}$$

$$aE = 2dr^2$$

$$r^2 = \frac{aE}{2d}$$

$$r = \sqrt{\frac{aE}{2d}}$$

$$\boxed{\begin{array}{l} F = \dfrac{mv^2}{r} \\ m = \dfrac{W}{g} \end{array}}$$

a) $a = \dfrac{F}{m}$

   $a = \dfrac{V^2}{r}$

b) $\dfrac{F}{m} = \dfrac{V^2}{r}$

c) $V = \sqrt{\dfrac{Fr}{m}}$, from:

   $V^2 = \dfrac{Fr}{m}$

68. At the left below, we eliminated E and then solved for I. Using the same steps, eliminate I from the same system and then solve for E.

$$\boxed{\begin{array}{l} E = IR \\ P = EI \end{array}}$$

$E = IR$ and $E = \dfrac{P}{I}$

$$IR = \dfrac{P}{I}$$

$$I^2R = P$$

$$I^2 = \dfrac{P}{R}$$

$$I = \sqrt{\dfrac{P}{R}}$$

$$\boxed{\begin{array}{l} E = IR \\ P = EI \end{array}}$$

$m = \dfrac{Fr}{v^2}$ and $m = \dfrac{W}{g}$

$$\dfrac{Fr}{v^2} = \dfrac{W}{g}$$

$$v = \sqrt{\dfrac{Fgr}{W}}$$

69. Sometimes when one common variable is eliminated, a second common variable is also eliminated. This is true for the system at the right in which both H and "t" are common variables.

We can use the equivalence principle to eliminate H. We get:

If we solve for "b" in the new formula, "t" is also eliminated because we can cancel to reduce to lowest terms. That is:

$$\boxed{\begin{array}{l} H = bt \\ H = \dfrac{GTt}{p^2} \end{array}}$$

$$bt = \dfrac{GTt}{p^2}$$

$$b = \dfrac{GT\cancel{t}}{p^2\cancel{t}} = \underline{\hspace{1cm}}$$

$I = \dfrac{E}{R}$ and $I = \dfrac{P}{E}$

$$\dfrac{E}{R} = \dfrac{P}{E}$$

$$E^2 = PR$$

$$E = \sqrt{PR}$$

70. Both S and T are common variables in the system at the right. If we eliminate S and solve for "d", we will also eliminate T.

a) Solve for S in each formula.

b) Use the equivalence principle.

c) Now solve for "d" and reduce the fraction to lowest terms.

$$\boxed{\begin{array}{l} Cd = \dfrac{pT}{S} \\ T = SV \end{array}}$$

$$\boxed{\begin{array}{l} S = \\[6pt] S = \end{array}}$$

$$\underline{\hspace{1cm}} = \underline{\hspace{1cm}}$$

$$b = \dfrac{GT}{p^2}$$

71. We eliminated F from the system below and then solved for "v". Notice that "m" was also eliminated. Eliminate "s" and then solve for V in the other system. Notice how "a" is also eliminated.

$$\boxed{\begin{array}{l} F = mg \\ Fr = mv^2 \end{array}}$$

$$F = mg \quad \text{and} \quad F = \frac{mv^2}{r}$$

$$mg = \frac{mv^2}{r}$$

$$mgr = mv^2$$

$$v^2 = \frac{\cancel{m}gr}{\cancel{m}}$$

$$v^2 = gr$$

$$v = \sqrt{gr}$$

$$\boxed{\begin{array}{l} K = mas \\ s = \dfrac{V^2}{2a} \end{array}}$$

a) $S = \dfrac{pT}{Cd}$

$S = \dfrac{T}{V}$

b) $\dfrac{pT}{Cd} = \dfrac{T}{V}$

c) $d = \dfrac{pV}{C}$, from:

$$pTV = CdT$$

$$d = \frac{p\cancel{T}V}{C\cancel{T}}$$

---

72. Sometimes we can eliminate two variables at one time. For example, we can eliminate both V and "p" at the same time from the system at the right. To do so, we solve for $\dfrac{V}{p}$ in both formulas.

Solving for $\dfrac{V}{p}$ in each, we get:

Using the equivalence principle, we get:

Solve for "t" in the new formula.

$$\boxed{\begin{array}{l} H = \dfrac{V}{p} \\ V = \dfrac{ABp}{t^2} \end{array}}$$

$$\boxed{\begin{array}{l} \dfrac{V}{p} = H \\ \dfrac{V}{p} = \dfrac{AB}{t^2} \end{array}}$$

$$H = \frac{AB}{t^2}$$

$V = \sqrt{\dfrac{2K}{m}}$, from:

$$\frac{K}{ma} = \frac{V^2}{2a}$$

$$V^2 = \frac{2\cancel{a}K}{m\cancel{a}}$$

---

73. Here is another system in which we can eliminate two variables at one time. To do so, we solve for "as" in both formulas.

a) Solve for "as" in both formulas.

b) Now use the equivalence principle and then solve for W.

$$\boxed{\begin{array}{l} W = mas \\ v_f^2 = v_0^2 + 2as \end{array}}$$

$$\boxed{\begin{array}{l} as = \\ \\ as = \end{array}}$$

$t = \sqrt{\dfrac{AB}{H}}$

74.   We eliminated E from the system below and then solved for G.   Notice that we had to factor by the distributive principle because E appears twice in the top formula.   Eliminate "b" from the other system and then solve for "t".

$$\boxed{\begin{array}{l} E = A(N - BE) \\ E = GN \end{array}}$$

$$\boxed{\begin{array}{l} b = ct \\ b = d(b + c) \end{array}}$$

$E = A(N - BE)$

$E = AN - ABE$

$E + ABE = AN$

$E(1 + AB) = AN$

$$E = \frac{AN}{1 + AB}$$

$$GN = \frac{AN}{1 + AB}$$

$$G = \frac{A\cancel{N}}{(1 + AB)\cancel{N}}$$

$$G = \frac{A}{1 + AB}$$

a) $as = \dfrac{W}{m}$

   $as = \dfrac{v_f^2 - v_o^2}{2}$

b) $W = \dfrac{m(v_f^2 - v_o^2)}{2}$

   or

   $W = \dfrac{mv_f^2 - mv_o^2}{2}$

   or

   $W = \dfrac{1}{2}mv_f^2 - \dfrac{1}{2}mv_o^2$

$$t = \frac{d}{1 - d}$$

10-11   THE SUBSTITUTION METHOD AND FORMULA DERIVATION

In this section, we will show how the substitution method can be used to perform formula derivations.

75.   When the variable we want to eliminate is already solved for in one formula, it is easy to use the substitution method.   An example is discussed.

A is already solved for in the top formula in the system at the right.

$$\boxed{\begin{array}{l} A = DT \\ M = AD \end{array}}$$

To eliminate A, we can substitute DT for A in the bottom formula.

$M = AD$

$M = (DT)D$

$M = D^2T$

Solve for T in the new formula.

$$T = \frac{M}{D^2}$$

76.   Since E is solved for in the top formula at the right, we can eliminate E by substituting IR for E in the bottom formula. The steps are shown.

$$E = IR$$
$$P = EI$$

$$P = EI$$
$$P = (IR)I$$
$$P = I^2R$$

Solve for I in the new formula.

---

77.   We eliminated X_L by the substitution method and then solved for L below. Use the same method to eliminate "b" from the other system and then solve for "t".

$$X = X_L - X_c$$
$$X_L = 2\pi fL$$

$$s = b + t$$
$$b = 2s + v$$

$$X = X_L - X_c$$
$$X = 2\pi fL - X_c$$
$$X + X_c = 2\pi fL$$
$$L = \frac{X + X_c}{2\pi f}$$

$$I = \sqrt{\frac{P}{R}}$$

---

78.   We used the substitution method to eliminate "a" below and then solved for V. Use the same method to eliminate "t" from the other formula and then solve for "r".

$$F = ma$$
$$a = \frac{V^2}{r}$$

$$E = dt$$
$$t = \frac{2r^2}{a}$$

$$F = ma$$
$$F = m\left(\frac{V^2}{r}\right)$$
$$F = \frac{mV^2}{r}$$
$$Fr = mV^2$$
$$V^2 = \frac{Fr}{m}$$
$$V = \sqrt{\frac{Fr}{m}}$$

$$t = -s - v, \text{ or}$$
$$t = -(s + v), \text{ from:}$$
$$s = (2s + v) + t$$

79. In the system below, we substituted to eliminate H and then solved for "t".
     Notice how we reduced the solution to lowest terms. Use substitution to
     eliminate "s" from the other system and then solve for V. Be sure to
     reduce to lowest terms.

$$r = \sqrt{\dfrac{aE}{2d}} \text{ , from:}$$

$$E = \dfrac{2dr^2}{a}$$

$$\boxed{\begin{array}{l} H = cd \\ Hp = ct^2 \end{array}}$$

$$\boxed{\begin{array}{l} K = mas \\ s = \dfrac{V^2}{2a} \end{array}}$$

$$Hp = ct^2$$

$$(cd)p = ct^2$$

$$t^2 = \dfrac{\cancel{c}dp}{\cancel{c}}$$

$$t^2 = dp$$

$$t = \sqrt{dp}$$

80. In the system at the right, we can
     eliminate both G and H at the same
     time since GH is solved for in the
     bottom formula.

$$\boxed{\begin{array}{l} K = GHv \\ R = GH \end{array}}$$

$$V = \sqrt{\dfrac{2K}{m}} \text{ , from:}$$

$$K = ma\left(\dfrac{V^2}{2a}\right)$$

$$K = \dfrac{m\cancel{a}V^2}{2\cancel{a}}$$

$$K = \dfrac{mV^2}{2}$$

   a) Substitute R for GH in the top
       formula.

   b) Solve for "v" in the new formula.

81. If the variable we want to eliminate is
     not already solved for in one of the
     formulas, we must solve for it first.
     For example, to eliminate D from the
     system at the right:

$$\boxed{\begin{array}{l} B = DT \\ S = BD \end{array}}$$

   a) K = GHv
       K = (R)v
       K = Rv

   b) $v = \dfrac{K}{R}$

   1. Solve for D in the top formula.

$$D = \dfrac{B}{T}$$

   2. Substitute that value for D in the
       bottom formula.

$$S = BD$$

$$S = B\left(\dfrac{B}{T}\right)$$

$$S = \dfrac{B^2}{T}$$

     Solve for T in the new formula.

82. Let's eliminate P by the substitution method and then solve for X.

$$Kd = \frac{aT}{P}$$
$$T = PX$$

$$T = \frac{B^2}{S}$$

   a) Solve for P in the top formula.

   b) Substitute that value for P in the bottom formula.

   c) Now solve for X in the new formula. Be sure to reduce to lowest terms.

---

83. Let's use substitution to eliminate "m" from this system and then solve for "v".

$$F = mg$$
$$Fr = mv^2$$

   a) Solve for "m" in the top formula.

a) $P = \dfrac{aT}{Kd}$

b) $T = \left(\dfrac{aT}{Kd}\right)X$

    $T = \dfrac{aTX}{Kd}$

   b) Substitute that value for "m" in the bottom formula.

c) $X = \dfrac{Kd\cancel{T}}{a\cancel{T}}$

    $X = \dfrac{Kd}{a}$

   c) Now solve for "v". Be sure to reduce to lowest terms.

84.   We used the substitution method to eliminate E below and then solved for
R. Use the same method to eliminate D and then solve for A in the
other formula. (Note: We began the solution. Complete it.)

$$i = \dfrac{E}{R + r}$$
$$E = IR$$

$$i = \dfrac{E}{R + r}$$

$$i = \dfrac{IR}{R + r}$$

$$i(R + r) = IR$$

$$iR + ir = IR$$

$$ir = IR - iR$$

$$ir = R(I - i)$$

$$R = \dfrac{ir}{I - i}$$

$$h = \dfrac{A}{D + t}$$
$$A = BD$$

$$D = \dfrac{A}{B}$$

$$h = \dfrac{A}{D + t}$$

$$h = \dfrac{A}{\dfrac{A}{B} + t}$$

$$h = \dfrac{A}{\dfrac{A + Bt}{B}}$$

$$h = \dfrac{AB}{A + Bt}$$

a) $m = \dfrac{F}{g}$

b) $Fr = \left(\dfrac{F}{g}\right)v^2$

   $Fr = \dfrac{Fv^2}{g}$

c) $v = \sqrt{gr}$, from:

   $v^2 = \dfrac{\cancel{F}gr}{\cancel{F}}$

---

85.   When using the substitution method, we can get a complex fraction that
is difficult to handle. For example, if we use substitution to eliminate
"w" from the system below, we get:

$$r = \dfrac{bw}{s}$$
$$w = \dfrac{r}{2s}$$

$$r = \dfrac{bw}{s}$$

$$r = \dfrac{b\left(\dfrac{r}{2s}\right)}{s}$$

$$r = \dfrac{\dfrac{br}{2s}}{s}$$

$$r = \dfrac{br}{2s(s)}$$

$$r = \dfrac{br}{2s^2}$$

$A = \dfrac{Bht}{B - h}$

Solve for "s" in the new formula.

86.   Though both the equivalence method and the substitution method can be
      used for formula derivations, the substitution method sometimes leads
      to a complex fraction. In such cases, it is easier to use the equiva-
      lence method. Use either method for the derivations below.

$$s = \sqrt{\frac{b}{2}}$$

Did you reduce to
lowest terms?

a) Eliminate PV and solve for E.   b) Eliminate H and solve for R.

$$H = E + PV$$

$$\frac{PV}{C(T_1 - T_2)} = R$$

$$\frac{HL}{t_2 - t_1} = AKT$$

$$\frac{H}{0.24} = IR$$

a) $E = H - CR(T_1 - T_2)$

b) $R = \dfrac{AKT(t_2 - t_1)}{0.24IL}$

<u>SELF-TEST</u> <u>35</u> (pages <u>437</u>–<u>449</u>)

1. Eliminate "d" and solve for "t".

$$d = ht$$
$$w = dh$$

2. Eliminate A and solve for "r".

$$S = \frac{F}{A}$$
$$A = \pi r^2$$

3. Eliminate B and solve for T.

$$K = \frac{B}{NT}$$
$$N = \frac{BP}{K}$$

4. Eliminate "v" and solve for "s".

$$p = v - ks$$
$$s = a + v$$

ANSWERS: 1. $t = \dfrac{w}{h^2}$ 2. $r = \sqrt{\dfrac{F}{\pi s}}$ 3. $T = \dfrac{1}{P}$ 4. $s = \dfrac{a + p}{1 - k}$

SUPPLEMENTARY PROBLEMS – CHAPTER 10

Assignment 33

Solve each system by the equivalence method.

1. $y = x + 6$
   $y = 4x$

2. $s = 20t$
   $s = 16t + 8$

3. $d = 2k - 7$
   $d = 3k - 10$

4. $x = 3 - y$
   $x = y + 9$

5. $a + b = 5$
   $a = 3b + 1$

6. $w = 4r - 2$
   $w - 3r = 1$

7. $p = 2h - 12$
   $h = p + 7$

8. $2E - 3 = R$
   $R + 5 = E$

9. $r = v - 2$
   $3r = 2v + 1$

10. $5y = 3x + 4$
    $y = 2x - 2$

11. $7 - 4G = 3P$
    $1 - G = P$

12. $3t + 2w = 8$
    $t - 11 = w$

13. $5P + 3Q = 1$
    $Q + 2P = 1$

14. $4r + s = 3$
    $3r + 2s = 11$

15. $5x - y = 13$
    $2y - x = 1$

16. $2B - 5A = 8$
    $3A - B = -5$

17. $4d_1 - 3d_2 = 13$
    $3d_1 + 2d_2 = 14$

18. $2p + 3t = 4$
    $3p - 5t = 25$

19. $4V_1 - 3V_2 = 0$
    $5V_2 - 2V_1 = 7$

20. $4h = 7 - 3k$
    $2k = 6h + 9$

Assignment 34

Solve each system by the substitution method.

1. $d = 3t$
   $2d + t = 14$

2. $E = 4R$
   $3E - 8R = 12$

3. $5x - y = 8$
   $y = 2x - 5$

4. $7p - 3v = 8$
   $v = 2p - 3$

5. $2b = 5m$
   $3m - b = 2$

6. $4y - 5x = 2$
   $4x = 3y$

7. $3F_1 - 2F_2 = 40$
   $4F_1 - 3F_2 = 50$

8. $7h + 3k = 12$
   $5h - 14 = 6k$

Solve each system by any method.

9. $7x - 3y = 34$
   $5x + 3y = 14$

10. $p + 4r = 5$
    $p + 5r = 7$

11. $s = 15 - 2w$
    $s = 4w$

12. $5a - 4b = 17$
    $2a - 5b = 17$

13. $\dfrac{b}{2} + \dfrac{d}{3} = 4$
    $\dfrac{3b}{4} + \dfrac{d}{6} = 4$

14. $\dfrac{5m}{3} - \dfrac{s}{2} = 21$
    $\dfrac{m}{4} + \dfrac{3s}{2} = 0$

15. $2x + \dfrac{y}{5} = 20$
    $5x - y = 5$

16. $\dfrac{a}{4} - 1 = \dfrac{b}{2}$
    $a - 4b = 3$

17. $h + 2(h + k) = 0$
    $3(h - k) - k = 18$

18. $5(2P + T) - 4P = 14$
    $3T - (P - T) = 17$

19. $0.1b + 0.2d = 3$
    $0.5b - 0.4d = 8$

20. $2.4r + 3.2 = w$
    $3(w - 4) = 4r$

Solve these applied problems.

21. To find the slope "m" and the y-intercept "b" of a straight line through (2, 5) and (6, 3), the system below was set up. Solve the system.

    $5 = 2m + b$
    $3 = 6m + b$

22. To solve a calculus problem involving integration by partial fractions, the system below was set up. Solve the system.

    $2A + 3B = 4$
    $A - 2B = 9$

23. To find currents i_1 and i_2 in an electric circuit, the system below was set up. Solve the system.

    $20(i_1 - i_2) + 4i_1 = 12$
    $8i_1 - 10i_2 = 0$

24. To find tensile forces T_1 and T_2 in two steel cables supporting a heavy load, the system below was set up. Solve the system.

    $0.9T_1 = 0.5T_2$
    $0.9T_1 + 0.5T_2 = 1,000$

Assignment 35

1. Eliminate "m" and solve for "f".

$$f = ma$$
$$m = kr$$

2. Eliminate W and solve for H.

$$A = LW$$
$$H = CW$$

3. Eliminate "v" and solve for "t".

$$d = vt$$
$$p = hv$$

4. Eliminate G and solve for R.

$$R = GN$$
$$G = NP$$

5. Eliminate "h" and solve for "s".

$$s = bh$$
$$b = hr$$

6. Eliminate T and solve for F.

$$K = FT$$
$$F = TV$$

7. Eliminate "a" and solve for "k".

$$a = ks$$
$$k = \frac{w}{a}$$

8. Eliminate H and solve for B.

$$D = BH$$
$$P = \frac{A}{H}$$

9. Eliminate "t" and solve for "p".

$$rt = pw$$
$$v = \frac{w}{t}$$

10. Eliminate "s" and solve for "f".

$$d - f = s$$
$$s + h = r$$

11. Eliminate P and solve for L.

$$M = L - P$$
$$P = B - L$$

12. Eliminate G and solve for R.

$$F = G + QR$$
$$R = K - G$$

13. Eliminate "t" and solve for "s".

$$r = \frac{p}{t}$$
$$p = \frac{t}{as}$$

14. Eliminate N and solve for A.

$$P = \frac{FN}{A}$$
$$N = \frac{P}{AR}$$

15. Eliminate "w" and solve for "t".

$$d = \frac{bw}{r}$$
$$w = \frac{d}{rt}$$

16. Eliminate G and solve for V.

$$G = \frac{V^2}{2}$$
$$B = GK$$

17. Eliminate C and solve for F.

$$K = C + 273$$
$$F = \frac{9C}{5} + 32$$

18. Eliminate "r" and solve for "t".

$$k = \frac{rt^2}{h}$$
$$h = rs$$

19. Eliminate "w" and solve for "p".

$$r = w - p$$
$$p = kw$$

20. Eliminate V and solve for "t".

$$V = at$$
$$V = T + t$$

21. Eliminate "k" and solve for "v".

$$v + k = r$$
$$mv = kr$$

ANSWERS FOR SUPPLEMENTARY PROBLEMS

CHAPTER 1 - SIGNED NUMBERS

Assignment 1
1. >   2. <   3. <   4. >   5. $\frac{3}{5}$ 6. 16   7. $\frac{8}{3}$ 8. 0   9. -8   10. 4   11. -2.6
12. -7.9   13. 13   14. -9   15. -1.29   16. -40   17. 0   18. 0   19. 6   20. 7   21. -14
22. -1   23. 0   24. 8   25. -11   26. -7   27. -8   28. -30   29. 7   30. -8   31. -35
32. 1.75   33. -5.5   34. -10.2   35. 8.6

Assignment 2
1. -48   2. -14   3. 35   4. 36   5. 0   6. 0   7. -4.8   8. 24.6   9. 60   10. -56
11. 0   12. 126   13. -6   14. -7   15. 1   16. 5   17. 0   18. -20   19. -12   20. 50
21. -8.6   22. 1   23. -7   24. 0   25. 20   26. -6   27. 53   28. -24   29. 0   30. -1
31. -6   32. 16   33. -21

Assignment 3
1. $-\frac{1}{15}$ 2. $-\frac{3}{4}$ 3. $\frac{10}{3}$ 4. $\frac{15}{8}$ 5. $-\frac{7}{4}$ 6. $-\frac{3}{4}$ 7. $\frac{9}{5}$ 8. 2   9. 0   10. $\frac{1}{4}$ 11. $-\frac{7}{6}$

12. -27   13. $-\frac{12}{27}$ 14. $-\frac{70}{40}$ 15. $\frac{100}{20}$ 16. $-\frac{18}{6}$ 17. $\frac{3}{10}$ 18. $-\frac{4}{3}$ 19. $-\frac{1}{3}$ 20. -5

21. $-\frac{1}{6}$ 22. $\frac{50}{9}$ 23. $-\frac{7}{4}$ 24. 6   25. $-\frac{4}{7}$ 26. 25   27. $\frac{2}{3}$ 28. -1

Assignment 4
1. 1   2. 0   3. $-\frac{7}{2}$ 4. -3   5. $-\frac{13}{20}$ 6. $\frac{8}{3}$ 7. $-\frac{5}{24}$ 8. $-\frac{7}{15}$ 9. $-\frac{7}{4}$ 10. $\frac{8}{3}$ 11. $-\frac{14}{5}$

12. $-\frac{37}{8}$ 13. $\frac{11}{9}$ 14. -1   15. -5   16. 0   17. $-\frac{39}{8}$ 18. $\frac{19}{5}$ 19. $-\frac{1}{30}$ 20. $-\frac{3}{20}$

21. $\frac{18}{7}$ 22. $-\frac{1}{3}$ 23. $-\frac{43}{8}$ 24. $-\frac{7}{16}$ 25. $-\frac{3}{10}$ 26. -2   27. $\frac{1}{6}$ 28. $-3\frac{1}{5}$ 29. $1\frac{1}{2}$

30. $-3\frac{3}{4}$

CHAPTER 2 - NON-FRACTIONAL EQUATIONS

Assignment 5
1. y = -4   2. t = 11   3. x = 41   4. n = 53   5. G = 1   6. d = -5   7. $p = \frac{4}{9}$ 8. $r = \frac{15}{8}$

9. F = -1   10. $y = \frac{1}{7}$ 11. $w = -\frac{5}{3}$ 12. h = 0   13. x = 4   14. r = -3   15. $m = -\frac{2}{3}$

16. $s = \frac{1}{4}$ 17. $H = -\frac{3}{2}$ 18. A = -1   19. $y = -\frac{7}{4}$ 20. $b = \frac{4}{5}$ 21. x = 1   22. $c = -\frac{1}{3}$
23. p = 0   24. v = 0

Assignment 6
1. t = 3   2. R = 0   3. $y = -\frac{13}{5}$ 4. $P = \frac{1}{3}$ 5. $x = -\frac{5}{2}$ 6. w = 0   7. $h = \frac{10}{3}$ 8. $F = -\frac{3}{4}$

9. s = -2   10. $m = \frac{4}{3}$ 11. b = -2   12. $x = -\frac{2}{3}$ 13. $r = \frac{5}{4}$ 14. $a = \frac{1}{2}$ 15. V = 1

16. $d = -\frac{3}{2}$ 17. $N = \frac{1}{3}$ 18. c = 0   19. E = 1   20. $t = -\frac{7}{4}$ 21. h = -12   22. y = 2

23. $w = -\frac{5}{2}$ 24. k = -1

Assignment 7

1. $x = -\frac{7}{4}$ 2. $w = -3$ 3. $y = 10$ 4. $t = \frac{4}{7}$ 5. $d = 2$ 6. $E = -\frac{3}{5}$ 7. $P = \frac{3}{2}$ 8. $r = \frac{2}{3}$

9. $F = \frac{5}{2}$ 10. $y = -30$ 11. $x = -2$ 12. $V = -5$ 13. $a = -\frac{1}{4}$ 14. $R = 4$ 15. $s = -\frac{1}{2}$

16. $x = -\frac{1}{6}$ 17. $w = 0$ 18. $b = \frac{5}{2}$ 19. $B = \frac{2}{3}$ 20. $K = 2$ 21. $G = -2$ 22. $h = -\frac{7}{2}$

23. $r = \frac{2}{5}$ 24. $t = \frac{1}{5}$

Assignment 8

1. $w = 5$ 2. $d = -\frac{1}{4}$ 3. $V = 1$ 4. $x = -\frac{15}{4}$ 5. $y = 15$ 6. $P = \frac{4}{5}$ 7. $E = -\frac{1}{2}$ 8. $r = 3$

9. $h = -3$ 10. $x = 1.1$ 11. $R = -3.7$ 12. $m = 40$ 13. $A = 10$ 14. $y = 20$ 15. $t = \frac{3}{2}$

16. $s = \frac{6}{5}$ 17. $E = \frac{1}{6}$ 18. $v = 96$ 19. $E = 72$ 20. $B = 70$ 21. $V = 400$ 22. $F = 68$

23. $I = 5$ 24. $h = 8$ 25. $N = 11$ 26. $W = 8$ 27. $t = 4$ 28. $A = 20$

CHAPTER 3 - FRACTIONAL EXPRESSIONS AND EQUATIONS

Assignment 9

1. $\frac{7x}{40}$ 2. $\frac{15}{4t}$ 3. $\frac{27y}{16}$ 4. $\frac{7}{6w}$ 5. $\frac{12p}{5}$ 6. $\frac{4r}{7}$ 7. $\frac{20}{3x}$ 8. $\frac{3}{R}$ 9. $\frac{4}{5}$ 10. 1 11. $\frac{2}{5}$

12. $\frac{3}{2}$ 13. $\frac{1}{2}$ 14. $3P$ 15. 7 16. $\frac{x}{3}$ 17. $\frac{3}{5}$ 18. $\frac{4}{3}$ 19. $12t$ 20. $\frac{1}{6}$ 21. 1 22. 1

23. 20 24. x 25. $\frac{7}{3}$ 26. 12 27. $\frac{3}{2}$ 28. 4 29. $3R$ 30. $21x$ 31. $8m$ 32. 1

Assignment 10

1. $w = 54$ 2. $x = 3$ 3. $y = 25$ 4. $R = 3$ 5. $G = \frac{12}{5}$ 6. $v = \frac{1}{10}$ 7. $d = \frac{7}{2}$ 8. $p = \frac{8}{3}$

9. $t = 0$ 10. $h = -3$ 11. $x = -1$ 12. $y = \frac{1}{2}$ 13. $w = -\frac{3}{4}$ 14. $F = 1$ 15. $a = 6$ 16. $k = \frac{3}{4}$

17. $x = -5$ 18. $t = \frac{3}{2}$ 19. $x = \frac{1}{4}$ 20. $H = 5$ 21. $v = -3$ 22. $P = -18$ 23. $y = -1$

24. $B = \frac{7}{5}$ 25. $t = -3$ 26. $k = \frac{9}{8}$ 27. $s = 0$

Assignment 11

1. $t = \frac{7}{2}$ 2. $y = \frac{3}{5}$ 3. $x = 1$ 4. $w = -6$ 5. $r = \frac{15}{8}$ 6. $d = \frac{11}{5}$ 7. $a = \frac{15}{4}$ 8. $h = -\frac{2}{7}$

9. $P = -\frac{5}{6}$ 10. $x = 8$ 11. $y = 6$ 12. $t = -3$ 13. $h = -\frac{9}{2}$ 14. $p = -6$ 15. $w = \frac{15}{28}$

16. $m = \frac{2}{3}$ 17. $k = 7$ 18. $r = \frac{24}{11}$ 19. $x = \frac{12}{5}$ 20. $b = -\frac{1}{4}$ 21. $E = -\frac{7}{8}$ 22. $t = \frac{13}{21}$

23. $R = 5$ 24. $y = -1$ 25. $F = \frac{3}{5}$ 26. $w = -18$ 27. $x = -\frac{15}{7}$

Assignment 12

1. $y = \frac{1}{7}$ 2. $R = \frac{1}{10}$ 3. $t = -1$ 4. $x = \frac{1}{2}$ 5. $w = \frac{3}{10}$ 6. $d = -\frac{1}{2}$ 7. $F = 1$ 8. $p = -\frac{1}{6}$

9. $R = \frac{10}{7}$ 10. $m = -\frac{20}{9}$ 11. $h = \frac{14}{13}$ 12. $y = \frac{21}{40}$ 13. $a = 5$ 14. $W = 12$ 15. $I = 4$

16. $f = 4$ 17. $W = 6$ 18. $v = 4$ 19. $i = 3$ 20. $t_2 = 8$ 21. $v_1 = 4$ 22. $r = 6$ 23. $r = 10$

24. $b = 1$ 25. $A = 15$ 26. $R_1 = 3$

CHAPTER 4 – SPECIAL PRODUCTS AND FACTORING

Assignment 13
1. $3y^2$ 2. $8P^2$ 3. $45x^2$ 4. $56r^2$ 5. $4E^2$ 6. $25w^2$ 7. $121h^2$ 8. $900t^2$ 9. $p^2 + p$
10. $7x^2 - 6x$ 11. $2t^2 - 2t$ 12. $24y^2 + 84y$ 13. $20b^2 + 5b$ 14. $54R^2 - 63R$ 15. (12)(5)
16. (6h)(2)   17. (8t)(4t)   18. (d)(8d)   19. $2(t + 3)$ 20. $5(y + 1)$ 21. $4(3r + 2)$ 22. $3(h - 1)$
23. $4(4m - 1)$ 24. $4(2x - 7)$ 25. $5(8d + 5)$ 26. $6(4p - 3)$ 27. $12(5w - 3)$ 28. $s(s + 1)$
29. $2y(y + 5)$ 30. $3r(2r + 1)$ 31. $x(3x - 1)$ 32. $2b(2b - 3)$ 33. $5w(3w - 2)$ 34. $4r(5r + 3)$
35. $h(8h - 15)$ 36. $16t(3t + 5)$

Assignment 14
1. $x^2 + 8x + 15$ 2. $r^2 + 4r + 3$ 3. $m^2 + 14m + 48$ 4. $2y^2 + 3y + 1$ 5. $4d^2 + 11d + 7$
6. $6t^2 + 13t + 6$ 7. $10a^2 + 57a + 54$ 8. $12w^2 + 47w + 45$ 9. $72x^2 + 25x + 2$ 10. $t^2 + t - 6$
11. $b^2 - 3b - 4$ 12. $y^2 - 8y + 15$ 13. $4r^2 + 3r - 1$ 14. $5h^2 - 8h - 4$ 15. $12p^2 - 19p + 5$
16. $4m^2 - 4m - 35$ 17. $20x^2 + 37x - 18$ 18. $35a^2 - 17a + 2$ 19. $a^2 - 49$ 20. $V^2 - 1$
21. $9h^2 - 4$ 22. $25t^2 - 1$ 23. $16y^2 - 9$ 24. $49x^2 - 144$ 25. $R^2 + 2R + 1$ 26. $w^2 - 12w + 36$
27. $4x^2 + 20x + 25$ 28. $36d^2 - 12d + 1$ 29. $9 - 6t + t^2$ 30. $1 + 8y + 16y^2$

Assignment 15
1. $(t + 7)(t - 7)$ 2. $(3y + 1)(3y - 1)$ 3. $(2F + 5)(2F - 5)$ 4. $(4x + 9)(4x - 9)$ 5. $m^2 + 6m + 5$
6. $w^2 - 4w - 12$ 7. $h^2 - 16h + 63$ 8. $P^2 - 5P - 24$ 9. $d^2 - 9d + 20$ 10. $r^2 + r - 90$
11. $(x + 1)(x + 2)$ 12. $(a + 2)(a + 5)$ 13. $(y + 3)(y + 3)$ 14. $(t - 1)(t - 3)$ 15. $(R - 2)(R - 4)$
16. $(w + 3)(w - 4)$ 17. $(h + 5)(h - 1)$ 18. $(d + 2)(d - 5)$ 19. $(m + 7)(m - 3)$ 20. $(2y + 1)(y + 1)$
21. $(3x + 1)(x + 2)$ 22. $(3p + 2)(p + 3)$ 23. $(2r - 1)(r - 2)$ 24. $(5t - 3)(t - 1)$ 25. $(3w - 4)(2w - 1)$
26. $(2E + 3)(2E - 1)$ 27. $(d + 2)(3d - 4)$ 28. $(7x + 3)(x - 1)$ 29. $(4m + 3)(2m - 3)$
30. $(2b - 1)(2b - 1)$ 31. $(3w + 2)(3w + 2)$

CHAPTER 5 – QUADRATIC EQUATIONS

Assignment 16
1. $\frac{1}{16}$ 2. $2,500$ 3. $\frac{9}{64}$ 4. 0.04 5. 45 6. -5 7. $d = 225$ 8. $M = 48$ 9. $r = 20$
10. $E = 6$ 11. $s = 144$ 12. $K = 4$ 13. $I = 50$ 14. $w = 200$ 15. -7 16. $\pm\frac{1}{5}$ 17. ± 3.87
18. ± 13.89 19. $x = \pm 1$ 20. $t = \pm\frac{1}{2}$ 21. $y = \pm\frac{10}{3}$ 22. $w = \pm\frac{5}{2}$ 23. $p = \pm\frac{4}{5}$ 24. $d = \pm\frac{2}{9}$
25. $y = \pm\frac{3}{4}$ 26. $d = \pm\frac{5}{6}$ 27. $r = \pm 4$ 28. $m = \pm 3.16$ 29. $y = \pm 8.94$ 30. $t = \pm 1.73$
31. $x = \pm 6.32$ 32. $r = \pm 12.25$ 33. $d = \pm 2.45$

Assignment 17
1. $t = -1$ 2. $w = \frac{1}{2}$ 3. $p = 0$ 4. $x = 0$ 5. $y = 0$ 6. $d = 0$ 7. $r = 0$ 8. $m = 0$
   $t = 7$ $w = \frac{5}{3}$ $p = -\frac{2}{5}$ $x = -1$ $y = \frac{3}{4}$ $d = \frac{1}{2}$ $r = -6$ $m = \frac{1}{3}$

9. $t = 0$ 10. $h = 0$ 11. $w = 0$ 12. $R = 0$ 13. $x = -1$ 14. $t = -2$ 15. $k = 1$
   $t = -\frac{2}{3}$ $h = -\frac{1}{5}$ $w = \frac{1}{4}$ $R = \frac{3}{2}$ $x = -3$ $t = 5$ $k = -7$

16. $y = -5$ 17. $r = 1$ 18. $x = 5$ 19. $P = 2$ 20. $F = 1$ 21. $h = 3$ 22. $m = 2$
    $y = -5$ $r = 1$ $x = -3$ $P = -3$ $F = 4$ $h = -4$ $m = -1$

23. $y = 4$ 24. $x = 1$ 25. $t = -2$ 26. $d = 1$ 27. $y = 3$ 28. $x = -1$ 29. $w = 4$
    $y = -2$ $x = -3$ $t = \frac{1}{2}$ $d = -\frac{1}{3}$ $y = \frac{1}{2}$ $x = -\frac{2}{3}$ $w = -\frac{1}{2}$

30. $x = \frac{1}{3}$ 31. $h = 2$ 32. $P = \frac{2}{5}$ 33. $m = 2$ 34. $V = -6$ 35. $y = 3$ 36. $x = \frac{2}{3}$
    $x = -\frac{3}{2}$ $h = -\frac{4}{3}$ $P = -\frac{1}{2}$ $m = \frac{3}{4}$ $V = \frac{3}{2}$ $y = \frac{5}{2}$ $x = -\frac{3}{5}$

Assignment 18

1. x = 4   2. t = $\frac{1}{2}$ or 0.5   3. w = 1   4. R = -4   5. y = -0.27
    x = -2   y = -3.73
              t = $-\frac{2}{3}$ or -0.67   w = $-\frac{3}{4}$ or -0.75   R = $\frac{5}{2}$ or 2.5

6. d = 1.85   7. p = 1.29   8. F = 1.20   9. h = 0.42   10. x = 1   11. y = -3
    d = -0.18   p = 0.31   F = -1.45   h = -2.42
                                                               x = $\frac{2}{3}$ or 0.67   y = $-\frac{2}{3}$ or -0.67

12. t = 5   13. r = -4   14. N = 1.62   15. w = 0.60   16. x = 3
    t = $-\frac{1}{2}$ or -0.5   r = $\frac{3}{2}$ or 1.5   N = -0.62   w = -2.10   x = -1

17. p = 2   18. y = $-\frac{1}{2}$ or -0.5   19. I = 2   20. r = 1.41   21. v = 12   22. R = 5.29
    p = $\frac{3}{2}$ or 1.5   y = $-\frac{1}{3}$ or -0.33

23. s = 2.83   24. t = $\frac{5}{2}$ or 2.5   25. Width = 2 meters   26. First leg = 8.94 cm
                                                      Length = 6 meters   Second leg = 17.88 cm

27. Side = 4 in   28. First leg = 2.63 cm   29. t = 5 seconds   30. N = 5.85
                         Second leg = 7.63 cm

CHAPTER 6 - GRAPHING

Assignment 19

1.

x	y
5	60
1	12
0	0
-2	-24
-4	-48

2.

x	y
8	8
3	3
0	0
-1	-1
-5	-5

3.

x	y
6	-8
2	0
0	4
-3	10
-7	18

4.

x	y
4	6
2	-2
0	-10
-2	-18
-4	-26

5.

x	y
8	-10
4	0
0	10
-2	15
-6	25

6.

x	y
10	9
5	6
0	3
-5	0
-10	-3

7. A: (3, 3)   B: (3, -2)   C: (-2, -1)   D: (-4, 2)   E: (2, 0)   F: (0, -3)

8. G: (2, 15)   H: (-4, 20)   I: (-6, -10)   J: (4, -15)   K: (-8, 0)   L: (0, 5)

9. a) Quadrant 2   b) Quadrant 4   c) Quadrant 1   d) Quadrant 3   e) Quadrant 2

10. b, c, and d   11. a, d, and e   12. origin   13. 9   14. -10   15. (17, -9)

Assignment 20

1. $\boxed{y = x}$

x	y
-5	-5
-2	-2
0	0
1	1
4	4

2. $\boxed{y = x^2 - 4}$

x	y	x	y
$\frac{1}{2}$	$-3\frac{3}{4}$	$-\frac{1}{2}$	$-3\frac{3}{4}$
1	-3	-1	-3
2	0	-2	0
3	5	-3	5
0	-4		

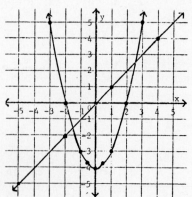

Assignment 20 (continued)

3. $\boxed{2x + y = 4}$

x	y
-3	10
-1	6
0	4
1	2
2	0
5	-6

4. $\boxed{y = 3x - x^2}$

x	y		x	y
-2	-10		2	2
-1	-4		3	0
0	0		4	-4
1	2		5	-10
$1\frac{1}{2}$	$2\frac{1}{4}$			

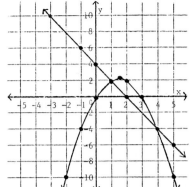

5. $\boxed{5y = 2x - 50}$

x	y
-25	-20
-15	-16
0	-10
10	-6
25	0

6. $\boxed{xy = 240}$

x	y		x	y
5	48		-5	-48
6	40		-6	-40
10	24		-10	-24
15	16		-15	-16
20	12		-20	-12
24	10		-24	-10
0	--		0	--

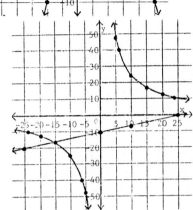

Assignment 21

1. $\boxed{w = 10p^2}$

p	w
0	0
1	10
2	40
3	90
4	160
5	250

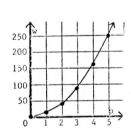

2. $\boxed{E + 2I = 20}$

E	I
0	10
4	8
8	6
12	4
16	2
20	0

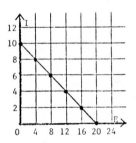

3. a) 3 centimeters
   b) 8 centimeters
   c) 12 centimeters
   d) 24 centimeters

4. a) 80 square millimeters
   b) 500 square millimeters
   c) 11 millimeters
   d) 18 millimeters

5. a) #1: <u>180</u> m   #2: <u>120</u> m   #3: <u>60</u> m

   b) #1: <u>4</u> sec   #2: <u>6</u> sec   #3: <u>12</u> sec

CHAPTER 7 - ALGEBRAIC FRACTIONS

Assignment 22

1. 1   2. $\frac{a}{2d}$ 3. $\frac{G}{3P}$ 4. $\frac{3}{4}$ 5. $\frac{1}{c}$ 6. 4T   7. $\frac{t}{b}$ 8. $\frac{1}{5r}$ 9. $\frac{h}{xy}$ 10. $\frac{2}{b}$ 11. $\frac{p^2t}{k}$

12. a   13. $\frac{E}{G}$ 14. $\frac{1}{t}$ 15. 2tx   16. $\frac{1}{k-1}$ 17. h + rw   18. $\frac{v}{x-y}$ 19. $\frac{b}{d}$ 20. $\frac{1}{x}$

21. 2P   22. $\frac{ks}{t}$ 23. $\frac{m}{2pr}$ 24. $\frac{F}{T(1-A)}$ 25. c + k   26. $\frac{4d}{m}$ 27. $\frac{1}{2(y+1)}$ 28. $\frac{1}{3}$

29. 2   30. 1

Assignment 23

1. $\dfrac{w}{2}$ 2. $\dfrac{v+w}{r}$ 3. $3ab$ 4. $\dfrac{3t+2}{w}$ 5. $\dfrac{F+H}{E-2}$ 6. $\dfrac{2h+1}{x+y}$ 7. $\dfrac{c+d+p}{h}$ 8. $\dfrac{3}{F}$ 9. $\dfrac{7r}{8}$

10. $\dfrac{7}{3G}$ 11. $\dfrac{5d+r}{5x}$ 12. $\dfrac{4a}{3t}$ 13. $\dfrac{2m+3}{6}$ 14. $\dfrac{3d+4r}{24}$ 15. $\dfrac{kv+pr}{rv}$ 16. $\dfrac{P+Q}{PQ}$ 17. $\dfrac{h+sw}{rw}$

18. $\dfrac{5m+3}{6m}$ 19. $\dfrac{2bw+at}{2ab}$ 20. $\dfrac{c+v}{cpv}$ 21. $\dfrac{E+2}{2}$ 22. $\dfrac{r+w}{r}$ 23. $\dfrac{2+3P}{3}$ 24. $\dfrac{5B+4}{B}$

25. $\dfrac{4t+1}{t}$ 26. $\dfrac{mk+w}{k}$ 27. $\dfrac{r+2adv}{av}$ 28. $\dfrac{1+x}{x}$

Assignment 24

1. $\dfrac{gh-d}{m}$ 2. $\dfrac{1}{2p}$ 3. 0 4. $2R$ 5. $\dfrac{1}{T}$ 6. $\dfrac{d+1}{a+b}$ 7. $\dfrac{P}{6}$ 8. $\dfrac{4r-t}{4t}$ 9. $\dfrac{3}{2x}$ 10. $\dfrac{a-d}{ad}$

11. $\dfrac{rw-st}{sw}$ 12. $\dfrac{2v-bp}{pw}$ 13. $\dfrac{ar-1}{r}$ 14. $\dfrac{h-6}{3}$ 15. $\dfrac{A-B}{A}$ 16. $\dfrac{m-tw}{t}$ 17. $\dfrac{1}{H}+\dfrac{1}{F}$

18. $\dfrac{2}{3}-\dfrac{1}{2r}$ 19. $\dfrac{B}{C}-\dfrac{1}{A}$ 20. $\dfrac{c}{r}-\dfrac{b}{t}$ 21. $\dfrac{r}{d}+\dfrac{s}{d}+\dfrac{t}{d}$ 22. $\dfrac{x}{y+1}+\dfrac{1}{y+1}$ 23. $\dfrac{a}{4}+2$ 24. $\dfrac{1}{r}-1$

25. True   26. False   27. False   28. $6t(3a+2)$ 29. $F(1-P)$ 30. $d(c+h)$ 31. Not possible

Assignment 25

1. $\dfrac{2x+4y}{3w}$ 2. $\dfrac{t}{p-2}$ 3. $\dfrac{d}{w+1}$ 4. $\dfrac{3s}{s-2v}$ 5. $\dfrac{p+a}{4}$ 6. $\dfrac{1-P}{R}$ 7. $\dfrac{y+1}{1-y}$ 8. $\dfrac{k-2}{2r-1}$

9. $\dfrac{2t+w}{p+3v}$ 10. $\dfrac{1}{A+1}$ 11. $\dfrac{1}{2y-3}$ 12. $b-1$ 13. True   14. True   15. False   16. $k(r-a)$

17. $\dfrac{bd}{t+w}$ 18. $\dfrac{x-y}{2pv}$ 19. $\dfrac{1}{D+H}$ 20. $\dfrac{x+1}{2xy}$ 21. $\dfrac{rt}{1-r}$ 22. $\dfrac{R}{R+2}$ 23. $\dfrac{w-1}{bw}$ 24. $\dfrac{3(x+y)}{t}$

25. $\dfrac{p-t}{c-k}$ 26. $\dfrac{R+1}{R-1}$ 27. $\dfrac{t(a-w)}{w(b+t)}$

CHAPTER 8 - FORMULA REARRANGEMENT

Assignment 26

1. $s=\dfrac{P}{4}$ 2. $L=\dfrac{A}{W}$ 3. $r=\dfrac{C}{2\pi}$ 4. $R=\dfrac{E^2}{P}$ 5. $h=\dfrac{V}{\pi r^2}$ 6. $v=\dfrac{d}{t_2-t_1}$ 7. $M=\dfrac{F(p-a)}{2m}$

8. $V_2=\dfrac{T_2V_1}{T_1}$ 9. $W=dF$ 10. $f=\dfrac{1}{t}$ 11. $c=\dfrac{Q}{mT}$ 12. $G=\dfrac{R^2W}{KN}$ 13. $R=\dfrac{E+e}{I}$ 14. $C=\dfrac{2s}{gt^2}$

15. $h=\dfrac{2dp}{b(1-a)}$ 16. $L=\dfrac{HK(b_1+b_2)}{A}$ 17. $A_2=\dfrac{A_1P_1}{P_2}$ 18. $R_2=\dfrac{(E_2)^2R_1}{(E_1)^2}$ 19. $d=\dfrac{hpt}{bw}$ 20. $a=\dfrac{r^2}{m}$

Assignment 27

1. $H=\sqrt{D+T}$ 2. $c=\sqrt{\dfrac{E}{m}}$ 3. $v=\sqrt{2as}$ 4. $t=\sqrt{\dfrac{d}{2k}}$ 5. $h=\sqrt{\dfrac{2M}{b}}$ 6. $P=\sqrt{GN}$ 7. $r=\sqrt{\dfrac{2cw}{a}}$

8. $V=\sqrt{\dfrac{FgR}{W}}$ 9. $F_t=F_1+F_2$ 10. $W=\dfrac{P-2L}{2}$ 11. $Z=\sqrt{R^2+X^2}$ 12. $E=e+ir$

13. $s_2=\dfrac{-s_1f_1}{f_2}$ 14. $h=\sqrt{r^2-v^2}$ 15. $P_1=P_2-EI$ 16. $k=\dfrac{ah-w}{s}$ 17. $d_2=d_1-vt$

18. $r=\sqrt{R^2-hN}$ 19. $G_1=PW-G_2$ 20. $c=\dfrac{1-2ad}{m}$

Assignment 28

1. $w = \dfrac{b - hd}{h}$ or $w = \dfrac{b}{h} - d$ 2. $t_1 = \dfrac{at_2 - v}{a}$ or $t_1 = t_2 - \dfrac{v}{a}$ 3. $d = \sqrt{\dfrac{\pi D^2 - 4A}{\pi}}$ or $d = \sqrt{D^2 - \dfrac{4A}{\pi}}$

4. $p = \sqrt{\dfrac{k + w}{k}}$ or $p = \sqrt{1 + \dfrac{w}{k}}$ 5. $d_2 = \dfrac{2V - bhd_1}{bh}$ or $d_2 = \dfrac{2V}{bh} - d_1$

6. $P_1 = \dfrac{SW + FTP_2}{FT}$ or $P_1 = \dfrac{SW}{FT} + P_2$ 7. $v = \sqrt{\dfrac{crt - ah^2}{a}}$ or $v = \sqrt{\dfrac{crt}{a} - h^2}$

8. $P_2 = \dfrac{dHP_1 - 3Fms}{dH}$ or $P_2 = P_1 - \dfrac{3Fms}{dH}$ 9. $P = \dfrac{F}{A_1 + A_2 + A_3}$ 10. $x = \dfrac{b}{m - 1}$

11. $f_1 = \dfrac{w + df_2}{d}$ or $f_1 = \dfrac{w}{d} + f_2$ 12. $T = \dfrac{AK}{P - 2}$ 13. $r = \dfrac{a}{h + v}$ 14. $A = \dfrac{G}{D - 1}$ 15. $t = \dfrac{sv}{s + v}$

16. $G_2 = \dfrac{G_1 G_t}{G_1 - G_t}$ 17. $d = \dfrac{b}{s + 1}$ 18. $H = \dfrac{FV}{F - V}$ 19. $N = \dfrac{L}{L + 1}$ 20. $v = \sqrt{\dfrac{p(e - k)}{m}}$

Assignment 29

1. $R = \dfrac{A}{1 - H}$ 2. $a = \dfrac{bc}{b - c}$ 3. $F_1 = \dfrac{F_2 F_t}{F_2 - F_t}$ 4. $k = \dfrac{mt}{m - h}$ 5. $P = \dfrac{M + T}{M - B}$ 6. $r = \dfrac{N}{1 + hN}$

7. $M = \dfrac{s - 1}{s + 1}$ 8. $t_1 = T - b(p - p_1)$ 9. $a = t - m(d_1 + d_2)$ 10. $k = \dfrac{v - h}{w - 1}$

11. $T = \dfrac{p_1 - p_2 + bt}{b}$ or $T = \dfrac{p_1 - p_2}{b} + t$ 12. $R_2 = \dfrac{R_t - dR_1}{d + 1}$ 13. $F = \dfrac{h + w}{t + w}$ 14. $r_1 = dG(p_1 - p_2) - r_2$

15. $H = \dfrac{AS(B_1 + B_2)}{V_2 - V_1}$ 16. $m = \dfrac{w(h - t)}{p(a - k)}$ 17. $e_2 = \dfrac{C_1 + C_2 - C_1 e_1 Q}{C_1 Q}$ or $e_2 = \dfrac{C_1 + C_2}{C_1 Q} - e_1$

18. $d_1 = \dfrac{P - W + Ad_2 P}{AP}$ or $d_1 = \dfrac{P - W}{AP} + d_2$ 19. $r_1 = \dfrac{r_2}{f(w_1 + w_2) + 1}$ 20. $d = \dfrac{r(1 - ah)}{1 + ah}$

CHAPTER 9 - AN INTRODUCTION TO SYSTEMS OF EQUATIONS

Assignment 30
1. $x = 5$, $y = 3$

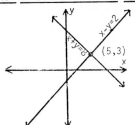

2. $x = 2$, $y = -4$

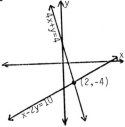

3. $x = 4$, $y = -1$ 4. $p = 5$, $d = 1$ 5. $h = 6$, $m = 3$ 6. $x = 3$, $y = -2$ 7. $t = 4$, $w = -4$
8. $x = 1$, $y = 1$ 9. $a = 2$, $b = -3$ 10. $r = 10$, $v = 3$ 11. $h = 7$, $k = 0$ 12. $s = 5$, $t = -2$
13. $b = 4$, $m = -8$ 14. $p = 20$, $r = 5$ 15. $p = 4$, $v = 1.2$ 16. $m = 2.5$, $n = 1.2$
17. $x = 3$, $y = -0.2$ 18. $r = 1.42$, $w = -2.06$

Assignment 31
1. $x = 5$, $y = 1$ 2. $x = 3$, $y = 2$ 3. $x = 4$, $y = -3$ 4. $x = 1$, $y = -1$ 5. $t = 2$, $w = 1$
6. $h = 6$, $k = 5$ 7. $a = 3$, $b = -3$ 8. $r = -1$, $s = 4$ 9. $m = 2$, $n = 0$ 10. $c = 1$, $d = -2$
11. $p = -4$, $t = 10$ 12. $v = 3$, $w = -1$ 13. $x = 2$, $y = 4$ 14. $a = 3$, $b = -3$ 15. $p = 1$, $r = -1$
16. $v = -2$, $w = -4$ 17. $h = 5$, $k = 1$ 18. $d = 10$, $h = -3$ 19. $x = -1$, $y = -6$ 20. $s = 8$, $t = 0$
21. $m = -2$, $n = -5$ 22. $x = 2$, $y = -1$ 23. $b = 3$, $d = -5$ 24. $t = 1$, $v = 2$

Assignment 32
1. $x = 5$, $y = 3$ 2. $a = 3$, $b = 9$ 3. $t = 6$, $w = 1$ 4. $r = 4$, $v = -1$ 5. $h = 20$, $m = 5$
6. $p = 1$, $s = 4$ 7. $d = 8$, $h = -2$ 8. $k = -3$, $m = 1$ 9. $r = 4$, $t = 1$ 10, $p = 2$, $w = 1$
11. $x = 2$, $y = 2$ 12. $d = 3$, $p = 4$
(Continued)

<u>Assignment</u> <u>32</u> (Continued)
13. 150 and 68   14. 48 and 16   15. 55 centimeters and 45 centimeters   16. camera costs $250;
calculator costs $170   17. length is 18 centimeters; width is 12 centimeters   18. 16 meters and 4 meters
19. 40 liters of solution A; 20 liters of solution B   20. 75 milliliters of 10% solution; 125 milliliters of
50% solution   21. 100 pounds of 8% alloy; 200 pounds of 5% alloy   22. 5 kilograms of pure zinc;
10 kilograms of 40% zinc

<u>CHAPTER</u> <u>10</u> – <u>SYSTEMS</u> <u>OF</u> <u>TWO</u> <u>EQUATIONS</u> <u>AND</u> <u>FORMULAS</u>

<u>Assignment</u> <u>33</u>
1. $x = 2$, $y = 8$ 2. $t = 2$, $s = 40$ 3. $d = -1$, $k = 3$ 4. $x = 6$, $y = -3$ 5. $a = 4$, $b = 1$
6. $r = 3$, $w = 10$ 7. $h = 5$, $p = -2$ 8. $E = -2$, $R = -7$ 9. $r = 5$, $v = 7$ 10. $x = 2$, $y = 2$
11. $G = 4$, $P = -3$ 12. $t = 6$, $w = -5$ 13. $P = 2$, $Q = -3$ 14. $r = -1$, $s = 7$ 15. $x = 3$, $y = 2$
16. $A = -2$, $B = -1$ 17. $d_1 = 4$, $d_2 = 1$ 18. $p = 5$, $t = -2$ 19. $V_1 = \frac{3}{2}$, $V_2 = 2$
20. $h = -\frac{1}{2}$, $k = 3$

<u>Assignment</u> <u>34</u>
1. $d = 6$, $t = 2$ 2. $E = 12$, $R = 3$ 3. $x = 1$, $y = -3$ 4. $p = -1$, $v = -5$ 5. $b = 10$, $m = 4$
6. $x = 6$, $y = 8$ 7. $F_1 = 20$, $F_2 = 10$ 8. $h = 2$, $k = -\frac{2}{3}$ 9. $x = 4$, $y = -2$ 10. $p = -3$, $r = 2$
11. $s = 10$, $w = \frac{5}{2}$ 12. $a = 1$, $b = -3$ 13. $b = 4$, $d = 6$ 14. $m = 12$, $s = -2$ 15. $x = 7$, $y = 30$
16. $a = 5$, $b = \frac{1}{2}$ 17. $h = 2$, $k = -3$ 18. $P = -1$, $T = 4$ 19. $b = 20$, $d = 5$ 20. $r = \frac{3}{4}$, $w = 5$
21. $b = 6$, $m = -\frac{1}{2}$ 22. $A = 5$, $B = -2$ 23. $i_1 = 1.5$, $i_2 = 1.2$ 24. $T_1 = 556$, $T_2 = 1,000$

<u>Assignment</u> <u>35</u>
1. $f = akr$ 2. $H = \dfrac{AC}{L}$ 3. $t = \dfrac{dh}{p}$ 4. $R = N^2 P$ 5. $s = \dfrac{b^2}{r}$ 6. $F = \sqrt{KV}$ 7. $k = \sqrt{\dfrac{w}{s}}$
8. $B = \dfrac{DP}{A}$ 9. $p = \dfrac{r}{v}$ 10. $f = d + h - r$ 11. $L = \dfrac{B + M}{2}$ 12. $R = \dfrac{F - K}{Q - 1}$ or $\dfrac{K - F}{1 - Q}$ 13. $s = \dfrac{1}{ar}$
14. $A = \sqrt{\dfrac{F}{R}}$ 15. $t = \dfrac{b}{r^2}$ 16. $V = \sqrt{\dfrac{2B}{K}}$ 17. $F = \dfrac{9(K - 273)}{5} + 32$ or $F = \dfrac{9K - 2297}{5}$ 18. $t = \sqrt{ks}$
19. $p = \dfrac{kr}{1 - k}$ 20. $t = \dfrac{T}{a - 1}$ 21. $v = \dfrac{r^2}{m + r}$

INDEX